MINI-WOOD COMMUNITY

Other Books by the Author

Insect Behavior

Backyard Exploration

Hand Book for the Curious

The Giant Cactus Forest and Its World

This World of Living Things

Photographer in the Rain-Forests

With William Beebe and George Inness Hartley
Tropical Wild Life in British Guiana

MINI-WOOD COMMUNITY

PAUL GRISWOLD HOWES

Director-Curator Emeritus
Bruce Museum of Natural History, History and Art
Greenwich, Connecticut

With 495 Photographs
and 24 Drawings
by the Author

Sylvanus, 1975

Published by
EXPOSITION PRESS, HICKSVILLE, NEW YORK
For
SYLVANUS BOOKS, Noroton, Connecticut

CONTENTS

PREFACE

The sight of our home, the planet Earth, as a breath-taking blue ball partly enshrouded in dense clouds, has now become a fixed object in the mind's eye of all of us.

In closely examining the truly superb photographs secured by the astronauts, we became suddenly more easily able to grasp the truth of how we exist upon the thin skin of a ball of mineral matter partly veiled by a veneer of green vegetation and all but surrounded by salt water, a small globe comparatively, isolated save for its one still tinier satellite, a pair endlessly traveling—and one whirling as well—through space.

But greatly as these photographs have enlightened us in many ways, how far we still are from comprehending how we intelligent beings and hundreds of thousands of organisms of lesser status came to evolve and to dwell successfully side by side on this mysterious blue ball!

Let us imagine for a minute some other sort of intelligent beings traveling comfortably in a spaceship and for the first time setting eyes upon our earth from not too many miles above it. Would it enter the imagination of these beings, we wonder, that down under those cloud blankets this blue sphere might be tenanted over portions of its surface by strange green plants and literally crawling with animal organisms, vast numbers of living things dependent upon this green wherewithal, and the whole enigmatic complex bound together in an interrelated, but invisible, web?

In all probability this would be beyond their capability to visualize with any degree of accuracy; possibly it would be even beyond their comprehension to imagine such a living whole.

Likewise, right here on earth, only a small proportion of its human inhabitants can comprehend or imagine the density of life which occurs here—swarms, sometimes, in as small an area of the green mantle as half an acre. Even fewer realize that in addition to the more familiar, often visible, organisms, as, for instance, birds and mammals, thousands—perhaps millions—of small to minute organisms also inhabit the humus and soil, actually, one might say, beneath our feet.

The title of the book refers to such a community of plants and animals; it is the story of life—macro and micro—in this bit of property which the author prefers to call a mini-wood.

For a time, the title favored was *The Little Survivors*, but on contemplation it was soon felt that many of those who oppose the environmentalists might misconstrue its intended meaning, and without a glance beyond such a title might take comfort in a snap judgment and interpretation, believing that the author was on their side of the fence.

The story of the mini-wood does reveal how numerous small creatures have managed to survive the pesticide ordeal and the present pollution of air and water while others have *not*. But this is not just another book pertaining to pollution. It is the story of what went on day by day and season by season during fifteen years within and on the borders of a half-acre of New England hardwood deciduous woodland possessed by the author.

Employing the scientific (Latin) as well as the common English names of the organisms throughout, but using other technical terms consistent only with accuracy and usefulness, usually with immediate explanations, the book covers the vastly interesting, often astonishing, train of events in a living community—the features about it as a whole, and the individual appearances, habits, smells, and interrelationships of its plants and animals. This is all that has been intended, and those who wish to learn more concerning the anatomies and the taxonomy of the said organisms than has been very briefly included here are referred to a library or to the list of volumes for further reading given at the end of this book.

Commencing with a brief description of the local climate and terrain, then the trees and shrubs of an association sprung from a rock-strewn (glaciated) woodland with an interesting soil, the chapters cover the climbing vines, the mosses and lichens, the slime molds and fungi, and the ferns and wildflowers, and in covering the area's animal life, the author has endeavored to bring home, with special emphasis, how numerous, how complex, and yet how little known to the average person are most of these medium-small, very small, and micro organisms which make up so important a part of such a mini-wood complex. Chapters X, XI, and XII might be said to form the keystone of the book, for in them, it will be clearly seen how all of these small animate things fit into the picture of the living woodland as a whole.

This is not to say that *your* bit of deciduous New England woods, be it larger or smaller than the author's, will duplicate just what occurred in this wood over a fifteen-year period of study, although much of it would be about the same in a similar soil supporting a matching botanical setup. Usually, any moist woodland humus will be found teeming with living minutiae.

With clear new pictures, this book has been designed to open up this amazingly interesting mini-wood world afresh—to show the astonishing goings on among persistently viable plant and animal organisms, regardless of how fragmentary their woodland habitat may be.

Happily, and notwithstanding the manner in which we have abused our world, the mini-wood may still have its higher animal inhabitants and transients also—amphibians and reptiles, birds and mammals, valiant survivors which will be discussed in later chapters.

Songbirds have of course been the great sufferers from our past stupidities. I share this guilt, for in years gone by, I, like so many others, unwittingly gave the "tree men"

carte blanche to do what they deemed necessary when caterpillar-explosion time came around (see chapters XIV and XV on the mini-wood insects). As so many others had done, we allowed unsupervised spraying with lethal chemicals and, sad to say, at that very period when the spring migration was at its height: when hordes of succulent hairless caterpillars, upon which the incoming migrants customarily fed, were most numerous in the trees.

Among these spring caterpillars there are many fat green prizes which we see the Orioles and Tanagers, the Warblers and Vireos gobble up, but the small leaf-rollers and the cankerworms form the bulk of this seasonal food.

To the dismay of the author and his wife, it was found that some of our most wanted and cherished birds were dying. Wood Thrushes and Robins, Warblers and Tanagers were picked up in a strange quivering state, dying soon afterward. Vireos disappeared from our woods and shade trees. So did the Wood Pewee the "Chebec" Flycatcher, and other small songbirds, notably the Yellow Warbler. Even now, many years after banning all dangerous sprays at our property, many of the species have not increased to their former numbers; some are still absent.

It was during this period of thoughtlessness, when we and neighbors on both sides of our property were uninformed about the full dangers of pesticides, that we noticed cessation of all activity around certain birdhouses, but knowing House Wrens to be fickle little birds anyway, we imagined that they had simply found sites better to their liking elsewhere. Thus, for that time, we thought little more about the matter, and, indeed, we were not even certain that any eggs had been laid. Later, however, soon after the nests had been deserted, we found two adult Wrens dead near the boxes: one, a mere bunch of bones and brown feathers; the other, freshly dead and lying on the ground with wings outspread in a position in which we had found other poisoned species. When, still later, the boxes were opened for fall cleaning, there lay the all-too-graphic evidence of what uncontrolled spraying was doing to our bird life.

In one nest, Wrens had evidently hatched successfully enough but had died, either from poisoned food or—what was more likely—from starvation, when the parents could no longer bring them food of any kind. Lying in the nest were their little skeletons, the bones white and clean, no doubt left thus by blowflies, following the nestlings' death (see plate 158). In another birdhouse that year was a set of seven unhatched eggs, their contents dried up within the unbroken shells.

We had had our graphic lesson. Sensing a "Silent Spring" in the making, we immediately ended our spraying with these lethal-to-all-wildlife solutions; then, convincing our nearby neighbors of what was taking place, all cooperated forthwith, long before DDT was officially banned in Connecticut under most circumstances. The skeletons' photograph (plate 158) seemed to the author to be a useful picture which might be widely used to symbolize the meaning of uncontrolled spraying. Enlarged, it was mailed to important newspapers and magazines, especially to those emphasizing the dangers birds were facing. Each copy was accompanied by a short description of what had been observed, and the statement that no remuneration was wanted.

But not *one* editor would publish this graphic picture! The reasons given were unconvincing: "Not clear as to meaning . . ." "Dramatic, but if published, the bones would not be distinguishable from the surrounding grass. . . ." Some did not reply at all. Others never even returned the prints, nor offered opinions of any kind. That the editors of these conservation-minded magazines turned the picture down makes one wonder, does it not? And as for the photograph's meaning being cryptic, I will let the reader judge for himself.

Baffled, the author gave up the endeavor, and the picture is here published for the first time.

Some readers may wonder why the illustrations were not printed in color, and some may even pass the volume by because of it, but the reason was of course that such a publication (with colored plates) would have had to sell at a price much higher than that which many who may want the book would be ready to pay or unable to lay out. Black and white was therefore decided upon, but a special effort was made by the author to obtain clear new photographs, and it was this which held up the book's printing for some time after its writing had been completed. The fifteen-year study covered in these chapters extended from 1955 through 1970, and it may be said in retrospect that things new to the min-wood are still being found therein!

Aside from our day-to-day observations and simple rec-

ords, there are the basic overall mysteries concerning any woodland and its plant and animal inhabitants, three of which come to mind whenever contemplating such a complex.

First, we have the puzzle of how, in each instance and in the first place, these organisms came to be the ones selected, with each minor or major plant or animal successfully unlocking many knotty problems in becoming a part of the living whole.

Second, there is the mystery of how each of these organisms found the others necessary to it in this living web.

And, third, there is the imponderable question of how many centuries or ages were required for the successful forging of the woodland's living chains as we now find them.

As the metric system, long used by scientists and naturalists, is to be generally adopted in America, measurements in millimeters in describing plant and animal forms have been liberally sprinkled throughout this book in order to familiarize the uninitiated with such dimensions, their approximations in inches being given in each case.

In regard to the scientific (Latin) names of the organisms included, it need only be said that in binomials, the first term designates the genus, and the second term the species. In trinomials, the third different term indicates that the plant or animal is a subspecies. Occasional abbreviated names placed *after* the scientific names have to do with which scientist or scientists originally named the organism in question.

As the Latin names of plants and animals are employed universally in scientific circles, they are of necessity used throughout this book for the information of the technically concerned reader. By means of these names, the exact species or subspecies referred to may be known by naturalists and scientists *anywhere,* regardless of differing English names that may be used for them in the vernacular of the country.

The author will be grateful to any naturalist or scientist who may find an error or errors in this book and report same to him promptly so that corrections may be made in any future edition or printing of the work.

Also, changes in nomenclature are not always easy to keep up with, and it is therefore hoped that the technical names, as used in this book, will be mostly up to date, and in any case will serve to inform the professional naturalist, anywhere, just what species or subspecies were encountered and illustrated. At the same time it is hoped that changes from earlier writings will not seriously confuse the less well versed reader who may have seen other names, technical or otherwise, used elsewhere for the same organisms.

South Norwalk, Connecticut
1974

PAUL GRISWOLD HOWES

ACKNOWLEDGMENTS

The author is deeply grateful to the many who, in various ways, have made this book's publication possible. Special thanks are extended to the following scientists and naturalists for identifications pertaining to organisms treated in these chapters and made over a considerable period, or for checking or verifying the author's own identifications in a few cases where this was felt to be necessary:

Professor Robert L. Hulbary, Chairman, Department of Botany, the University of Iowa, Iowa City, for identifications of moss specimens; Dr. Alexander E. Smith, University of Michigan Herbarium, Ann Arbor, Michigan, for suggestions regarding certain fungi and fungus anomalies; Dr. Karl V. Krombein, Chairman, Department of Entomology, National Museum of Natural History, Smithsonian Institution, Washington, D.C., for information on pollinia of *Asclepias syriaca;* Dr. Richard H. Eyde, Curator, Department of Botany, National Museum of Natural History, Smithsonian Institution, Washington, D.C., for correspondence pertaining to pepperidge trees; Dr. Clark T. Rogerson, Senior Curator, Cryptogamic Botany, the New York Botanical Garden, New York City, for checking identification of mini-fungus of the genus *Marasmius;* Dr. John A. L. Cooke, Department of Entomology, the American Museum of Natural History, New York City, for information on millipede literature; Dr. Richard Hoffman, Radford College, Radford, Virginia, authority on millipedes, for technical information on these organisms and millipede literature; Dr. Frederick H. Rindge, Curator, Lepidoptera, the American Museum of Natural History, New York City, for checking identification of the moth *Ennomos subsignarius;* Dr. D. C. Ferguson, National Museum of Natural History, Smithsonian Institution, Washington, D.C., for identification of the "fern ball" moth, *Macrobotys aeglialis;* and Dr. William D. Field, Supervisor, Lepidoptera and Diptera, Department of Entomology, National Museum of Natural History, for assistance in obtaining this identification important to this book, and also for checking and correcting the spelling of the name of the moth *Archips argyrespilus.*

PART ONE

BOTANICAL

CHAPTER I

THE MINI-WOOD AREA
—ITS TREES AND SHRUBS

To one casually admiring a bit of deciduous New England woodland thriving in terrain once widely glaciated, an established and seasoned wood with its mature and aging trees, the overall impression received is that this is indeed a static scene. In its serene beauty and staunchness it may seem to the uninitiated beholder that it must have remained like this for years, or decades, or perhaps for a century, showing little or nothing in the way of change. What is so hard to realize and visualize is that such a wooded area, be it less than an acre or covering a great many acres, was once barren land, swept clean of vegetation by slowly moving glacial ice, hundreds to thousands of feet in thickness, which, as it melted during its tedious retreat thousands of years ago, left its overlying blankets of sand and clay and gravel, and the familiar embedded rocks and surface boulders that characterize so many parts of New England today.

No one knows how many centuries were required for the post-glacial pioneer forests to reclothe the land. But we do know that most of these were destroyed by man, and that our woodlands here in New England today are mostly second or third stands, grown since the pioneer fellings, or since later cuttings by our forefathers.

As the ice retreated, leaving behind the surface layers of sand and gravel, the fragmented rocks and boulders, the pioneer plants which required little else were the first organic arrivals. The first pioneers were no doubt partnerships between algae and fungi in the form of those curious plants we call lichens, some of which indeed may have already been established on the boulders when they were dumped by the glacier.

The following chapters will endeavor to describe and illustrate the complex community which evolved on one half-acre of such glaciated land as it was found and studied over the fifteen-year period between 1955 and 1970 at South Norwalk, Connecticut.

The lichens must have been followed by the mosses and ferns in ways that will be described in more detail later. Next, creeping in from territory south of the glacier's farthest advance, came a succession of larger, partly arboreal forms, and, maybe last of all, the herbaceous species suitable to conditions on the evolving woodland's floor.

Probably no one can say exactly how many decades have been required by *any* forest in evolving a so-called climax association or how many years passed before pioneer plants, alone with the elements, prepared a soil congenial to the first sun-tolerant trees, shrubs, and herbs. Nor can anyone state, with complete assurance of accuracy, just how many years must have elapsed subsequently before such an early association of plants was eliminated by excess shade, cast by the coming of still larger dominant trees and the somewhat lesser associate species which gradually moved in with them.

Looking into the mini-wood (which is the subject of this book) on a quiet sunny morning in May or June, the brilliancy of the new season's climax foliage, with the sun, filtered by the newly expanded delicate green leaves and casting an ethereal glow on others beneath (see top picture, plate 35), brought always a special delight to the author. Such a scene might indeed seem static at such times, but its calm beauty is an illusion, of course, for actually there is constant change or activity of some sort going on among all of the woodland organisms. Even long before winter is over, for instance, fluids are restless in tree trunks, limbs, and branches, in bulbs and corms and rootstocks. At the tips of twigs snapped off by winter storms, opaque sap icicles dangle from the bleeding ends, revealing the accelerated flow and signaling to the initiate that the vernal awakening is already here (top left photo, plate 4).

When at length the warm spring days are with us once more, we may see with surprise that the previous autumn's thick dead-leaf blanket has shrunk to a mere fraction of its late fall depth, reminding us sharply that wherever this vegetable blanket is left undisturbed, it never grows any deeper from year to year—and why this is, we shall soon see.

Added to this annual leaf-fall laid down by trees and other plants, there is a gradual year-round "shower" of other dead matter—twigs and bark fragments, limbs and branches infiltrated with fungi, and other dead and living discards, all part of a natural pruning process important to the economy of the woodland in many ways. People who remove all such debris from the woods are really interfering with nature and are, to repeat a statement for emphasis, removing raw materials needed in one way or another in the overall economy of the woodland.

In viewing this so-called waste and the heavy annual leaf-fall in particular, we should be reminded that, not too long before, this was alive and green and, with the aid of that complicated compound within it which we condense into the single word "chlorophyll," was making its contribution to the foods upon which all animals and the plants themselves depend. This process—photosynthesis—will be briefly touched upon elsewhere in this book, but let us not be diverted in this direction here, for now we are concerned with glaciation and the particular terrain of which the mini-wood forms a fragment.

The glacier's movements and its enormous pressure cracked and ground rocks to pieces and to powder, and also left behind thousands of less altered examples in its retreat. A short distance to the west of the author's property lies a swamp, and to the south of this bit of wetland, an outwash area of gravels, laid down by the melting ice, is now mostly hidden beneath numerous modern homes. Half-buried and fully exposed boulders of many sizes are perhaps our best reminders of the Ice Age in New England (see plate 2), and surely the most easily read evidence.

Carried in the ice and on its surface were vast quantities of lesser-abrading materials representing the churning turmoil of the ages. Plunging from glacial vents all along the ice front came the finest stuff of all, to form beds of clay which, mixed with gravel and larger stones, laid down the strata we now call "hardpan." Deposited in the glacial debris were also occasional stone objects which the author likes to call "glacial eggs." These are fragments so smoothed and shaped by fine abrasive mineral matter in churning water that they do resemble large ova, even to their dull eggshell sheen (bottom photo, plate 3).

Glacial signs are everywhere in this part of New England, yet many pass them by without a second glance or thought. Ponds and swamps, rounded hills and rocky ridges, caves in tumbled rock, planed and gouged mineral surfaces, boulders and debris—all bear witness to the passage of mountains of ice forming the Pleistocene glacier known as the Wisconsin.

Such an enormous frozen mass, whose thickness no one knows, passed over all the land that is now Connecticut. It had come from the north, moving southwestward, and right here—at the site of the mini-wood—neared the end of its journey when changes in the climate spelled fini. Surprising as it may seem, the mini-wood soil, tediously accumulated over the centuries since, is now very rich in organic material, and in places is deep enough to have partly or wholly buried some of these glacially deposited rocks and boulders.

Many *cultivated* Connecticut soils analyze only 4 to 10 percent organic matter, but, as was ascertained by the loss-through-ignition analytical procedure, samples from the mini-wood tested as high as 21.1 percent by comparison. As was to be expected (a fact which will be better understood upon reading of the arboreal makeup of this bit of woodland), its soil tested very acid, or pH 4.5. Other tests, made at Storrs laboratory,* showed it to be low in calcium and higher in magnesium, phosphorus, and potassium. Color samples of the soil were dark gray to brownish gray (according, of course, to depth of soil and where the specimens were taken within the mini-wood), while the humus layer was generally jet black. All samples showed good water-holding capacity.

A forty-four-year record of rainfall at Norwalk, Connecticut, recorded at an elevation of 116 feet above sea

---

*Cooperative Extension Service, College of Agriculture and Natural Resources, the University of Connecticut, Storrs, Connecticut.

level, averaged 44.69 inches for the whole period, the highest precipitation being in 1934, when 58.06 inches were recorded. The average snowfall for the same period averaged 33.09 inches, with a high of 78.3 inches in the winter of 1914-1915.* This forty-four-year record for Norwalk is of particular interest when it is compared with the almost identical average rainfall of 44.85 inches over a fifteen-year period in the nineteen-fifties and -sixties recorded by the weather station of the Stamford Museum and Nature Center, situated only a few miles south of the mini-wood's location.

Severe winters, mild winters, hot and dry or hot and wet summers come and go in New England, as we who live here well know, but the average monthly temperatures for the forty-four-year period given above were anything but extreme: January, 27.4°; February, 26.6°; March, 36.3°; April, 49.9°; May, 58.5°; June, 67.2°; July, 72.2°; August, 69.8°; September, 63.4°; October, 52.3°; November, 49.8°; December, 30.1° (all readings of course being Farenheit).

Happily, where the mini-wood is situated, the temperature has seldom passed ninety degrees, for the locality in summer is frequently tempered and cooled by southwesterly breezes from Long Island Sound, but three miles away to the south.

Comparatively few people realize how many forms of small to minute animal life may be present in even an area as small as that under scrutiny in this book. We are referring in this instance particularly to the animals of the soil, organisms so diverse, so numerous, and so important in the ecology of every forest, no matter what its size, that three long chapters have been devoted to them in this volume.

Moisture loving, moisture *requiring*, the small fry in question dwell within and under the dead-leaf and debris blanket on the woodland floor, and in the humus layer just below it, wherein its *microclimates* in summer remain from eight to ten degrees cooler than in the more open woodland habitats above.

We will find that every niche in a forest has its inhabitants in one or many forms, into scores of whose lives and habits, whether they be plants or animals, we will forthwith pry. First, however, we must examine and know the *macro* vegetation of the mini-wood under observation, the trees and shrubs, in other words, of this association—an arboreal group which came into being in this particular climate and in the particular soil suitable to their special needs, and in which for a long time now most of them have been standing and growing. Many others have aged and died, as in any wood, but always leaving behind their offspring, either as young trees and saplings, or as seeds just sprouting.

As has already been stated, we know that all woodlands differ, some in almost all ways, while others may be very much the same, but the purpose of this particular mini-wood study has been, simply, to point out how very rich in organisms *any* woodland area may be.

Although the arboreal aspects of the area under examination have been but briefly described in this chapter, they should serve to familiarize the reader with the twenty-five species of trees and shrubs which constituted this part of the mini-wood proper, that is to say, the trees and shrubs which through many long years became established in this particular bit of land which was no doubt once a part of a much more extensive forest. These were the species tolerant of one another in root competition for the available nutrients and the moisture in the humus and soil which they themselves helped to build—the species tolerant of the varying amounts of shade and sunlight also regulated by the members of the association themselves.

Although but briefly treated here, many a species will enter the picture again in various interrelationships with other plants or with animal organisms, great numbers of which also dwelt in the mini-wood and which are covered in later chapters. But first we must look further into the botanical makeup of the community, observe its climbing vines, its wild flowers and ferns, its few mosses and lichens, its astonishing macro fungi, and some plants that had tried for membership in the mini-wood but failed to attain it.

Doubtless for the most part and in many ways a botanical group of long standing, as it was found to be during the fifteen years of this study, the tight little association turned thumbs down on two newcomers during this period, and doubtless had done the same with others beforehand.

Seeds of many other species must have blown into the mini-wood over the years or have been deposited there in the excrement of birds or as acorns of some species of oaks buried therein by squirrels. Some of these undoubtedly sprouted and found temporary footing, living for a time on food still available in their seeds, only to succumb soon to uncongenial company, soil, and general habitat. As in an exclusive, snobbish club, the mini-wood's long-established

---

*State of Connecticut. State Geological and Natural History Survey, Bulletin no. 61: *The Weather and Climate of Connecticut*, 1939.

founders managed to maintain the purity of its membership, only on the rarest occasions accepting new applicants as additions to the exclusive roll.

Let us never cease to revere trees and all other green plants upon whose chlorophyll all life directly or indirectly depends for food. Nor let us ever forget that oxygen, upon which organisms also depend for existence, also comes from green plants, probably the only possible source capable of maintaining the fabulous supply of the element necessary to life on this planet.

Keep these two main facts in mind. Give them first priority and thought in depth before ever again unnecessarily cutting down a tree.

When a mature tree succumbs naturally, even if to the gusts of a gale or a hurricane, it is a sad day for its owner and its habitat, but, having no control over such events, one may find solace and compensation, if one is inclined toward the study of nature, in a woodpile of logs obtained from such a tree, for if stacked and left *uncovered* in the woods or on its border, over the years it will yield many intensely interesting things, as we shall see in many parts of this book.

Let us now examine the mini-wood trees and shrubs—the major arboreal species—some briefly, others in more detail.

Dominating this fragment of forested land which the author affectionately calls the mini-wood were the stately black oaks, *Quercus velutina*, sixty- to eighty-foot examples that had stood on the gently sloping upper portion of the area through all of the summer and winter storms, and even two or three hurricanes, for up to, and, in one case, well over a hundred years. Estimating this tree's age by comparing its trunk with a thirty-six-inch cross section of a black oak blown down in the hurricane of 1938 (and on view at the Bruce Museum in Greenwich, Connecticut), a tree that was 130 years old when felled, the specimen in the mini-wood must have sprouted in the horse-and-buggy era, about 1821, long before the first steam locomotive puffed its way into the village of Norwalk. (See plates 1, 5, 6, and 7.)

The bark of this species is rough, over an inch thick in a mature tree, and almost black; the inner portion, however, is more or less orange in color, and rich in tannin. The leaves measure five to six inches in length, have rounded deep indentations in most examples, are dark glossy green on the upper surface, and *bristle-tipped*, and have long *yellow* stems.

The pollen-bearing catkins (see plate 15, top) and also the pistillate ones which produce the acorns are borne on the same tree, the former in such profusion that fertilization of the flowers is easily accomplished by wind, and indeed, so much pollen is shed that the ground beneath a big tree is sometimes turned yellow. In May, no species of tree in the mini-wood was more popular with migrant Warblers, Vireos and Orioles, and several other species of summer visitors and resident birds, all of which fed mostly among the trees' catkins, removing from them large numbers of minute caterpillars and eating scores of others from the young leaves.

It should be stated here that the half-acre area that became the subject of this study originally included a small strip of adjoining woodland north of the author's property line. But a few yards in width, this strip included one very tall tulip-tree and several mature beeches of particular beauty. Now on the north side of a high anchor-post fence erected by the author to protect the property from intruders from a public road on the east, the strip always remained as a part of the half-acre investigated.

Occasional large limbs blown down by various storms over the years of this study (some from near the tops of these tall oaks), showed that the acorns of the species varied considerably in size from tree to tree and from acorn-producing year to year (bottom photograph, plate 15). After the caterpillar (elm looper) explosion of 1970,* the acorns were unusually small. It seemed also that the black oak, like some other species in the mini-wood, produced a crop of mast only every other year.

The largest black oak in the mini-wood grew up with a large portion of its butt surrounding and growing over a considerable area of a huge glacial boulder. This did not seem to inhibit the tree in any way, nor did seepage of sap ever occur at the junction of rock and wood, as it sometimes did in the case of white oaks overgrowing rocks in the woods and apparently due to irritation caused by the slight motion of the butts on the rough mineral matter during heavy winds.

Four other species of oaks, some of which no doubt owed their existence in or on the borders of the mini-wood to long-dead gray squirrels that had buried their acorns for winter use (but never found them again) were: the white oak, *Quercus alba;* the swamp white oak, *Quercus bicolor;* the northern red oak, *Quercus borealis;* and the densely

---

*See chapter XIV.

twigged pin oak, *Quercus palustris.* Thus, with the majestic black species just described as the dominant tree, the little mini-wood boasted five splendid species of this genus.

The acorn-burying habit of the gray squirrel is very advantageous to both trees and mammals. Each squirrel of the local population where there are oaks unwittingly helps in the winter maintenance of the others; that is to say, it buries so many acorns each fall that "everybody" recovers a share of the collective hoard during the cold months, with each squirrel having the privilege of sniffing out, digging up, and consuming any acorn that it may locate, regardless of "who" may have taken the trouble to bury it. All of the squirrels play by the same fair rule of chance. One con-sequence of this natural arrangement is that enough unre-covered acorns remain safely in the ground, thus ensuring the sprouting of many oak seedlings each spring. Young oaks from a few inches to a foot in height were frequent in the mini-wood; many, of course, sprang up even where squirrels could not have buried the acorns.

The white oaks of the area grew very tall and slender and had narrow canopies, due to their having sprouted too close to other trees of their kind. By mid-May their pollen-bearing catkins dangled from the twigs in long clusters, with their less obvious pistillate or female flowers appearing in somewhat smaller catkins (plates 8 and 9). The bright yellow pollen was shed in enormous quantities some years, and the species is doubtless chiefly wind-pollinated. Spent catkins that dropped from the twigs high above soon be-came dry and brown and were then sometimes used for nesting material by the Crested Flycatcher, the Tufted Titmouse, and probably too by other small birds such as the House Sparrow.

Superficially, the catkins of the five species of oaks in the mini-wood area were very similar in appearance, but when magnified, differences were visible as will be seen by examining the photograph in plate 15. The acorns of these oaks differed greatly however, both in shape and size, and in the form and texture of the cups, and they varied in size from year to year. (See plate 15.)

The single pin oak within the mini-wood, with its roots penetrating one of the wettest portions of the area, had sprouted and grown tall. Nearby, a tiny natural pool was much beloved by birds of many species, for here, almost hidden in spring and summer, was a dependable bathing and drinking spot with cool, shallow water which dried up for a time only during the hottest summers. Surrounded by vigorous skunk cabbages and here and there a jack-in-the-pulpit, it was a favorite haunt of the Maryland Yellow-throat when that was present in the spring, and once a pair nested not far from the pool.

This pin oak, a rather straggly specimen, was tall and irregular, with its myriad twigs curiously involved, and its every leaf the species' autograph (see plate 10). Staminate and pistillate catkins appeared on new growth of wood, but seldom on the lowest, down-growing branches of this monoecious species. The brown, tassellike staminate cat-kins occurred in early spring, and the young leaves ap-peared in May. Where these new leaflets joined the main stems, the somewhat smaller pistillate flowers appeared. The species' *paired* acorns require two years to reach matu-rity, and as can be seen in the illustration (plate 15), were small and squatty and held in narrow-rimmed little cups.

One who studies the various oak species closely will find that there is much variation in the size and other physi-cal aspects of their acorns, sometimes even among those from individual trees, as well from crop to crop. In years when caterpillar explosions caused serious defoliation, some of the acorns were found to be much stunted.

Where they were more or less isolated on the mini-wood borders and had room to expand, specimens of the swamp white oak grew to sixty or seventy feet, with wide-spreading branches and foliage of great beauty (plate 11).

Flowering from the middle to about the end of May, the staminate and pistillate flowers appeared on the same trees (plate 12), the male catkins being three or four inches long at the pollen-shedding stage, and producing the yellow dust in great quantities. The pistillate catkins were more furry than these others, and the acorns in their wide-rimmed cups—and sometimes in pairs—were produced on distinct stems rather than directly against the twigs.

A northern red oak, *Q. borealis,* grew on the woodland border. It was a baffling "individual" for the reason that it produced leaves varying greatly in their contours and in depth and form of the lobes, some being scarcely lobed at all—more like leaves of the swamp white oak—and some leaves bristle-tipped and some not. If there is such a thing as a hybrid oak, this indeed might be it! (See plates 13 and 14, photo at right.) This tree held its leaves, which all turned a dull brown, until almost the end of November, when they all came down within three or four days. It flowered around the latter part of May, with some spent catkins already falling by the twenty-fifth of the month. Its ovoid, rather sharply pointed acorns were green at first, light brown when dry, and held at the ends of short stems (plate 15).

Every two or three years the leaves of this oak were subject to the attack of a tiny gall wasp, *Neuroterus pernotus*, which deposited its eggs in the leaf tissues. The resulting galls, often several to a leaf, were soft rounded growths on the leaf upper surfaces, about a half-inch in diameter, and resembled yellowish wool. The material was mysteriously produced by the plant's cells under the irritations set up by the feeding grubs of this gall wasp.

The American beech trees, *Fagus grandifolia*, splendid in full leaf in spring and summer, lent an even greater touch of beauty to the area in winter, what with their massive gray trunks and fully exposed silvery gray and often oddly bent limbs and branches (plate 16). One aging beech measured seven feet nine inches in circumference, and thirty-five inches in diameter at waist level. Years before the author acquired the land, this tree had been disfigured by people's initials crudely and deeply cut into the bark and cambium or outer growing layer of wood, which, as this book was being written, the old tree had not yet been able to obliterate.

Doubtless well over one hundred years of age, some of its large limbs were gone or already stubs in which woodpeckers had drilled their nesting holes; and from time to time, during windstorms, fragments had been heard falling from it. Ice, gales, and hurricanes had taken their toll over the years, and the center of the aging tree was hollow for many feet upward from a large, roughly triangular opening at the base, a cavity possibly initiated by lightning which had left a long scar on the trunk years ago. (Top photos, plates 17 and 19. The canopy of a mature beech is shown in plate 20.)

Whatever the future might bring to this parent tree, its chromosomes and genes would survive in the form of fifty or more shoots and saplings arisen from the old tree's shallower roots. Some of the saplings had reached twenty feet in height as this was being written, and one day one or two of these will survive the competition when the old beech finally dies in the midst of its family (plate 19, top picture).

When freshly expanded in May, the bright green glossy beech leaves are especially beautiful as the sun shines through them, casting an ethereal glow on other leaves and even on the ocher and sepia leaf carpet below (see plate 35, top picture), already alluded to.

This species is *monoecious*, which means it has staminate and pistillate flowers (or male and female organs) on the same tree. The male flower "balls," well under an inch in diameter, are produced on long stems (plate 18), while the pistillate or female blossoms appear in pairs, borne on hairy, shorter stems.

Toward the end of summer the prickly little burs of the beech split open, revealing the two or three triangular, glossy, sweetish but extremely hard nuts within (plate 18). These were relished by gray squirrels, Crows, and Blue Jays, the latter birds gobbling them up whole despite their hardness and, one would imagine, tastelessness in this form.

As a rule, the mini-wood beeches suffered hardly at all from foliage-eating insects, their leaves usually remaining whole throughout the summers, but during the great elm looper caterpillar explosion of 1970, old trees and their saplings suffered great damage. Many of all ages were completely denuded, but the trees recovered remarkably well. While very few new leaves were put forth that summer, they succeeded in forming strong new buds for the following spring, and in 1971 the foliage was normal once more.

Rivaling some of the tallest oaks in the mini-wood were a few sassafras trees, *Sassafras albidum*. The largest of these, some sixty-five feet in height, measured four feet six inches in circumference, and twenty-six inches in diameter at waist level.

Where these trees unfolded their interestingly colored yellow-green new wood and highest twigs—and there produced their heaviest foliage—rather dense canopies sometimes resulted, especially where two trees had grown up close together (plate 23).

Sometimes blossoming as early as April seventh, but normally by the tenth or fifteenth of that month, the staminate and pistillate flowers occurred on different trees. The former were borne at the ends of stout yellowish green twigs on short stems, above the first unfolding leaves. In the mini-wood, as many as seven flowers made up these pollen-bearing groups, each flower having six petals and nine stamens, and all about the same color. Fully expanded, they measured around a third of an inch in width, and the small new leaves opening just below or between the flower clusters quickly expanded to half their mature length before the last spent blossoms had dropped to the ground (plate 21).

The fruit, which ripened in late September or early October, was obling, dark blue, and berrylike, and about a half an inch in length. Borne singly at the ends of slender stems, they dropped off, leaving the stems behind, still attached to the crimson calyxes of the flowers. These remained on the trees long after the dropped fruits had disintegrated or had been eaten. Some woodland birds and mammals probably also swallowed the single, pointed seeds; by passing these pits in their excrement, they helped in the distribution of the sassafras trees.

Unique, and identifying this species of tree even as very young specimens, are the yellowish green leaves of three different shapes, which may occur even on a single branch. The most distinctive of these are old-fashioned mittenlike in outline, with the "thumb" plain to be seen, while the other leaves may be plain ovate in shape, or three-lobed. Thus we have in the sassafras another tree whose leaves are its autographs (plate 22).

Not everyone who owns a New England woodlot is aware of the possibility that large sassafras trees and their saplings may well be present therein in numbers, some of them forming part of the forest canopy along with oaks and beeches, maples, white ash, and other mature species. As the young trees (of a size whose top branches a man can still reach) do not blossom, it is even less likely that the average mini-wood proprietor has ever seen the sassafras flowers, most of which open early in the spring high above the ground, and most profusely in the canopy (plate 21).

Crumble the sassafras leaves or bruise the twigs, and a delicious aromatic odor is emitted. As children we often sought the saplings in the woods and chewed their green-colored twigs for their pleasant sarsaparillalike flavor. It might be that this odd organic secretion of the plant—essential oil, or whatever it may be chemically—acts as a repellent agent against insects, for this tree is notably free from caterpillar attacks by most species, even when other trees are being nearly denuded. During the devastating ravages of the elm looper caterpillars described in chapter XIV, sassafras trees in the mini-wood remained unharmed throughout the two seasons of this caterpillar explosion.

In times gone by, sassafras had many uses. There was a spring tonic called "sassafras tea." Oil distilled from the roots went into perfumes and patent medicines, and it is still used in soaps.

It is a persistent species. If a sassafras is blown down or uprooted, a family of new sprouts may arise from its roots in a single season, due to the food released by the parent and to the increase of available sunlight. Forty such shoots followed the removal of a mini-wood parent tree, which very suddenly left a space wide open to the sky. The sprouts shot up *five and six feet* within a single year!

Heart rot is a serious enemy of the sassafras. Resulting from the intrusion of a fungus and bacterial action, it may not be apparent for years, but then, perhaps even in a moderate wind, an exterior crack may occur, sometimes several feet above the ground. In later winds the trunk may give way and suddenly break off, its demise, however, giving almost immediate birth to offshoots as mentioned above.

Such was the *audible* fate of a forty-five-year-old sassafras tree in the author's mini-wood one windy April night—a tree in which a horizontal crack had already been observed. Opened up, it was found that where once had been sound heart wood, now there was only soft dark brown and black material resembling wet sawdust. In the laboratory, the microscope revealed this rot to be well populated with minute animal consumers. How they got into this dark decaying world of the tree's interior was of course obvious yet no less surprising. Among these creatures were glossy, dark brown, eight-legged mites, and larger species with long legs and oval yellowish bodies, while crawling all through the stuff were scores of ultra-minute animals called *collembolans* or *springtails*. Full details about these various arthropods will be found with illustrations in later chapters, along with many others.

Later, from some of this black rot enclosed in an observation box, came a strange little plant. Sprouting after a few days, it grew with vigor, anchoring its long, pure white stem in the highly acid goo with hundreds of silvery root hairs. At its top, the yellow beanlike seed from which it had sprouted remained straddled, but no first leaves appeared. Becoming too dry, it withered and died, like some animal forms when unable to escape from their egg or cocoon, but the little plant helped to confirm the assertion that nothing or next to nothing goes to waste in nature.* Here was a seed that had somehow become lodged in the heart rot of a tree, was buried therein for who may say how long but which, upon release and under the benign influ-

---

*Some scientists dispute this concept, pointing to the beds of peat and coal as negating examples, an absurd contention, however, what with the animal, man, being a greedy consumer of these natural products of ancient plant life.

ence of light, was able to germinate and grow, only to be frustrated by a straddled seed.

The American basswood, *Tilia americana*, was present in the mini-wood in small size, the tallest had reached about thirty-five feet as this was being written. The seeds of this species, few of which seem to become naturally planted, produced very slow growing saplings. This tree, often known as the linden, is shown in plate 26, in the left photo. Its heart-shaped leaves measured five to six inches in length, and from four to five in width. By mid-July the clusters of blossoms, some white, but mostly pale yellow-ocher (plate 26) opened below a single long slender stem, each of which was attached to a leaflike bract above it (plate 27). The round, extremely hard seeds which follow blossoming remained on the trees until late summer or early fall.

Basswood leaves were popular with many consumer insects. During some springs, hordes of cankerworms (caterpillars) perforated most of the foliage and were often eaten themselves by the birds; elm loopers ravaged the trees in other seasons; and, periodically, real crowds of the larvae of the pretty little moth *Archips argyrespilus* and other larger caterpillars such as one fat green species with a horn on its tail end (which, if it escaped bird predators, later became a sphinx or hummingbird moth) fed on basswood leaves. But in spite of all these and still other avid foliage consumers, the basswoods bloomed on and on, and produced their buckshotlike seeds year after year.

And what a perfume is that which emanates from these delicately formed and tinted blossoms! Broadcast in July, some years as early as the fourth of the mouth, it velvets the nights with an indescribably subtle yet alluring scent, an odor which blends well with moonlight nights, bringing to at least one naturalist an immediate sensation of delight at being alive, able to enjoy nature's endless offerings all around one.

Some insect species seemed equally affected. At night came little female moths to partake of basswood nectar or to linger and mate with arriving males. By day came bees and wasps of a dozen species, likewise to sip nectar or gather the pollen, cross-pollinating the flowers.

Thrips—minute insects most gardeners hate—also gathered at the basswood blossoms. Clinging to the petals by their hind legs, they eagerly stretched far forward, their four tiny wings expanded and rapidly vibrating, thus flash-ing signals of their whereabouts and mating readiness in the ambient perfume of the flowers.

Years ago, these trees were lumbered for carriage wheels and bodies, and for neat boxes and plaques in which designs were burned with a red-hot needle when the pyrography fad was at its height. In the open, unencumbered by other trees, basswoods grow into massive specimens, and crowds of visiting honeybees may then be heard at some distance from the lavishly blossoming trees. Basswood honey, by the way, was always one of the best to be had.

The red maple, *Acer rubrum*, and the sugar maple, *Acer saccharum*, were represented in the mini-wood by numerous examples (plate 28). While the former preferred the wetter portions of the area, it also seemed to do as well at times in drier situations, in which most of the sugar maples were located in the dense shade of other, larger species and under the influence of whose shade they grew very slowly.

A. rubrum put forth its rich red flower clusters in March and early April (plate 29), before the leaves expanded. Staminate and pistillate blossoms, either on the same tree, or segregated, appeared to be usual for this species. Being most easily wind fertilized, but also fertilized through the agency of early honeybees, the winged seeds or "keys" were produced some years in truly fabulous numbers (plate 30), and soon after these seeds were wind-distributed, tiny seedlings could be seen in paths and driveways, in the woodland and on its borders, and even in the gutters of buildings or other unsuitable places, but here most of these latter soon died.

The red maples add greatly to the fall scene. Brilliant scarlet, yellow, and orange colors occur in the leaves in endless combinations and, due to the crystalline pigment carotene, become visible after the fading of the green chlorophyll.

The sugar maple—the species from whose sap the syrup and sugar are boiled down—was present in the mini-wood only as moderately tall trees of the understory, and, as already stated, were all slow growing in the deep shade cast by the beeches and oaks, as the dominant species.

The flowers of the sugar maples appeared in late April or early May, as the leaf buds were opening. Drooping at the ends of long pedicles, both the staminate and pistillate blossoms occurred on the same tree. The winged seeds, produced in pairs and in large numbers in some years, or few or none in other years, were similar to those of the red maple, but a close observer would find some differences by comparing them when they were newly formed. The leaves of these trees, however, could never be mistaken for those of the red maple. Compare the leaves of the sugar maple in plate 31 with those of the red maple in plate 30.

Little brown ants (genus *Lasius*) sometimes placed herds of aphids on the sugar maple leaves. As we know, upon imbibing sap, these tiny bugs secrete a very sweet sugar substance which is eliminated in their excretia. To feed heavily upon this sugar, as we also know, is the sole reason why ants of so many species take such tender care of these insects. It occurred to the author, when thus observing the *Lasius* ants and their aphid flock, that possibly, when feeding on sugar maple sap, the latter may exude an even sweeter and more desirable product which this species of ant possibly has long "known" about and learned to appreciate (see plate 137). In one instance, *Lasius* ants were found tending a herd of aphids late in November on a sugar maple leaf that was still green.

The young of many trees produce what might seem to be abnormally large leaves, but this is nothing extraordinary. Sugar maple saplings in the mini-wood, for instance, produced some with long wine-colored stems and leaves alone ten inches in length and seven inches in width, whereas those on the older trees were much smaller, and shorter than wide, or about the same in both length and width. Oversize leaves found in the saplings of many species are probably the result of comparatively few leaves receiving all of the nutrients which in older specimens must be spread among hundreds or thousands of individual leaves.

In the fall, the foliage of mature sugar maples may be as colorful as that of the red maples, yet within the mini-wood, this was never the case, all leaves turning a solid bright yellow.

Two splendid tulip-trees, *Liriodendron tulipifera*, seventy or eighty feet tall, matured on the mini-wood borders, doubtless having been well *within* more extensive forest when young. A third, even taller specimen, just beyond the mini-wood's north border, was already taller than the mature beeches when this was written. (Plate 32.)

Tulip-tree leaves are distinctive, being squarish, with four or more points, and oddly notched, clearly showing leaf form on young specimens (plate 33, bottom photo).

Glossy and beautifully green when young, in times when aphids are overabundant, the insects' sugary exudations on these leaves support a smothering *black* fungus or mold which soon coats their entire surfaces.

No flowers of the woodlands are more beautiful than that of this tree of the magnolia family. They open about the middle of June. From three to four inches in width, the large pale green and orange petals cradle the fat pistil and a whorl of stoutly formed stamens (plate 34).

Tulip-tree flowers, in the author's opinion, seem as if they might be misplaced water lilies taking to an aerial form of existence! Although they are produced in large numbers some years, how many of us have stopped to secure one and hold it in the hand, wondering, perhaps a little sadly (while conscious, of course, that reproduction is the blossom's purpose), why such exquisite creations must so *quickly* wither in favor of the seed?

Following the fading of the flowers come conical fruits, each consisting of large numbers of closely packed, single-winged seeds (plate 35). These remain on the trees until November, when, at varying dates, they are set free to be wind distributed or, on calm days to lodge in great numbers on the ground beneath or not far from the parent tree. Cardinals, Towhees, White-throated Sparrows and House Sparrows are among the resident and transient birds that have learned to extract and eat the single small hard seeds. A group of these seeds is shown, actual size, in plate 79.

As with the sugar maple, the tulip-tree's leaves turn a solid yellow in the fall, and it is at this time of year that twigs bearing the seed "cones" are often blown down.

Known also as "whitewood," lumber produced from the tulip-tree has many commercial uses, an important one being for the core of plywood veneer.

A more numerous species in the mini-wood and about its borders was the white ash, *Fraxinus americana*. As can be seen in plate 37, the leaves of this tree, which usually grow seven to nine in a group, arise from a stalk or twig which, in the autumn, after turning yellow, itself falls to the ground not long after the leaves. At such times these tough, conspicuous, but spent stalks may be found in early October littering the ground beneath the trees. Upon abscission (separation from twigs and stems by disorganization of a natural separation layer), the leaves leave deep notches on the twigs from which they have fallen, which could be easily matched with the stem ends of the leaves.

Measuring from three to six inches in length and from two to three inches in width, white ash leaves have wavy edges or borders (plate 37).

The big buds open about April 15 and the leaves expand rapidly into the first week in May, but the staminate and pistillate flowers (on different trees) precede this young foliage, the dark reddish purple male blossoms appearing in tufts close in by the twigs. The pistillate flowers produce groups of drooping single-seeded wings up to two inches in length which are wind distributed or sometimes gathered and eaten by deer mice and, less often, by birds, including the Cardinal, Towhee, White-throated Sparrow, and Fox Sparrow.

A white ash which reached fifty feet in height in a very wet situation (sometimes remaining flooded for days after heavy rains) produced stout buttresses or upgrowths from the roots. Whether or not buttresses are "anchors" developed for extra support is a debatable question, but during a heavy windstorm which toppled a mature beech tree on the edge of this habitat, the buttressed ash stood firm throughout the strongest gusts. (Butts and buttresses of trees in the mini-wood are illustrated in plates 24 and 25.)

Unlike fallen leaves of the oaks, the beech, and some others, those of the ash are soft and easily decomposed, usually disappearing completely within a single year, much of their substance becoming food for earthworms and many other small animals of the soil army as described in detail in chapters X, XI, and XII.

Ash is one of the trees which is able to make a rather rapid recovery after serious or even complete denudation in years of zenith caterpillar infestations. In a number of cases in the author's woodland, where the trees had been eaten bare of leaves at the end of May, they were putting forth new foliage by June 21, and by the middle of July had actually reclothed themselves very respectably that same year. Although the new leaves never seem to attain their normal mature dimensions under such circumstances, they are able to manufacture sufficient food to tide the trees over until the following spring.

Still another species represented in the mini-wood by only one mature specimen was the pignut hickory, *Carya glabra,* but there were probably other specimens in woodlands nearby.

Unlike the hickory which bears edible and delicious nuts, the pignut hickory's bark is comparatively smooth rather than in shaggy slabs, as in the former tree, which

have given it its common name of "shagbark." The leaf arrangement and the shape of the leaflets, however, is about the same in both species. (Plate 38.)*

The staminate flowers open in drooping catkins in May, and the insignificant-looking pistillate blossoms some distance from these pollen bearers, but on the same tree. The specimen in question, which became planted in very wet soil where skunk cabbages grew luxuriantly beneath it, produced nuts but rarely, and then they were few in number. They formed within fig-shaped pods or fruits, sometimes in pairs. When ripe, these split along four well-marked sutures, revealing their brownish contents. The nuts sometimes dropped out before the pods themselves fell, much as our native sweet chestnuts fell from their spiny burs after the first heavy frosts.

Undistinguished in other ways, this pignut did once prove attractive to a pair of Scarlet Tanagers, whose nest of twigs and tendrils was constructed among the foliage on an almost horizontal branch twenty-five feet above the ground.

A tree represented by two mature specimens with trunks nine and seven inches in diameter at waist level, and a family of ten saplings from three to fifteen feet in height, was the tupelo, *Nyssa aquatica,* otherwise known as the "pepperidge" (plate 39). These two specimens grew up among white oaks, red maples, and other rather heavily foliaged species. One of the tupelos on the woodland border produced many almost horizontal branches that stuck straight out, away from the dense summer shade at its back, as illustrated in the plate; it shows the effects of crowding. Not shown, however, are the bunched-up canopies of the two mature trees squeezed between these species. Male and female flowers occur on separate trees in this species, and as the mature specimens were both staminate individuals, the saplings were doubtless sprouts or offshoots from the adults' roots.

Tupelo leaves are arranged in clusters of four or five, with the many staminate flowers growing in dense little greenish white clusters more or less centered, as shown, life-size, in plate 39. The larger pistillate or female flowers are borne on longer stems, and grow singly. The fruits, which occur in drooping clusters, look like elongated plums, and are said to ripen in September.

------

*The leaflets of the pignut are often very much larger than those of the white ash, for which it is sometimes mistaken. Giant leaves occur on sapling pignuts.

Here, the author wishes to urge all who are nature students or who are simply interested in our woodland trees to make their own investigations as well as studying books, for pleasing surprises may be in store, and things may be found at home that may not perchance be found in the book or nature guide at hand. In certain of these works, the reader is advised to take special notice of the *ranges* given, thereby possibly eliminating a number of species during one's efforts at identification.

But this is not always good policy. In the case of the tupelo, for instance, in two well-known tree books, its range is given in one as: "swamps of the southern coastal plain, se. Virginia to Florida and e. Texas, and north in the Mississippi Valley to s. Indiana, s. Illinois, and se. Missouri"; in the other book the range is similarly written as: "the coast region from southeastern Virginia to northern Florida, and through the Gulf States to Texas."

It is a long trek from these areas to southern Connecticut, yet here is the tupelo, so to speak, in the author's own backyard! There is still another mature tupelo tree just beyond the mini-wood's borders, while on the edges of what was once a large swamp, a short distance away, there were, even quite recently, other tupelos growing, all of these trees with their roots in wet soil.

From the human point of view, the tree is of interest for the reason that famous breads and other baked foods are prepared (not too far from the author's habitat) on land that was once a farm, and where still grows a "pepperidge" specimen much older and much larger than those of the mini-wood, and from which the old farm and these now universally known foods took their name. Few people seem to know that the tupelo grows in Connecticut as well as in the southern states, and most abundantly in large swampy areas.

Standing just outside of the mini-wood there was also a single ironwood tree, *Carpinus caroliniana*, a specimen twenty feet high, with almost bluish gray to dark gray curiously smooth bark. Being tolerant of heavy shade, it doubtless could grow very well within the mini-wood proper; indeed, when it became established some years ago, it was, in all probability, in denser shade.

In the spring, the curious pollen-bearing catkins, about an inch and a quarter in length, form compactly, while the pistillate, semi-erect female flowers occur at the ends of the twigs. (Plates 40 and 41 [top two pictures].)

The fruit is a very small brown nut borne at the end of leaflike bracts, many of which are crowded together on stems five to six inches in length. These turn brown and remain on the trees until late in the fall (plates 40 and 41).

Wild chokecherry trees, *Prunus virginiana*, were scattered through the mini-wood and on its borders. Probably mostly planted by birds, they sprout readily from excreted pits. Woodland deer mice also unwittingly plant other viable cherry pits which occasionally work through and drop out of the nests of shredded bark and other loose material in which the seeds have been hoarded.

The oval, dark green, rather sharply pointed leaves of this cherry often hang down from the twigs. Toward the end of May, drooping clusters, consisting of dozens of small flowers with five white petals, reach their full beauty. Such a *raceme* or flowering arrangement, in which the individual blossoms are borne on short stalks along an axis which continues to grow and on which flowers continue to open, may reach five or six inches in length (plate 42).

From these flower clusters come a peculiar and characteristic scent unlike that which emanates from any other *inflorescence* (an arrangement of flowers on an axis) known to the author except other species of cherries. The odor is really indescribable, and might be said to border between pleasant and unpleasant, from the human point of view, but evidently alluring enough to many winged insects.

Birds do not always wait for the cherries which result from these crowded blossoms to ripen into juicy black berries, often plucking them when they are still partly green (photo D, plate 42).

Most of us who have lived in the country have ourselves made or at least have imbided a glass or two of delectable "cherry bounce" perhaps made by a good friend. This is wine prepared from a mix of purple wild cherry pulp and sugar, and named appropriately for its potency.

The wild cherry has long been favored as a food plant by many kinds of insects, the tent caterpillar in particular, but it is a tree with an extraordinary ability to reclothe itself with new foliage even though its first suit of young leaves may have been completely consumed. This whole interesting story of the chokecherry–tent caterpillar association will be found in Chapter XIV. (See also plate 41.)

Around the third week in April, within the area under examination and in many areas of New England's woodlands as well, isolated patches of yellow haze reveal the whereabouts of the spicebush, *Lindera benzoin*, blooming among the other low-growing, later-flowering shrubs.

With the sun climbing higher each day, the blooming of the species signals the surging of the forest sap. At the same time, such sights beget a fresh surge of the naturalist's blood, bringing an inimitable sensation of vernal uplift after the drabness of long winter days which make one cognizant anew and thankful for this subtle interrelationship among sunlight, plants, and man.

Opening close to the twigs and branches, the delicate yellow floral groups, their stamens rich with pollen, would be inconspicuous in the woods except for their numbers, which collectively produce this yellow haze in the April woodlands (plate 43).

If either the leaves or twigs of this species are broken or bruised, an odor at once pleasant to the human senses is emitted. As children, we often chewed the twigs for their delightful flavor, as we likewise munched on those of the sassafras saplings.

Gum benzoin, a fragrant substance obtained from a tree native to Nubia and Somaliland and named *Boswellia carteri,* is used in the preparation of frankincense. Someone must have compared the odor of the secretion from the spicebush's leaves and twigs and roots favorably with frankincense and the exotic tree, hence the term *benzoin* in the Latin binomial of the American shrub!

While very pleasant to the senses of a human being, it is possible that the spicebush is protected from some foliage consumers of the woodland by its odoriferous secretion. As in the case of sassafras leaves, the author's experience was that those of the spicebush in the mini-wood were scarcely touched by caterpillars, only an occasional larva or two of the spicebush swallowtail butterfly, *Papilio troilus,* being found on them, and, formerly, occasional caterpillars of the moth *Callosamia promethea,* which, however, much preferred the wild cherry, on whose twigs the prometheas' silken cocoons, wrapped in a dried leaf, could often be seen dangling all winter.

Why this fine big moth and the still larger *cecropia, polyphemus, regalis,* and *imperialis* moths disappeared from the author's general locality, and from many other parts of the country as well, along with the truly beautiful pale green, long-tailed *luna* moth, we do not know for certain, but very probably DDT had much to do with their decline, as their big fat caterpillars were not very numerous at any time. Their loss is one of the sad developments within many ecosystems. (See chapter XIII and plate 131.)

* * * * *

The mini-wood soil proved suitable for the native red mulberry, *Morus rubra.* One tree (twenty-five feet in height at this writing) and another one somewhat smaller, became established on the sloping upper portion of the area where the soil had a tendency to dry out rather rapidly after each rain.

Despite their small sizes, both bore fruit, but their actual ages cannot be stated. Mulberry leaves (in this species) were dark green above, lighter below, and had a slightly rough or "sandpapery" feel when gently rubbed from tip to abscission point. With toothed edges, these leaves also had sharp points called "drip tips," which may or may not exist for the "purpose" of draining the leaf surfaces during extended rainy periods. Most of the leaves were of moderate size, but individual examples were found that measured from nine to twelve inches in length, and from seven to nine inches in width! Broadly heart shaped, occasionally lobed, they might easily be mistaken by the inexperienced for leaves of the basswood.

In May, cylindrical catkins were produced on moderately long stems, the staminate flower clusters being somewhat longer than the pistillate examples. These were followed (the anthers having forcibly ejected their pollen) by inch-long mulberries, at first green but later turning red, then almost black when ripe, around July 15 (plate 44). Like blackberries, the fruits are made up of many smaller juicy units, each containing a small seed. In the mini-wood, these were sought by Catbirds, Thrushes (including the Robin), by Brown Thrashers and squirrels. Both trees in the mini-wood supported heavy growths of the common frost grape, *Vitis vulpina,* described and illustrated in chapter II.

In a few places just within or on the border of the mini-wood, the sweet pepperbushes, *Clethra alnifolia,* grew twelve to fifteen feet in height. These were shrubs with dark green, somewhat wedge-shaped leaves with finely toothed borders. By August 1, the densely clustered racemes of white blossoms were broadcasting their delicate and extraordinarily delicious perfume, a scent unlike any other known to the author and rivaled in allure, subtlety, and interest only by emanations from the honeysuckle, the native frost wild grape, and basswood flowers (plate 45).

The buds continued to open along the lower ends of the axes for many days, the racemes reaching their fully expanded beauty about the fifteenth of the month. Individual blossoms averaged eleven millimeters (seven-sixteenths of an inch) in width, their five petals and ten white stamens with brownish yellow anthers surrounding the tall pistil with its three-lobed stigma (plate 45).

Their perfume proved alluring to a great variety of bees, wasps, and flower flies, that gathered at the racemes for nectar in return for their pollinating services. Another visitor for nectar was the diminutive azure butterfly *Lycaenopsis argiolus pseudargiolus* form *neglecta*. With touches of violet in its wings (dark-bordered in many specimens), these little butterflies represented the second or summer brood, some of them also feeding later at the flowers of the white snakeroot within the mini-wood.

First revealed to the author during his observation of the blooming pepperbushes was a remarkable interrelationship between certain large solitary wasps and *milkweed flowers*. One such wasp was the powerful *Chlorion ichneumonea*, with an inch-long body, all black except for anterior yellowish orange portions of its abdomen and its legs. A strong flier with translucent brown wings, it was a frequent visitor at the pepperbush blooms (plate 47). The other wasp involved in this odd association was the all-black duskywinged *Sphex pennsylvanicus,* a larger insect. Both were solitary species, living alone, and digging and provisioning nests by themselves (a subject explained more fully in chapter XV).

As with all arthropod material taken by the author for this study, wasps were carefully examined microscopically. At once, odd-looking objects were observed (on both species) *impaled* upon and protruding from various spines on the insects' tarsi or lower leg joints. Each of the objects was attached to the spine by a black, bean-shaped swelling which in turn was joined by an angular stalk to a pair of translucent "wings" (plate 46). In some cases, two or more of these things were linked together, but always with the lowest or terminal beanlike object impaled on a tarsal or other spine on some one of the wasp's six legs. Highly magnified, these "wings" had a cellular appearance or pattern. They were definitely flexible, whereas the beanlike objects were hard and glossy.

At first these things were mistaken for animal organisms, but unlike anything the author had ever encountered before. They suggested parasites of some sort which by this means might reach the wasps' burrows for their stores, but somewhat later, the author, aided by botanists, learned that these odd black objects with their attached yellowish "wings" were the *pollinia* from the flowers of the familiar northeastern milkweed, *Asclepias syriaca* (plate 47).

The milkweed blossoms are numerous, fragrant, and individually long stemmed. In the cryptic language of technical botany, we might read in regard to them that in this form of pollen-bearing organs, "the lateral margins of each anther are cartilaginous, separated by a narrow slit, and usually project radially from the gymnostegium into a low triangle with its salient angle near the base. An insect visiting a flower may get its foot into the open basal end of the slit, draw it up to the distal end, and upon pulling it loose bring with it the united pair of pollen-masses suspended from the translator."*

In language more easily understood, let us say that, within each five-lobed blossom surrounding a central tube, are five cups, collectively constituting what is called the *carona* or *crown*. From each of these receptacles a curved horn arises, its point turned inward toward the tip of the pistil (or stigma). Concealed are five pairs of the flexible pollen-bearing "wings" attached to the slit oval black portions of the device (as illustrated in plate 46). Actually, these bean-shaped cleft objects are glands into which visiting insects, such as the two large wasps described, get their tarsal spines firmly caught, as shown in the micrographs. As these wasps take off, they are powerful enough to pull out and carry the pollen-bearing "wings" with them, whereas weaker insects die trapped on the flowers!

In view of the milkweed's alluring pink to purplish pink flowers and their delightful fragrance, such elaboration to ensure pollination in the very numerous individual blossoms would seem excessive—or even wasteful, in a measure—as even the few seedpods which eventually result on each thick-leaved rugged plant are more than sufficient to make the milkweed's continuance secure, each pod containing great numbers of tightly packed seeds, each with an attached silken parachutelike device by which they are carried far and wide by wind, once the pods dry and split open. These flowers may *kill* more insect visitors than escape bearing the pollinia!

Another shrub of the mini-wood area whose fruit was eaten by some birds (and their hard seeds by occasional chipmunks) was the arrowwood, *Viburnum recognitum,* an escaped (Asian?) species growing ten to fifteen feet in height.

------

*Henry A. Gleason, *The New Britton and Brown Illustrated Flora of the Northeastern United States and Adjacent Canada,* vol. 3 (Lancaster, Pa.: The Lancaster Press, published for The New York Botanical Garden, 1958), p. 74.

With rather coarsely toothed, ribbed leaves, and cymes of closely growing small white flowers forming what looked from a distance like solid white platforms, this shrub in bloom might easily have been mistaken for an elderberry, especially when both species were growing in the same wet habitat on the mini-wood border (plate 48).

Opening during the first week in June, these flowers were followed by drupes, or one-seeded berries. Very gradually, these became bluish black as they ripened during the first ten days of August, and while some birds liked them, to the author's taste they seemed astringent and acrid. The hard stones with five lengthwise shallow grooves were occasionally found in late summer excreted in the bird-baths (plate 53).

The common or sweet elder, *Sambucus canadensis*, just referred to, also produced what from a distance looked like solid white platforms when blooming, but which upon close inspection were seen to be compound cymes made up of numerous separate and tiny white flowers. Cymes and the sharply toothed leaves are shown in plate 49.

In late summer and well into September, the shrub's clusters of round shiny black berries were eagerly sought by Brown Thrashers and Catbirds, Robins and Swainson's Thrushes. They are even today gathered by some human beings who dwell in areas still sufficiently rural, and who still know what to do with both the fruits and the flowers, or "blows," to produce elderberry wine in all its potency.

In an area of the mini-wood's western edge where the soil remained wet for a good part of the spring and summer, a habitat where tall jewelweeds flourished, the highbush blueberry, *Vaccinium corymbosum*, also became naturally established (plate 50).

Its dark green untoothed slightly glossy green leaves had a tendency to broaden considerably with age, in some instances. About May 15, the white to pinkish white blossoms, like narrowly formed little bells, were produced some years in profusion.

The blue to black berries, however, seldom followed the flowers in equal numbers. Ripening in July and August, many of them bore a whitish bloom. At their upper surfaces, the remains of the calyx (petallike sepals of the flower) was retained, and as such may be seen, as it is on other familiar fruits, such as the apple.

The winterberry, *Ilex verticillata*, often called the "black elder," although it belongs to the holly family, grew sparingly about the mini-wood area, mostly in fairly wet portions of its soil, the shrubs reaching fifteen to twenty feet in height.

In its better years (and all shrubs and trees seem to have their off years where blooming is concerned), the winterberry produced great numbers of small staminate and pistillate flowers, singly or in clusters, along the twigs. With white petals, the stamens were so heavy with bright yellow pollen at times that from a distance the blossoms seemed to be yellow also. Pollen was so abundant that much of it fell off or was pushed off by bees or was seen adhering to the dark green leaves (plate 51). Fully opened, the flowers measured only about five millimeters (three-sixteenths of an inch) in width. Under these loads of yellow pollen, the six stamens surrounding the shorter green pistil were white. Many small species of bees, honeybees, and flower flies with yellow-marked bodies visited these flowers every day during their blooming period of a week or longer, yet, notwithstanding these crowds of pollinators, there were years when no berries resulted. At such times, the barren branches of these shrubs were particularly noticeable when pepperbushes were nearby upon which every flower seemed to have made seeds which remained on the shrubs all winter.

In good years, the copious bright red winterberries reached the size of jumbo peas. Within their yellow pulp, the yellowish seeds, three to five in number, were shaped like those of pole beans in miniature. Berries were ripe and red by mid-September. Normally, these remained on the twigs long after the leaves had fallen, in some seasons, while in others they were soon spotted by such hungry migrating birds as Swainson's Thrushes, which annually pass through Connecticut in large numbers, flying southward by night and feeding in our woodlands and thickets during the days, especially when berries are ripe. Late Catbirds and Robins, and possibly Wood Thrushes, may have fed on winterberries at times, but the most avid devourers of these rather tasteless fruits were the Cedar Waxwings, when present here. After the nesting season, Waxwings travel in small flocks, and when such a group of twenty-five or so discovered a well-loaded winterberry shrub, they simply settled down on the spot until every last fruit had been consumed.

As we so well know, many kinds of trees and other plants owe their wide distribution, at least in part, to the birds which swallow their seeds along with the fruit and then pass these seeds, unharmed, in their excrement.

Wild cherry trees growing in a line beneath telephone wires tell this story very plainly at times. One summer, in the author's garden, wild cherry and grape seedlings sprouted directly beneath the long parallel slats of a tomato trellis which had become favorite perches for Robins and Catbirds.

As mentioned before, seed dispersal by birds may be conveniently studied by frequently scrutinizing a bird-bath in late summer, when many wild berries are ripening. A shaded bath constructed on top of a basinlike glacial boulder proved particularly popular with many kinds of birds during the summers, and often late into the fall, when the Thrush migration was in full swing. Uncaring about the state of their drinking water, the happy birds imbibed, bathed, and, as the spirit moved them, eliminated, thus leaving a variety of ejected seeds and pits in the bath as a welcome aid to the author's investigations (plate 53).

Single specimens of the gray birch, *Betula populifolia,* and the black birch, *Betula lenta,* became seeded on the edge of the mini-wood, probably five or six years before this study was begun. Both trees, growing with only a little more than a yard between them, had reached about thirty feet in height as this book was being completed. The trunk of the gray birch then measured three and a half inches in diameter at waist level, and that of the black birch five and seven-eighths inches (plate 52).

The otherwise quite smooth bark of these two trees was notable for the hundreds of slightly raised lines or ridges of various lengths running parallel around the trunks. On the gray species these ridges were mostly less than an inch in length, with an eighth to a quarter of an inch separating the parallel measurements. On this black birch the ridges measured from five-eighths of an inch to an inch and a quarter in length, and were somewhat farther apart than in the gray species. Both trunks bore the familiar wide black *chevron* marks below the butts of the branches and scattered elsewhere on the trunks, features characteristic of these birches.

The slow growth of these two trees was doubtless due to a soil often remaining saturated for too long a time and probably not well suited in other respects, as further attested to by the fact that no other trees of these species appeared, although both made abundant catkins and seed.

Pollen-bearing catkins of the black birch reached two and a half to three inches in length as they matured during the first week in May (plate 52). If the tree was then shaken (or merely touched, in some cases), copious showers of yellow pollen could be seen falling and being carried off on the slightest breeze, hence the ease with which the female catkins are fertilized.

Likewise, the gray birch produced catkins which released pollen grains by the million, while the thousands of seeds which formed in the female catkins were later also wind dispersed and distributed. In outline, these little seeds resembled miniature birds in flight, as shown in plate 52; because of their winged structure, they might move miles through the air when once released.

The author remembers an instance in which a barren hillside, denuded by clearing and burning, was quickly repopulated within two years by gray birch saplings. All being trees of equal size, the seeds must have been airborne together, and all must have landed at about the same time. Finding the soil and exposure just right, within eight years this hillside was clothed with a fine young birch forest which developed rapidly in a dry soil exposed to full sunlight.

In passing, it might be mentioned that the black birch is another of our trees which secretes a pleasant-smelling and -tasting oil with the flavor of wintergreen. It was another of the species whose twigs we chewed as children, along with those of the sassafras and spicebush saplings. In the soft-drink industry, birch beer has long been a favorite, and its flavor was originally obtained from the black birch bark.

CHAPTER II

THE PASSING OF THE CHESTNUT TREES; THE MINI-WOOD'S CLIMBING VINES

In addition to the species of trees and shrubs described in the previous chapter, so many lesser forms of vegetation were to be found in the mini-wood that separate chapters have had to be allotted to them. While the climbing vines make up this chapter for the most part, some material has been inserted here to recall the tragedy which overtook one of the finest species of trees that formerly flourished in our woodlands, the American chestnut, *Castanea dentata,* all but exterminated not so long ago by the Asiatic fungus blight *Endothia parasitica.* The organism attacked the bark and was thought by many to have been carried and spread on the feet of birds.

Perfectly suited to life in our rocky Connecticut forests, *C. dentata* yielded the sweetest small brown chestnuts for roasting that man had ever been blessed with. Our woods were prolific of chestnut trees, and it was a must to get out early, after the first good frosts had opened the burs, if one hoped to beat the red squirrels to the feast. In the early nineteen hundreds, when the author was a schoolboy, we gathered these chestnuts by the hatful.

Classified by specialists in forestry according to a region's dominant trees, most of Connecticut is said to lie in what is thus technically known as the "sprout-hardwoods region" for the reason that most if not all of the dominant species of the area (following the original or several cuttings) often arose as sprouts from the stumps or roots of older trees and less often from seeds dropped naturally, or brought in and buried or lost by animals. Many young chestnuts which thus appeared grew to the sapling stage but never reached the fruiting stage. Occasionally, however, a few did produce nuts of a sort.

The other forested Connecticut areas, or "belts," have been designated as the "spruce region," the "northern hardwoods region," and the "white pine region." With all due respect to the authorities, however, it would seem to this author that the mini-wood per se might best be described as a remnant of a mixed *mesophytic* hardwoods association, the italicized term referring to the species adapted to a moderately wet environment, the woodland fragment in question having supported twenty-five species of trees and shrubs over the years, many with their roots in very wet soil for months at a time.

Long before this study was initiated, a few sapling chestnuts were found in the mini-wood area (about 1945), but all of these gradually died out without producing burs. Where their forebears had flourished and these saplings had arisen from still viable old roots, the whole area was strewn with rocks left by the retreating glacier—fragments of basalt, hornblende-schist, and granitoid gneiss, of hornblende-syenite, quartz, and iron-stained granite, of red sandstone, and bits of rock smoothed and polished by sand and pebbles in churning glacial waters. Samples of these rocks now in the walls of the author's garage present a study in glaciation in themselves (plate 55).

Here it was instructive to see how many kinds of plants had become adapted to a highly acid soil.

In soil testing, a scale is employed which is marked with numbers from 0 to 14, and in which 7 (the value for pure water at 25° C) represents neutrality. These numbers designate the hydrogen ion (abbreviated as pH) concentration value, or degree of acidity or alkalinity of the sample being tested. Tests are made with indicator dyes whose color change in a specific dye; matched with a color in a prepared (printed) scale or chart, thus gives the approximate pH value of a soil sample extract. Most good garden soils for vegetables or crops test between 6 and 7.5. The lower the number (reading down from 7) on the scale, the more acid is the sample.

Directly below the largest black oak in the mini-wood, a two-inch layer of black humus or so-called mull tested pH 4.5, a surprisingly acid reading even where taken, beneath disintegrating oak leaves with a high tannin content. Under this black humus layer, a medium-brown-colored soil occurred which gradually lightened in shade as one dug downward toward the purely mineral strata.

Beneath a mixed growth of trees and shrubs, and close to the large angular glacial boulder shown in plate 2, a so-called podzolic soil occurred beneath highly acid humus, clearly differentiated therefrom by its medium-gray color, and containing little organic material.

In the mini-wood, as in any ecological system, *edaphic* (inherent) factors affecting the soil, such as acidity and alkalinity, drainage, and climatic conditions (as a few examples in a complexity of others) had all played a part in the community, as found and outlined in this book. How long these soils had remained as found, who might say with certainty? Long enough, however, to nourish and bring to maturity trees that sprouted well over a hundred years ago.

For trees, like the American chestnuts, that once flourished in the area, Connecticut soil must have been ideal. It was only the appearance of this exotic blight that doomed them soon after 1900.

The mini-wood and its borders proved congenial to six species of climbing vines. These were the Virginia creeper or woodbine, *Parthenocissus quinquefolia;* poison ivy, *Rhus radicans;* greenbrier or catbrier, *Smilax rotundifolia;* the sweet-scented grape or frost grape,* *Vitis vulpina,* the Japanese honeysuckle, *Lonicera Japonica;* and the wild yam, *Dioscorea villosa.* Two other plants, one a honeysuckle but not a climbing vine which appeared in the mini-wood, and a common species of bramble, have also been included here.

Species of plants which use others only as supports are called *epiphytes.* Often confused with *parasitic* species, the former take no nourishment from their hosts, merely ascending trees and other supports by various means, such as tendrils, adventitious roots, and "holdfasts." None of the mini-wood vines were parasitic.

True parasitic plants have evolved what are known botanically as *haustoria,* which actually penetrate the host's tissues, taking nourishment from it through these suckerlike modifications of roots. Wherever encountered, haustoria are interesting objects for study. A familiar plant which thus lives on the foods made by other species is the mistletoe. In different parts of the country it parasitizes many kinds of trees, its haustoria of course being specialized organs for the absorption of liquid nutrients directly from the host plant.

The Virginia creeper bears from two to five, and occasionally seven, compound leaflets arising from long reddish stalks or branches from the main stems. The leaflets are partly or completely toothed, and the vine's total foliage is so dense that many of the delicate little flower clusters, which will presently be described in detail for those who have never seen them, are very often high up or otherwise hidden from view.

The Virginia creeper flourished in many situations in, and on, the borders of the mini-wood. A species with great expanding ability, it covered many of the woodland rocks, while old vines that had long since reached to the tops of host trees rivaled the tropical lianas which everywhere bind the trees with their stems (plate 56). Like these lianas, it creeps over the ground in persistent search of a supporting host, and in its travels overruns even such noxious species as the poison ivy. It is tenacious, its smallest pulled-up segment of stem rooting and growing anew when tossed aside. It creates microclimates with its leaves, cooling the rocks and the highest branches alike, and affording good shelter for the nests of Robins, Tanagers, and other birds among the latter. No woodland plant is more dedicated in its climbing habit, and once it has grasped a tree with its many holdfasts, its blossoms are thereby assured of eventual sunlight high above the ground. Close examination of these holdfasts, growing from the stipules at intervals on the stems, will reveal numerous tiny disklike objects which adhere to whatever object is to be climbed. In their eagerness to thus serve the main stems, holdfasts may occasionally grasp bits of dead wood or even pebbles, hanging on to them with great tenacity (plate 56).

---

*"Frost grape" is the proper name, although it is referred to in this book as the "sweet-scented grape," for reasons that will be explained.

The mini-wood in the dead of winter. In the center is the woodland's oldest beech, surrounded by its "family" of silvery gray saplings. On each side are black oaks of varying
ages.   (Page 6.)

PLATE 1

A portion of the mini-wood in winter garb. Here, because of their proximity to one another, white oaks, red maples, and other trees have grown up tall and spindly in the rock-strewn land. Mounds of snow hide these glacial deposits. Birdbox housed three flying squirrels.

An angular or "erratic" boulder, left by the glacier, rests alone in the mini-wood. (Page 4.)

Top: Some of the crusty lichens growing on this boulder may very well have been descended from species brought here by the glacier. (Page 3.) *Bottom:* Glacial "eggs" dug out of mini-wood soil. Largest one, at right, a solid fragment of white quartz, had been so smoothed by abrasion in glacial times that it possessed a distinct eggshell sheen. (Page 4.)

PLATE 3

Above: Opaque sap icicles dangling from winter-damaged twigs are sure signs that spring is not far away. (Page 4.) *Right:* The entrance to the mini-wood, and the bloom of spring leafage. Icicle, February 12; woodland, May 10.

The layer of rich black humus lying just below the blanket of dead leaves and woody debris in the mini-wood is shown as it looks when exposed. The humus supports vast numbers of living animal and plant organisms bound together in complex relationships.

Plate 4

The largest black oak in the mini-wood. Estimated to have sprouted about 1821, it has grown up alongside, and partly grown over, the large glacial boulder, as shown at left. (Page 6.)

A pair of black oaks in their prime rise above young red and sugar maples. An ancient beech shows in the background. (Page 6.)

Black oak. *Upper left*: Staminate flowers, shown $\times$ 2½. They produce pollen in such quantities that the ground under the tree is often turned yellow. (Page 6.) *Above*: Branch and leaf arràngements. *Left*: Yellow-stemmed leaves and acorn in situ. (Page 6.)

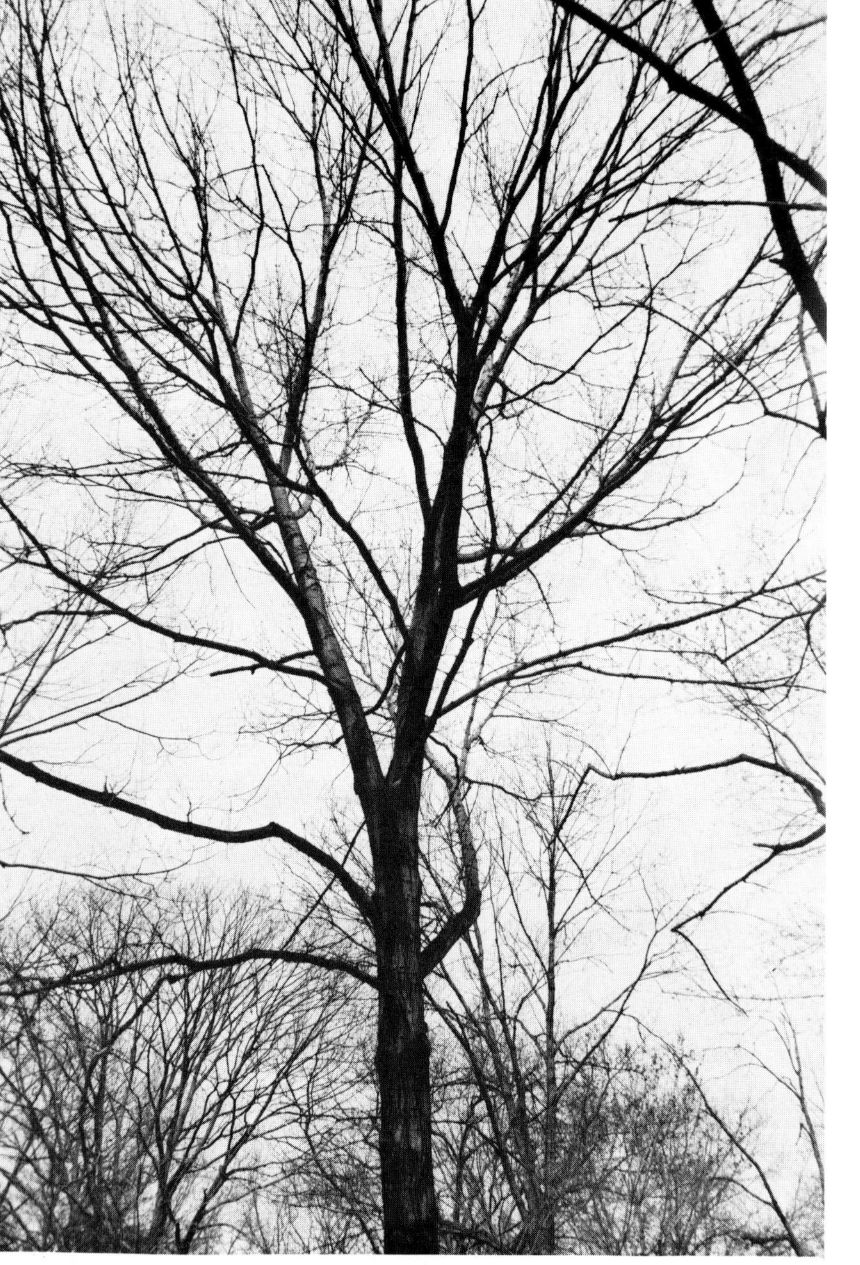

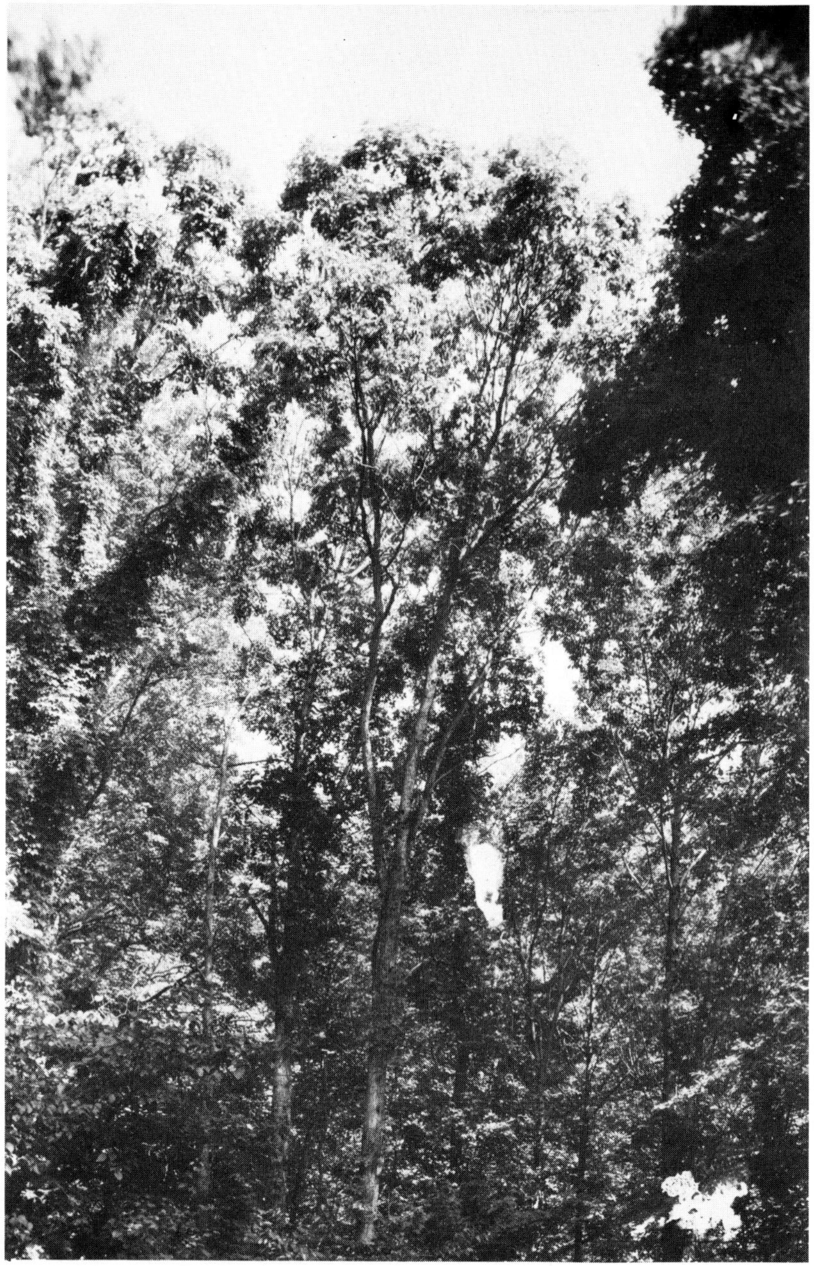

Left: A white oak, uninhibited by other trees nearby, may take a broad form, as shown here. *Right:* When white oaks grow up close together, they become tall and spindly. The two at left and center show heavy growths of Virginia creeper vines on their oblique limbs. (Page 8.)

PLATE 8

Staminate catkins of the white oak, shown $\times$ 2½, and a branch showing leaf form and arrangement. (Page 7.)

A tall and irregularly branched pin oak is shown in the center, with other oaks and beeches at either side and in the background. (Page 7.)

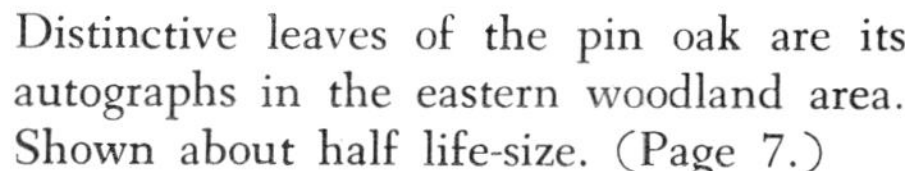

Distinctive leaves of the pin oak are its autographs in the eastern woodland area. Shown about half life-size. (Page 7.)

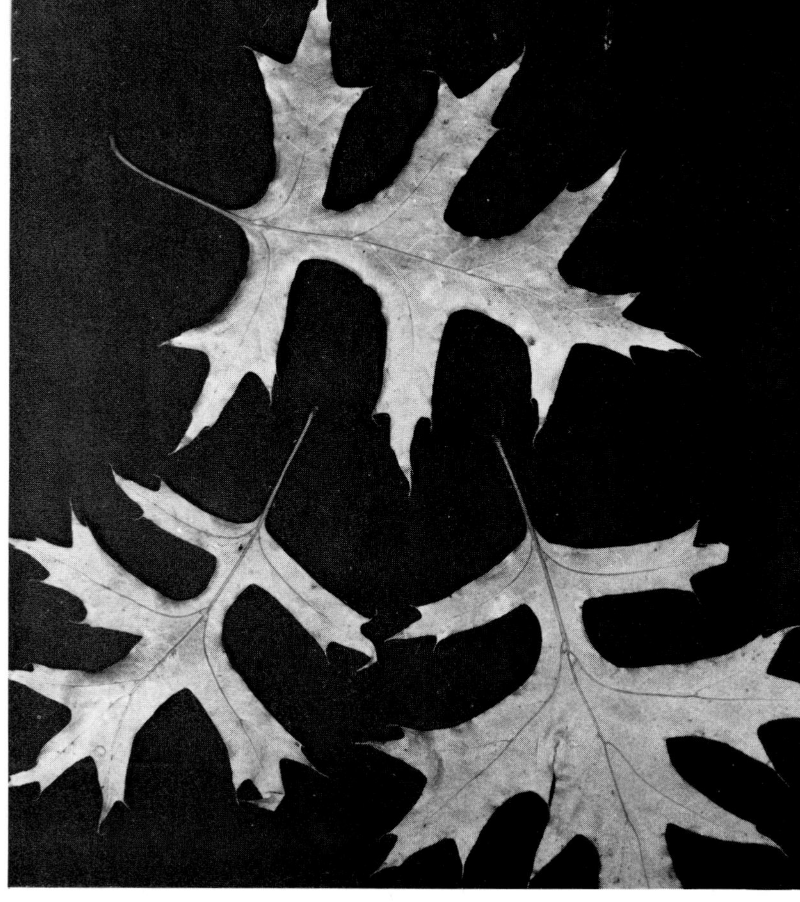

Plate 10

Left: A fine specimen of the swamp white oak which grew where it was uninhibited by other trees. *Right*: The same tree in full and splendid leaf, photographed June 27. (Page 7.)

Plate 11

Above, and top right: Catkins and unfolding leaves of the swamp white oak, photographed May .13, life-size, and May 19, about half life-size. *Right*: Typical leaf shape and branch arrangement. Date: June 30. One-third life-size. (Page 7.)

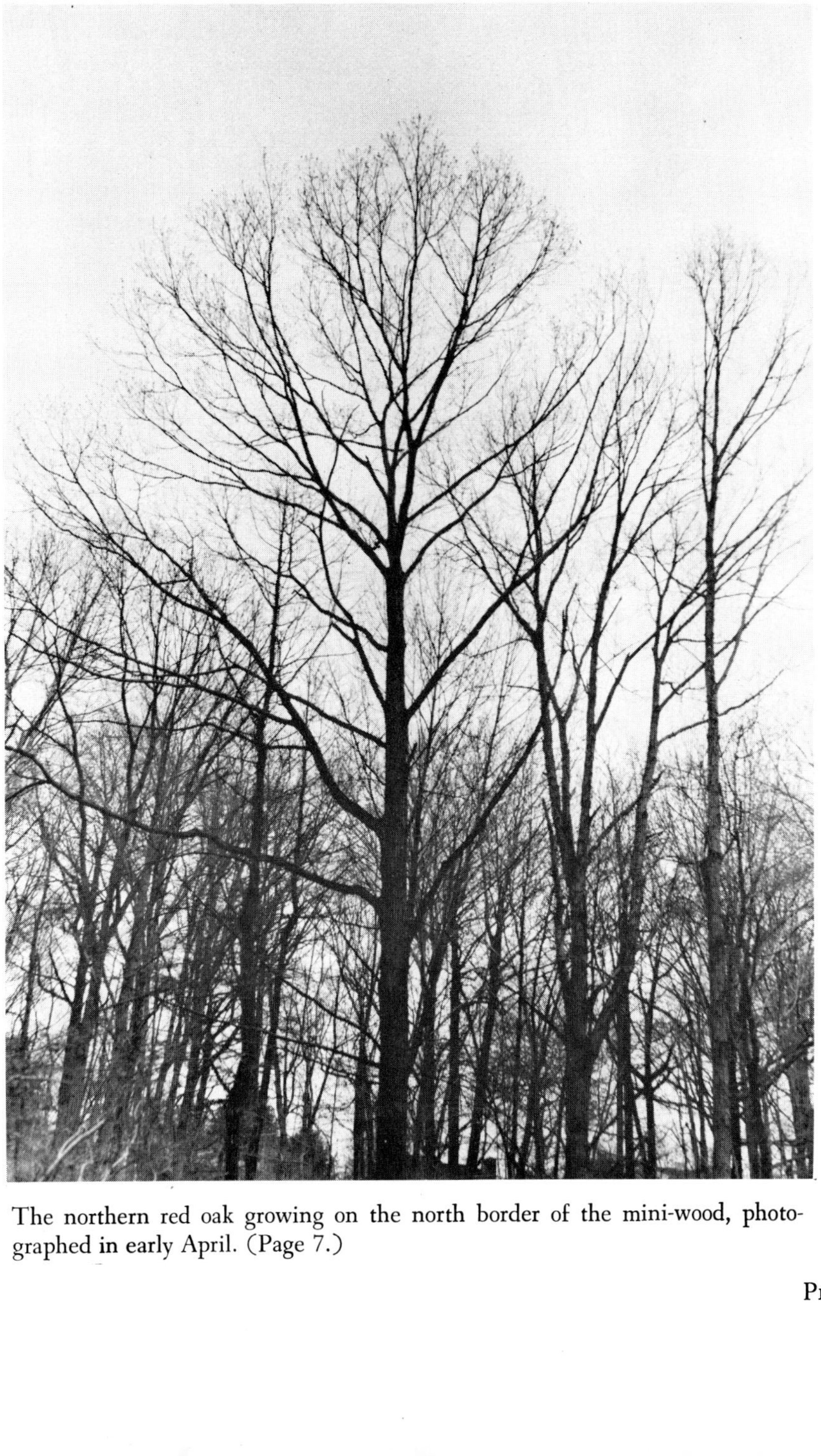

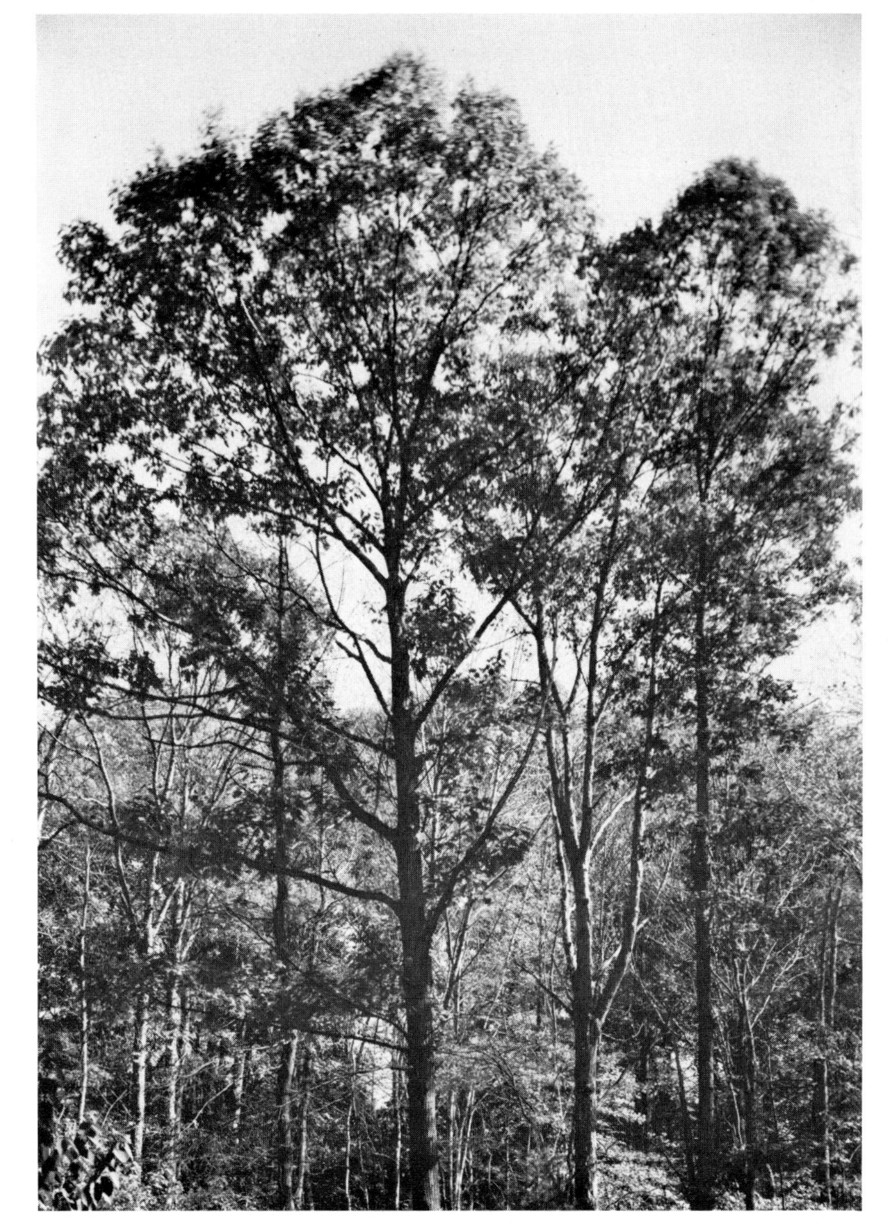

The northern red oak growing on the north border of the mini-wood, photographed in early April. (Page 7.)

The same tree in fall, photographed in late November, some time before it shed its leaves of various shapes. (Page 7.)

PLATE 13

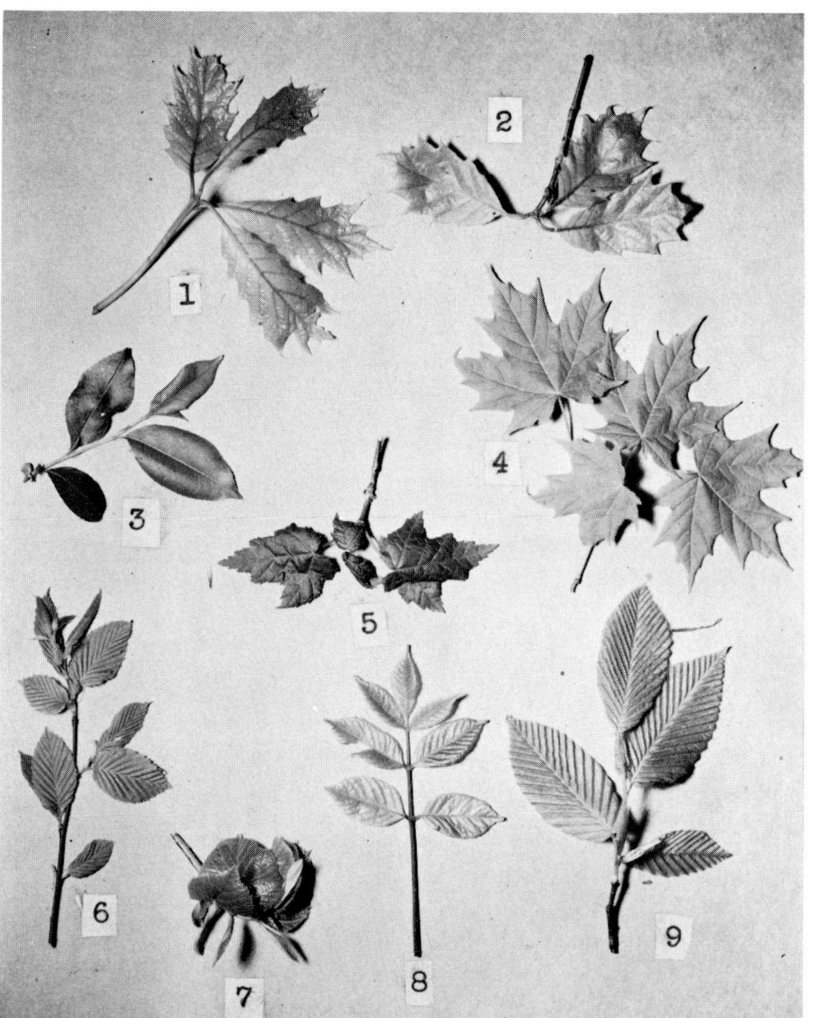

Above: Tender young leaves from the mini-wood, photographed May 10, and shown about half life-size. (1) Black oak. (2) Same, but from a different tree. (3) Wild cherry. (4) Sugar maple. (5) Red maple. (6) Ironwood. (7) Basswood. (8) White ash. (9) Beech.

Below: A sugar maple leaf (*top*) and two examples of one of the odd forms which fell on November 29 from the oak shown in plate 13. One-third life-size. (Page 7.)

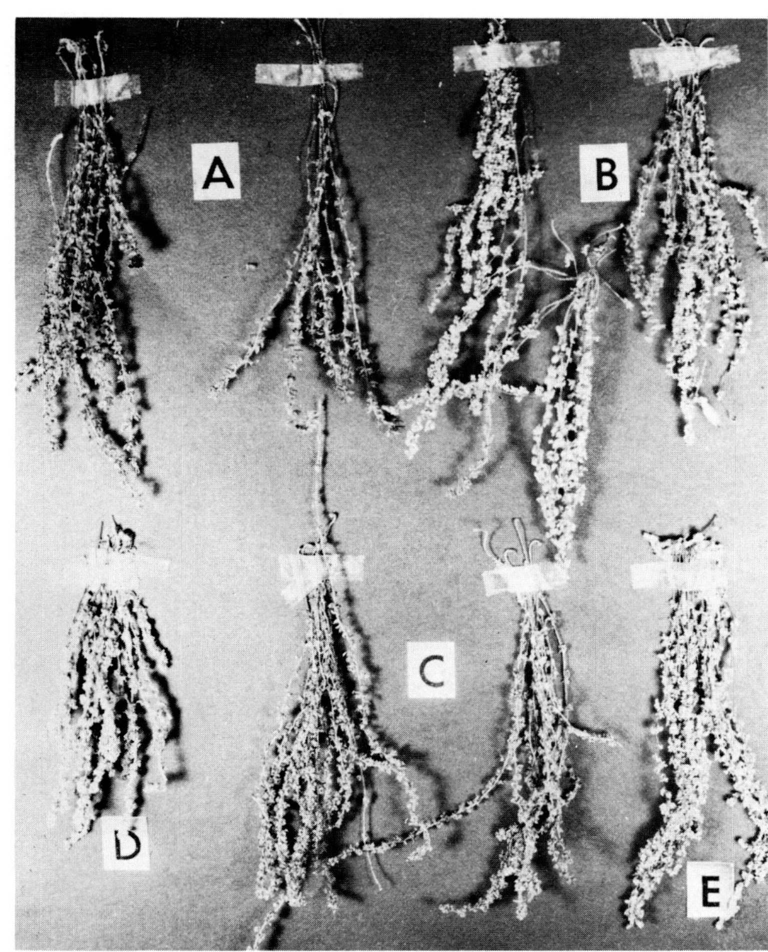

Left: Freshly dropped oak catkins, shown two-thirds natural size. (A) Black oak. (B) Northern red oak. (C) White oak. (D) Swamp white oak. (E) Pin oak. Date: May 30. Note similarity. *Below:* Acorns vary considerably in different trees of the same species. Those shown are: (A) Black oak. (B) White oak. (C) Swamp white oak. (D) Pin oak. (E) Northern red oak. All shown about two-thirds life-size. (Pages 6, 7.)

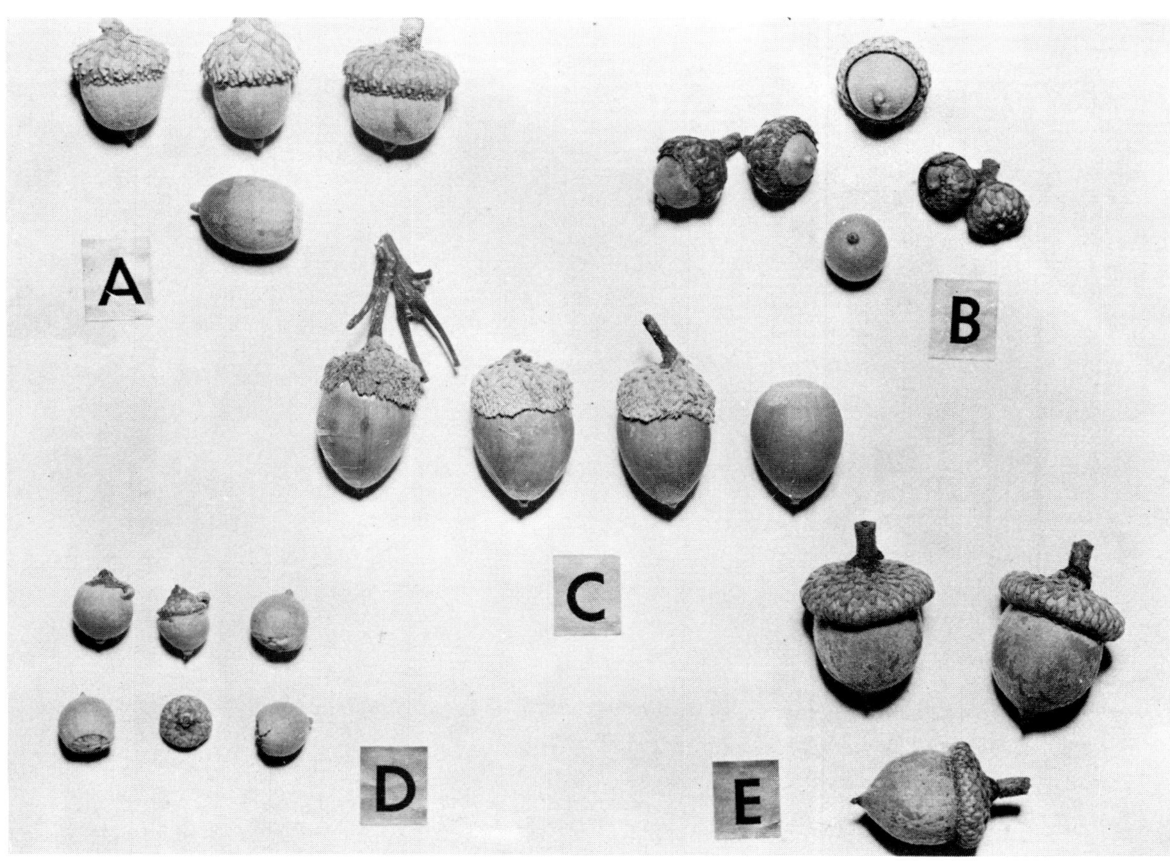

PLATE 15

A mature and perfect American beech of the mini-wood, photographed in early April in all its silvery gray beauty. (Page 8.)

PLATE 16

Beech tree well over a hundred years old, showing a large cavity at the bottom, probably initiated by fire (lightning bolt?) many years ago. (Page 8.)

The woodpile resulting from a mature beech, blown down on the woodland border two years before the picture was taken. The fine young beech just behind it, growing as a shoot from the old tree's roots, will now make better progress in sunlight. Date: May 15.

Plate 17

Left: Staminate beech flowers, $\times$ 3. Date: May 10. *Above:* Newly opened beech burs, August 28. *Below:* Leaves and closed burs, August 4. Natural size. (Page 8.)

Plate 18

Top: An ancient beech and its "family" of saplings, early morning, May 10, in the mini-wood. Note that young beech leaves are still limp (drooping) at this date. *Below*: Young beeches, skunk cabbages, mature tulip-tree (foreground), and early morning sunlight, May 18. (Page 8.)

PLATE 19

The massive trunk, major limbs, and part of the canopy of a ninety-year-old beech. With its roots in very swampy ground, it finally succumbed and came down in the storm of August 27, 1971. Its remains, in the form of a woodpile, are shown in plate 17.

Plate 20

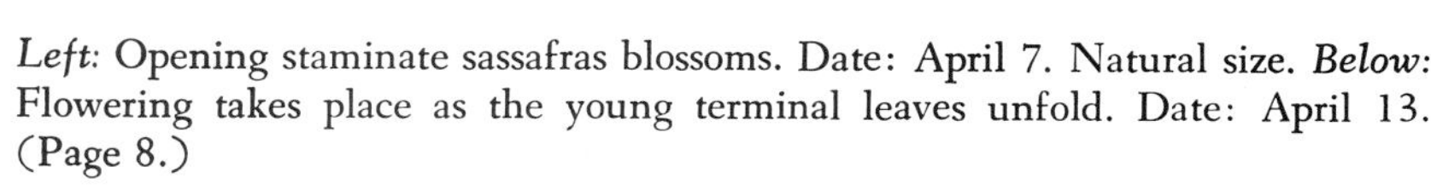

Left: Opening staminate sassafras blossoms. Date: April 7. Natural size. *Below:* Flowering takes place as the young terminal leaves unfold. Date: April 13. (Page 8.)

Here the leaves have expanded to half their mature length, and now the blossoms have permanently closed. Date: April 21. Shown somewhat reduced. (Page 8.)

Leaf forms of the sassafras are its autographs. The plain ovate, the three-lobed, and the mitten shapes are shown at left, about half life-size. Date: June 1. (Page 9.)

Above: A sapling sassafras. Date: September 29. *Left*: Typical young leaves of the sassafras tree, shown much reduced. Date: May 25. (Page 9.)

Plate 22

The sassafras canopy shown was quite dense where two trees had grown up together. In late September, an occasional dark blue berrylike fruit tumbled to the ground, but most were probably eaten from the trees by birds and squirrels. (Page 8.)

PLATE 23

Many trees which sprouted in the mini-wood found it necessary to grow up over, or around, glacially deposited rocks. *Top left*: White oak. *Above*: White oak which seems to be grasping a rock. *Bottom left*: Swamp white oak. (Page 12.)

Some of the oldest vines in and out of the mini-wood area reached an inch and a half in diameter, their aerial foliage in such examples spreading thickly along limbs forty and fifty feet above the ground, and upon these higher portions of the plants their berries were produced in greatest numbers.

As in the case of the sweet-scented grape and also some other familiar plants, one finds little or nothing about their flowers in most of the older as well as recent nature field guides; perhaps because some of these, including the flowers of the Virginia creeper are small and inconspicuously colored, they have been dismissed with such statements as: "Flowers small, greenish; clustered." Or "Reddish or greenish small flowers grow in cymes."

The smaller flowers may appear at first sight, the more closely they should be examined. In the case of the Virginia creeper's inflorescence, three or more branched short stems arise from single stems from one to two inches in length. The flower buds are at the ends of these short stems; therefore, when the buds open, the blossoms form compact little clusters botanically called *cymes*. The buds are pale green; their stems, pink and green. When open, the five pale green petals are seen to be rather thick and fleshy, and shaped exactly like little boats, their square "sterns" nearest the ovary, and their higher, canoelike "prows" turned backward. Wide-open flowers measured up to six millimeters (somewhat under a quarter of an inch) in width, and displayed the onion-shaped pale green pistil surrounded by five white stamens and anthers (plate 57).

By late September or early October, many of the fertilized flowers produced globular dark blue or black berries, their two divisions each containing two seeds.

In the fall, when the berries are ripening and bird migration is again at its height, among scores of other species, Swainson's Thrushes are then moving leisurely southward in large numbers. Annually, some of these stop off in the mini-wood to rest and then feed up during the daytime on the berries of this vine, continuing their stops well into October.

During these cool fall nights, the Thrushes' distinctive, far-carrying calls and whistles to one another far overhead may well fill the listener with wonder at their innate navigational abilities, as well as with wishful thinking—flavored with a bit of envy, perhaps—that one can never know the ecstacy of freedom to go anywhere in the world one might care to go, on one's own wings, and with one's anatomically built-in compass and radar system.

At this time of year, every available berry was stripped from the vines, Robins, Catbirds, Jays, and even Doves and Woodpeckers joining in the feast along with the Thrushes.

That this creeper is often bird distributed was evidenced by numerous seeds and occasional unripe whole berries recovered from birdbaths established on the borders of the mini-wood; the food had been dropped there or passed in excrement. Remembering that the berries only carry up to four seeds apiece, the wide distribution of the vine could hardly be accounted for if it were not for the birds that utilize and consume the soft parts of these fruits, ejecting the viable seeds wherever they may happen to be.

At one time, in a very rocky portion of the mini-wood, poison ivy had almost completely taken over the ground. It was mostly eradicated by a mild herbicide solution carefully and individually applied to the stems. Two years later, this area had been covered by Virginia creepers, and even though poison ivy sprouts arose anew from the old roots, they were mostly smothered by heavy growths of this other, somewhat beneficial vine.

Few who live in the country are unfamiliar with the appearance of poison ivy, technically known as *Rhus radicans,* yet occasionally one still hears of someone picking bunches of the foliage upon seeing and admiring the bright red, yellow, and intergrading shades of the leaves in October.

Distinguished for its *glossy green* compound foliage, with twigs ending in three leaves, it should be recognized even by a casual observer. Where it has climbed high into trees, the stems in old vines may be an inch or more in diameter and be held to the supporting tree trunk by hundreds of threadlike black outgrowths sometimes called adventitious roots (plate 58).

The inflorescence of this ivy appears in the form of panicles, or little open groups of white to greenish white flowers, with five petals and five stamens at the ends of short stems; thus, the round, grayish white berries which follow appear as bunches, each berry measuring no more than six millimeters (less than a quarter of an inch) in diameter. Their pale color and glossy green leaves grouped in threes are diagnostic features which will easily distinguish this vine from the equally abundant climbing Virginia creeper often growing in the same habitats.

The poison ivy berries were eaten by a number of birds, although the fruit may become hard and dry. Jays, and Crows in particular, consumed them, and probably helped in the plant's wide distribution.

It is fortunate that there are not many species of plants which are capable of causing the kind of distress which some human beings suffer from poison ivy. A severe infection and the itching it brings on, and its tendency to spread as small blisters are scratched and ruptured, is nothing to make light of. While some people are easily and even seriously affected by it, many others, including the author, are only mildly discomforted after coming in contact with the foliage or perhaps just dry vines of the species.

The itching and other effects are caused by an oily secretion whose chemical formula (in case some reader may wish to know) is $C_{15}H_{27} C_6H_3 (OH)_2$, which is a phenolic compound also occurring in the sap of our native poison sumac, the poison oak, and in certain oriental lacquer trees. Its name: *urushiol.*

In the absence of commercial remedies, a baking soda paste gently applied to the rash may be helpful. In serious cases, corticosteroid drugs are applied. Washing with hot water and Ivory soap soon after contact has negated trouble for the author on several occasions, but as the secretion will follow the tiniest channels in the skin, it is obvious that the less a victim scratches, the better off he will be, and the less the rash will be spread over his limbs or body. Zinc ointment is an efficient commercial remedy which the author used successfully when infected by an itch-producing plant in South America.

A low climbing vine which grew here and there in the mini-wood was the greenbrier or catbrier, *Smilax rotundifolia.* Usually more luxuriant as a plant of more open situations—for instance, beside stone walls in fields, in fence corners, and other sunny locations—it became established in the mini-wood about the butts of oaks and other tall trees, probably being bird planted. (Plate 59.)

Its leaves are broadly heart shaped, and the plant has main stems which may remain green until cold weather sets in or, in some instances, all winter.

Long, twisting tendrils, which are organs modified from leaves and arising from either side of the leaf stems, twine about other plants and their branches for their own support. Such tendrils occur in many other species also, often reaching out toward the nearest possible supports, almost as if they might have "eyes" (plate 59).

Greenbrier stems bear sharp *prickles,* and while these may look like thorns, they are simply outgrowths of stem tissues. *Thorns* and *prickles* are easily confused; prickles are branch or stem modifications, whereas thorns arise from the axil (angle between the leaf and the stem on upper side) or somewhat above it, and prickles are more haphazardly placed, as in the blackberry stems shown in plate 58. Still other kinds of "thorny" plants are armed with *spines,* which botanically are leaves which have become modified into slender, conical, hard, and sharply pointed projections, as in the barberry (plate 58).

Greenbrier flowers, which are small and yellowish green, with three-celled ovaries, are produced in clusters in varying numbers, on short stems, and are followed in September by globular blue black berries containing up to three hard seeds each. Their skins acquire a lighter bloom as the berries ripen, and they remain on the vines only until discovered by woodland hunting birds, although they persist for considerable periods in situ if not so discovered. Succulent and sweetish, these berries are favorites of migrating Swainson's Thrushes, while the large dense clumps of the plant which grow in open situations have long been favorite nesting places for the Brown Thrasher, a bird which instinctively chooses the protection afforded by the greenbrier's wealth of sharp prickles (see plate 166).

Another of the woodland vines, a long-lived species and a powerful climber by means of long tendrils, a plant whose easily stripped shreds of bark enter into the nests of many birds, gray and flying squirrels, and deer mice, was the common frost grape, *Vitis vulpina* (plates 60 and 61).

Within the mini-wood, one aged vine climbed a red mulberry tree, spreading its medium-green, occasionally three-lobed leaves in massive array over and among the much larger mulberry leaves, the two species forming an especially dense canopy. Another example many years old reached and climbed a dead pin oak yards beyond its sprouting point, while a third vine "chose" a white oak on the mini-wood border for its support.

A few of this species' leaves grew to nine inches in length, and almost as wide, while the great majority were very much smaller. All had long green stems, and their undersides were considerably lighter than the upper surfaces (plate 60).

The tendrils by which this vigorous vine ascends its host trees arise from the points where the normal leaves join the vine. Twisting grotesquely and growing rapidly, they reach out and grasp twigs and branches in all directions, holding with great strength once they are coiled about any such support.

In the mini-wood, the long main stems measured up to an inch and a half in diameter at ground level. These stems, or "trunk" portions of the vines, often with loose shreds of bark hanging from them, may undulate for yards before turning upward into a host tree where the foliage, blossoms, and fruit are produced high above.

As for the flowers of this species, it must be admitted that they are small and plainly colored and therefore very inconspicuous, which may account for their dismissal in certain reference books with descriptions such as, "flowers greenish," or no description at all!

While the flowers did as a rule grow high up, where it was difficult to climb and locate them, these odd little clusters deserve at least a few more words in their favor. This is a species which the author prefers to call the *sweet-scented* wild grape, for a reason that might become obvious at once to anyone near the inflorescence.

The crowded little blossoms seem anything but inconspicuous when slightly magnified. With five petals, and five white, widely spread stamens with bright yellow anthers, they are delicately beautiful. Oddly enough, some flowers had greener and fatter pistils than their neighbors, while some also had much shorter stamens than others (plate 60). In various ways these blossoms were individuals! Some flowers were cuplike, with greenish rather than plain white petals. And a single flower with its stamens could spread up to six millimeters (under a quarter of an inch) in width.

From the whole clusters came a truly delicious scent, but a perfume too delicate, subtle, and enchanting to be adequately described by the author. To know it and to experience the pleasure of its presence, one must seek the flower clusters in situ in their lofty habitats during the first few days in July.

After opening, the blossoms fall quickly, and in the mini-wood, the grapes themselves were often few in number. Rainy spells during the blossoms' brief life-span seemed to inhibit pollination, and many of the fruits dried and shriveled on the vines. Normal grapes were sweetish, the skins covered with a bluish bloom, and some matured as late as the middle of September (plate 61).

They are doubltless planted by birds, for many species regularly eat the small, sparsely produced fruits, and then discard the hard little seeds in their excrement. The new sprouts from these seeds, by chance finding the proper surroundings, soil, and moisture, creep lianalike over the ground until they reach and grasp some suitable support with their young tendrils. This assured, the new grapevines may grow for many years, eventually all but overwhelming some of their host trees with their epiphytic tangles.

The fox grape and other wild native species might also well have found the mini-wood soil congenial enough and to thrive therein, but up to the writing of this book none had been become established there.

As mentioned before, grapevines yield important commodities to many animal forms of the woodland and elsewhere. Employed in nest building by many species of birds, by deer mice, flying squirrels, and sometimes by gray squirrels, the vines give up their bark in easily removed strips in many convenient widths which wind readily into nest linings. In addition, the species offer sweet fruit to these woodland inhabitants, and here again the beneficial mutual exchange is evident, with the birds, and possibly individual mice and squirrels, aiding in the distribution of viable seeds of the vines in return for favors received.

Wild grapes make excellent jelly. When as youths we camped along the Mianus River, a small stream in southern Connecticut, we gathered the wild grapes which then were to be found in profusion along its banks. Boiling them down with plenty of sugar, we produced a crude sort of jelly or half jam which we thought outclassed all of our "wild" camp foods, which then included certain fungi, wild berries, and snapping turtle stew!

The most tangled climbing vine, which grew mostly on the mini-wood borders rather than in its heavier shaded areas, was the Japanese honeysuckle, *Lonicera japonica,* a long-naturalized species.

Found abundantly in New England, its dense mats covering walls, fences, bushes, and small trees, or else creeping slowly over the ground, it is known to most everyone. Its stems climb and bind without tendrils or holdfasts, their twisting and spiraling forming dense tangles on host trees and other objects (plate 62).

Its dark green leaves seemed to have a built-in antifreeze, some withstanding heavy frosts and even hard freezes. Blossoming in mid-June, nothing within or outside of the mini-wood was more exquisitely and delicately fragrant than this honeysuckle's *paired,* white, or ocher-buff tubular flowers, with their distinctive protruding stamens and slender pistil (plate 62).

While the deep tubular construction of these blossoms may guard their delectable nectar from larger, maybe undesirable insects in some cases, tiny useful pollinators are admitted in numbers.

Nectar—"food for the gods"—is secreted lavishly by the blossoms of this species. You may taste it for yourself if you wish, using this procedure:

With your fingernails, very carefully cut through the bulbous lower end of the flower tube, but without severing the slender white shaft of the pistil within. Next, draw back on this severed end, very slowly and carefully, thus pulling the pistil through the tube with the stigma still attached. This organ will act as a piston, forcing back the nectar, which will then appear as a clear glistening droplet at the opening you have made.

Put the tip of your tongue to this precious sweet manna —the *pure nectar*—not honey, which is what nectar is converted to after processing by highly subtle chemistry in the stomach of the bee, to be regurgitated into the honeycomb in the form in which we later buy it in the stores.

In such simple tests with honeysuckle blossoms, you will have experienced a taste sensation which bees and butterflies, bugs and bats, hummingbirds—and the gods, as well—have known throughout the ages. The vine also supplies juicy black berries which sometimes remain undiscovered until the Pine Siskins, Redpolls, and Purple Finches arrive for the winter.

Stray seeds somehow arriving in the mini-wood, perhaps as long as twenty years back, or perhaps about five years before this particular study was begun, resulted in scattered examples of the honeysuckle *Lonicera Morrowi*, another naturalized foreign species. Although not a climbing vine, it has been placed here in this chapter simply to bring the species of *Lonicera* together.

Specimens of *Morrowi* in and on the edge of the mini-wood grew into shrubs four and five feet in height, with obvate to oblong leaves, oppositely arranged, and with up to twenty or more pairs on the new wood each spring. The form of the flowers—difficult to describe with clarity— will best be understood by studying the photograph in plate 63. Petals that were white or light buff occurred side by side, the buff ones being doubtless simply faded white ones, while all of the delicate stamens were of a uniform yellow. During the fifteen-year study, blossoms appeared

only three or four times, and no berries were found by the author, birds or small mammals probably finding them first.

One of the shrubs, which came up on the edge of the woods and partly through a brush pile, was selected by a pair of Catbirds, their nest being constructed in this plant, as shown in plate 164.

The author has often wished that more people were cognizant of wild perfumes. Four in particular come instantly to mind. Really indescribable, more subtle than any man-made scents, these emanate from the basswood and sweet pepperbush, the Japanese honeysuckle, and the wild frost grape. Suddenly floating to the senses in the form of odorous zephyrs, they happily inform us that summer with all its delights is really with us once again.

Continuing to spring surprises, the mini-wood and its borders came through again while this chapter was being written, and just as this portion of the book was being given its final touches. The cause was an organism nicely illustrating species change and succession.

The newcomer appeared in the form of a climbing vine never encountered before by the author, and for what it may be worth, later in the chapter a possible explanation as to its manner of becoming established will be suggested, but first follows a somewhat detailed description of this interesting and sudden arrival.

The plant was the wild yam, *Dioscorea villosa*, a climber without tendrils, ascending by firmly encircling its supporting host, and with more than one stem arising from a single rhizome (rootstock) and, by thus doubling up, greatly increasing its hold.

When first observed, it had appeared above a dense bed of sensitive ferns, brambles, and weeds interspersed with wild cherry saplings and spicebushes. The loose tips of the double stem seemed to be "feeling out" the situation in search of a host, but there being no support nearby, the author came to the rescue with a discarded long-handled, iron-ringed crab-catcher (minus the net) stuck into the ground in an upright position beside the plant in mid-May.

Sensing the support almost at once in the uncanny manner of many other climbing vines, the two stems encircled the net handle, and in eight weeks reached the iron ring, encircled two-thirds of the wire at a slower pace, and finally came into flower (plate 63).

At this stage of its growth the vine's foliage had almost hidden the supporting handle from view. Its earliest leaflets were green and white, the four furrows on each side of the midrib and in which the veins were located giving the young foliage a striped appearance, but in all older and mature leaves the color was a solid, rather dark green. Mature leaves measured up to three inches in length and two and a half inches in width. All were heart shaped, but with very long, sharp tips (plate 63).

Flowering commenced about July 15, when the vine had grown six feet high. From the last twenty-four inches of the stems, somewhat zigzag flower spikes were borne at intervals of two inches or less, each with from five to ten tiny blossoms, the six minute petals, clearly shown in the photographs, topping curious spindle-shaped ovaries measuring six millimeters (about a quarter of an inch) in length. As this yam is a *dioecious* species (bearing staminate and pistillate flowers on separate plants), no description of the staminate form was possible, for only this one vine was found.

Both the flower petals and their fusiform ovaries were greenish white, the latter with three lengthwise paler ridges which soon developed into the three thin wings of the fruit capsules (plate 65). Early rapid development of these capsules slowed down considerably before they reached full growth, about September 1. At this date, the three seeds, one in each wing, were white and soft, three millimeters in width, more or less lenticular in form, and each centered in a thin, papery surrounding wing with an irregular border and, roughly, six millimeters in diameter. By late October the capsules had turned brown. They remained on the vine until late in November.

The interesting flowers are shown with the leaves in plate 64; the rootstock and seed capsules in plate 65.

Minute as the flowers of this vine were, they were very fragrant, and having appeared in considerable numbers at the same time, their pleasing scent was noticeable at once as one neared the vine. When stems bearing these flowers were plucked and at once put in water in the laboratory, their fragrance, a natural perfume quite new to the author, seemed to be enhanced, and added to the pleasure of writing these words.

But alas, removing them thus from the growing vine was a bad experiment, for the flowers wilted all too rapidly, bringing up the mystery again of why the stamina of *cut* flowers varies so widely from species to species. Jewelweed flowers and foliage, for instance, wilt even quicker than a wild yam's, while spikes from the sweet pepperbush crowded with blossoms remain fresh and fragrant for days

if the stems are put in water. Fertilized flowers, it might be said, usually fade more quickly than unfertilized ones, but this was not true of the pepperbush. The subject would bear more study.

A possible way by which this wild yam may have become established on the woodland border is as follows: A Virginia creeper vine of great age annually spread its foliage and fruit on the limbs of a supporting white ash tree overhanging the thicketlike patch in which the yam was found. Each year in the fall, Swainson's Thrushes and other berry-eating birds fed regularly upon this old creeper's offerings, at the same time dropping seedy excrement into the habitat below. Thus, no doubt, had the young wild cherry, spicebush, and additional Virginia creepers been established therein, and also perhaps this wild yam, although only the Cardinal had been observed actually munching its seed capsules.

Another species, vinelike in some cases but usually considered a woody shrub, and one which occurred mostly on the mini-wood borders rather than in its deep shade and often with its roots in very wet soil, was a species of blackberry and a member of the crowded genus *Rubus*.

Strange as it may seem to the uninitiate in plant complexities, there are over two hundred species and subspecies of blackberries now recognized, and these plants thus constitute a group many of which may only be identified with certainty by botanical authorities.

The author is nevertheless chancing the extrusion of his neck (although hoping not too far) in stating that the plant found in the actual area of this study was the species which New Englanders have long called the "highbush" blackberry, *Rubus allegheniensis* (plate 66).

With its rugged green and reddish stems often gracefully bowed, this plant grows ten feet or more in length, and with many side branches. Most of the stems are ridged lengthwise so that a cross section yields a neat five-pointed, odd starlike figure. If such a section be cut with the aid of a microtome, so thin as to be transparent, and is then properly prepared, with its cell layers stained in contrasting colors, the specimen becomes highly instructive, and beautiful when viewed through a microscope (plate 66).

In the leaf arrangement on the fruit-laden branch at the lower right in plate 66, it might be noted that some leaves have shorter stems than others, the longer terminating in three-leaf groups. The white flowers of these blackberries have five sepals and five petals, with numerous stamens and numerous pistils in the center. Most of the mini-wood plants were in bloom around June 15, and fully formed green and ripening berries were numerous on these stems by July 21.

A blackberry, usually called a "blackcap" in country lingo, will be seen on careful examination to be made up of many small, complete fruits resulting from the numerous pistils, each of which contains seeds. At the upper end of each berry, the remains of the calyx may be seen, and the clusters of fruits (plate 66) are called *drupelets*.

Close examination of the stems will show that the prickles with which they are armed have arisen in no particular order. Some may be close together, others sparsely distributed and far apart. Some are straight, others curved backward, but always they are present in adequate numbers collectively, as anyone well knows who has unwittingly grasped a blackberry stem. (See plate 58.)

The pulpy cushion upon which the little individual fruits of each berry mature, and to which they cling even when the berry falls or is plucked, is called the *torus*. In raspberries, on the other hand, which also are made up of many small individual fruits, the white central receptacle remains on the plant and the berry with all its parts comes off hollow, but whole. When consuming blackberries, we, or the birds, or the other mammals, eat the fruit, torus and all.

Seldom does a single blackberry escape the birds. In and around the mini-wood, they were quickly gobbled up by Catbirds, Thrushes, late-lingering Thrashers, and occasionally Robins. The blackberry seeds being small, possibly some of them are digested and rendered non-viable, which might be one reason why there were not more of these plants in the area described, and why the species' population remained fairly static there during the fifteen years.

No one, the author believes, in contemplating a living association in which its arboreal dominants have grown a hundred or more annual rings, will fail to realize that such a community is a representative of nature's supreme achievement and triumph over many adversities.

Your woodland, whether only a half-acre or a great deal larger, may indeed, like this mini-wood of the author's, vary greatly in their soils, their organisms, and other features, yet every such forest or fragment of forest will prove equally interesting when minutely scrutinized. One will do well—and profit handsomely in mental uplift—through such an examination with newly opened eyes.

In this book, a beginning has now been made at pointing out the possibilities that await one, including some of the simpler plant and animal interrelationships, to which a great deal more attention will be paid and more details given in the nineteen chapters which follow.

CHAPTER III

THE HERBACEOUS ANGIOSPERMS OR WILDFLOWERS OF THE MINI-WOOD

Thousands of plant species which dominate the vegetation of the everyday world around us, and which through thousands of years and the tedious processes of evolution now reproduce themselves by means of seeds contained in an ovary and developed by the flower, are known botanically as *angiosperms,* and the term *herbaceous* in the title of the chapter signifies the *non-woody* plant species whose *above*ground portions die down completely each fall.

The other great plant group, whose seeds are not enclosed in an ovary and whose members are botanically known as the *gymnosperms,* includes the pines and other conifers whose seeds are produced on the surface of cone scales. No gymnosperms occurred in the mini-wood area under examination (ornamental species excepted).

The angiosperms, which includes the trees and shrubs of this study as well as the herbs, are botanically divided into *monocotyledonous* and *dicotyledonous* species, and for the purposes of this book it will be enough to know that the former's embryos produce but a single seed leaf, while those of the latter produce two seed leaves. The differences may be seen and understood at once if seeds about to sprout are cut open. A kernel of corn, for instance, will demonstrate the single seed leaf; a lima bean, the twin seed leaves. Henceforth, in this book, and as is customary, the two forms will be referred to simply as *monocots* and/or *dicots.* Major examples of the latter angiosperms were the shrubs, the largest trees in the half-acre woodland, including such species as the sassafras, tulip-tree, and tupelo, the ash and beech, and the oaks and maples. Characteristic monocots, on the other hand, if present, would be any species of grass such as is brought in and used in the nests of birds, or sweet and field corn, perhaps growing not far away.

The climbing vines covered in chapter II were all angiosperms, while the numerous herbaceous species we call wildflowers, as already briefly mentioned, were those with little or no woody tissue in their stems which, after making their annual foliage, flowers, and seed, at the end of a single season died down completely or to ground level only, leaving their living bulbs, corms, or rootstocks safely buried for future seasons' growth and flowering.

Many species of such wildflowers were firmly established in the mini-wood or on its borders, having found suitable soil in which to carry on their life functions. Some appeared as typical annuals, growing vigorously from seeds distributed mechanically or otherwise during the previous year. Some of these grew very tall and rugged in appearance, but withered and died before the first frosts arrived. Perennials arose each spring or later from the underground stems or rhizomes, bulbs, or corms, as mentioned earlier, all being food-storing devices for following years' growths and flowerings. All of these flowers had their particular soil and moisture requirements, their light or shade tolerances. All will be illustrated, and, it is hoped, more lucidly described in this chapter. But in covering the established species, they have not been placed in their order of natural classification in relation to each other; rather they have been placed according to their time of appearance and flowering. As has been emphasized before, this book has been designed only to point out *what goes on* and how much there is to see even within so small a woodland community as the author's half-acre. As has also been stated before, a reader's woodlot or forest or other owned and/or studied area may prove to be about the same as the mini-wood in organisms present, or again, it may be much richer or poorer in actual species, but the book as a whole should act as a guide, in many ways informing the uninitiate what to look for, and in some measure how to go about it.

Earliest herb to flower, in swamps and wet woodlands as well as in wet meadows and along the borders of brooks and ponds, is the skunk cabbage, *Symplocarpus foetidus* (plates 67 and 68). Unknown to many is the fact that its shoots for the following spring are already aboveground, although usually under the dead-leaf cover, as early as September 15. These shoots may be four to five inches in length, and may appear sometimes as slim and conical twins, or as single shoots with down-curving tips (plate 67).

Sometimes before all of the snow has melted in March, or even earlier, the purplish red, reddish mahogany, or streaked and spotted yellowish green flower spathes have come into view. Within the spathe, the more or less globular thick-stemmed spadix is housed, its whole surface interestingly sprinkled or dotted with minute perfect flowers (having both pistil and stamens). As the season advances, quantities of pollen are produced, cross-pollination being accomplished by small flies which crawl all over the spadix, traveling from one spathe to another on sunny days. Honeybees likewise help in cross-pollination, visiting the skunk cabbages' offerings eagerly as early as March 22, at a time when there is little or no other pollen available.

Following the flowering period, when both the odor and the pollen become so alluring to various insects, the softening spadix becomes covered with rounded soft berries which seem to be partly embedded in its tissues but which do not seem to be eaten by the small animal life of the habitat.

If one digs out the shoots before they have expanded into awkward plants, their rootstocks will be seen to be quite long, growing erect in the humus or soil, with their bases surrounded by whorls of fleshy roots (plate 67), reminding one somewhat of the exposed bases of corn stalks, or the supports aboveground which surround the trunks of tropical stilt palms.

The skunk cabbage plants which grew in the wettest portion of the author's woodland sometimes reached over three feet in height. Their huge leaves spread out as they aged, and were then quite uncabbagelike. Numbers of the plants came up so close together that, when fully expanded in May and June, they formed solid dark green masses, some of the big leaves overlapping (bottom photo, plate 68).

For all their apparent toughness and vigor at this time, they shrivel as summer advances, then blacken and completely disappear so that, by the middle of September—or earlier, sometimes—there is not a sign of them upon the dead-leaf carpet where their massed numbers shaded all below in June.

Apparently they have no caterpillar enemies, but the disappearing blackened remains of the leaves and stems of course mean that the micro flora and fauna of the forest floor—doubtless including many of the organisms which will be met in other parts of this book—are hard at work upon them.

Are these plants useful at all? Well, we have seen that the honeybees gather and use their pollen. Small flies no doubt enjoy their nectar, which no doubt is sweet despite the odor emanating from other parts of the cabbages. The author enjoys seeing their green vigor in spring as he welcomes all other natural comers in his woodland fragment, and once a pair of Maryland Yellow-throats placed their nest deep in the center of one of the plants close to the tiny natural pool where all kinds of birds come to drink and bathe in this wood in spring and early summer. (Plate 163 shows a nest of this species.)

Many plants, as we know, succumb rather rapidly when conditions in their natural environment are suddenly altered. Some are sensitive in the extreme, and others may be tolerant for a somewhat longer period before dying out, but the skunk cabbage seems to be the most tenacious under adverse circumstances. As an example, where a house with an open cellar had been built over once swampy ground, skunk cabbages continued to come up for ten years in semi-darkness. Their shoots and dwindling leaves turned white and soon withered, but the plants kept up the fight until utterly exhausted by a decade of endeavor to survive in the sterile fill in darkness.

During this fifteen-year study, it was noted that twenty-three species of herbaceous plants (wildflowers) blossomed, and they will undoubtedly continue to do so, perhaps with one or two exceptions. Year after year, all but two sudden newcomers came up and flowered at about the same dates each spring, summer, or fall, with dependable regularity; however, it should be remembered that the dates given for first flowerings will vary elsewhere, according to the type of soil, precipitation, temperature, and other less obvious factors. As recorded, they are solely dates for the locality of the mini-wood.

First to bloom after the skunk cabbages when their pollen-laden spadices had made their annual offerings to the insects, were the trout lilies, *Erythronium americanum.* These delicately beautiful, slender-petaled yellow blossoms, borne on long, pale-colored stems, bobbed and nodded conspicuously in every breeze (plate 69).

Authorities continue to call the species by different names—dogtooth violet, adder's-tongue, and trout lily (now make your own choice)—but since the plant is certainly not a violet, and since it certainly bears little resemblance to an adder's tongue but belongs definitely to the lily family, the name used in this chapter seemed the logical one to retain.

April 22 was the average date of first blooms in the mini-wood area. The woodland wanderer in April will sometimes see extensive beds of the species' long, narrow, grayish mottled green leaves, with only a few flowers showing, one reason being that flowering bulbs are produced sparingly, and from these, after considerable time has passed, much smaller bulbs result. These seem capable of sending up only single leaves for several seasons before producing *paired* leaves and blossoms. The great preponderance of foliage in the beds may also be due in part to seed which is occasionally produced, the few that sprout only adding to the leafy congestion. These trout lily bulbs, which have hairlike white roots, lie a few inches below the surface, in places forming veritable tangles, although the individual specimens may be tiny (plate 69).

Like so many other early spring flowers that are intolerant of the continued deep shade cast by the fully expanded arboreal foliage in May, the trout lilies arise and bloom quickly, after which, in a few weeks, even the leaves die down to ground level and are not seen again until the following spring. The leafage does not just disappear of itself, disintegration being aided and the leaf substances partly consumed by bacteria, fungi, and many tiny animals of the soil army, which will be described and illustrated in later chapters.

As mentioned before, by mid- or late summer, there remain no signs on the woodland floor of even such earlier massed foliage as that of the skunk cabbages. One would not then suspect that in May the woodland carpet had been entirely hidden beneath their massed broad leaves, or that the army of the soil would soon consume them.

Upon examining the trout lilies' insignificant-looking little bulbs, it is hard to imagine them as units for adequate food storage from which eventually come petals, pistils, stamens, and other intricate flower parts, and long-stemmed paired leaves for each single blossom.

Within the trout lily corolla, two circles of stamens surround the club-shaped style or stem of the pistil, which is tipped with rather gummy stigma ridges. These receive the pollen transported from other blossoms of the same kind on the hairy legs and bodies of bees and other visiting insects which thus accomplish their all important missions in return for pollen and nectar.

Here, in briefest outline, a few refresher notes on the subject of cross-pollination will serve for the wildflower species and other seed plants described in this book, and may thus become a useful reference for the uninitiated reader.

Fertilization and the production of seed occur only upon the union of the male and female sex cells, called *gametes.* Union of the two results in a fertilized egg cell, called the *zygote.*

Within the flower, the pistil and stamens are therefore the essential organs concerned with this process. At the apex of each stamen is an *anther,* which produces pollen grains which in turn give rise to male gametes or sperm cells. On the lower, enlarged, or often bulbous portion of the pistil, the *ovules* are situated, each of which contains an egg. When a proper species of pollen grain is carried and left on the sticky stigma of the pistil (either by an insect, or by the wind) a *pollen tube* grows out from the pollen grain and down through the style, carrying the male gamete to the egg cell, which, upon entering it, accomplishes fertilization, assuring a seed and embryo in that ovule.

Many plants are also reproduced vegetatively, as from cuttings, or even from leaves, in some cases, but sexual reproduction, as just outlined, when studied in all its complex details, is the most beautiful plant method, the result of numberless experiments tried out by nature during millions of years of evolution, and the retention of those most successful.

Contemporary as a blooming species (with the trout

lily) and most often found on the woodland border, was the spring beauty *Claytonia virginica,* a plant arising from a perennial root growing quite deeply. Plants on the mini-wood border were about six inches in height, with narrow, quite pointed leaves, and characterized by white, pink-veined blossoms in terminal racemes (plate 69).

Around April 15, when their habitat was still well bathed with sunlight, flowers were produced in profusion, their distinctly pink anthers tipping the yellow stamens adding greatly to their delicate beauty.

Blooming was quickly over and the occasional seeds reproduced in little capsules, and the entire aboveground portions of the plants vanished from the habitat by the first part of June.

For no discernible reason, a small, vigorous patch of spring beauties which bloomed for several seasons beneath a spicebush failed to appear one spring, and had not been seen there again up to the completion of the fifteen-year study period. This is the way it is with some plants, as will be learned by the close student of their idiosyncrasies. Some, like the skunk cabbages, seem to grow on, maybe even longer than a human life-span. Others come and are gone after a season or two, and we cannot always say why.

Among these early bloomers within and on the borders of the mini-wood, two species of violets, one of them in two color forms so different from each other that the uninitiate might easily mistake them for separate species, have occurred with dependable regularity every year since the author has lived in the area.

The common blue violet, *Viola papilionacea,* sends its numerous blossoms up among abundant heart-shaped leaves in late April and early May. The two forms in and about the mini-wood were (1) that with deep violet-colored petals, with the inner part of the lower one white streaked with violet, (2) and that with white petals having much violet on their inner portions, and with numerous darker violet lines extending outward, almost to their margins, in some flowers (plate 70). While the deep violet and partly streaked white-throated form is by far the commonest, and the white and violet-streaked phase the next most frequently seen, this species is very variable, and may be found in intermediate combinations as well. All violets are exquisite flowers even when casually observed at a distance; close up, their true delicate beauty and perfection of construction are best seen with the aid of a modern hand glass of about ten-power magnification (plate 70).

Violets have five petals. The central or lower one pro-vides a convenient landing stage for cross-pollinating insects. The two side petals are called *wings,* and above them, the fourth and fifth petals are often turned backward to a considerable extent, and in some cases twisted at their tips.

The inner bases of the side petals are provided with hairy tufts called the *beards.* Seen under the microscope, in this species of *Viola,* these appear like oval scrubbing brushes with white "bristles" which in reality are soft and delicate hairs.

Pushing these aside, as an insect crawls into the flower, the pollinator reaches the greenish pistil surrounded by five leaflike stamens with yellow-colored tips, the lower two of which secrete rewarding nectar. Close observation reveals that the beards sometimes become pollen dusted as they brush against hairy bees of small size, and as these beards are close to the pistil, they doubtless aid in pollination when pushed against that part of the stigma at the tip of the pistil, where the pollen grains adhere, and which, if of the proper species, may germinate.

The flowers thus fertilized produce tripartitioned capsules, each compartment of which contain a group of seeds. As these capsules open easily when ripe, these common violets "plant" themselves abundantly, which explains the annually expanding beds, particularly those of the usual commonest form, most often noticed.

The white, violet-streaked form of this species is impartial in its choice of wet or dry soil. Large plants (self-seeded) have appeared in the author's flower garden and also at the very edge of a trickling streamlet, with the roots in saturated soil, where it bloomed profusely. Later, more appeared in very wet soil and thereafter continued to thrive in this habitat.

In the upper, sloping portion of the mini-wood and close to a stone wall elsewhere on the author's land, a few plants of the sweet white violet, *Viola blanda,* appeared during this fifteen-year study, but disappeared before they could be photographed. So named because its little flowers emit a delicate, very faint perfume, it was alluring enough to attract many tiny winged insects through whose ultrasensitive antennae the messages no doubt came in clear and strong. Only half an inch in width (measured across the sparsely streaked lateral petals when wide open) the flowers seemed suited only to tiny bees and flower flies, the lower petal, or *landing stage,* being very frail, although alluringly veined with rich brown on a white background.

The pistil was pale green and surrounded by yellow-tipped pollen-bearing organs. If these oddly shaped parts were carefully lifted with the tip of a dissecting needle and placed under the microscope, individual pollen grains could then be seen adhering to the metal instrument.

The leaves of this species, as is usually the case with violets, were much more abundant than the flowers. Unusually small, some measured up to seven-eighths of an inch in length and five-eighths of an inch in width, but probably grew somewhat larger after flowering time. All were deeply notched at the stems.

One must be an expert botanist to correctly name many of our violets. As stated, they vary greatly, and hybridization adds to the difficulties. One will find, even in very up-to-date flower books and guides, that the descriptions are not always too helpful. The only way to be certain about a doubtful species or variety is to get the help of a specialist, supplying him with living, or properly prepared dried, specimens and full data.

Two other small violets occurred within the mini-wood. One of these, which, after some study, the author believes to have been *Viola conspersa,* produced delicate lateral petals spreading about three quarters of an inch. Of a medium shade—a soft but not pale violet—these lateral petals, in a plant examined as this was being written, were white toward their inner ends, with a few dark veins plainly showing, and with rather conspicuous beards. The lower petal or landing stage was also partly white but streaked with a few darker veins. The flower stems, which seemed quite long as the buds were opening, were greatly outdistanced by the stems of the thickly bunched heart-shaped leaves before blossoming was over. The largest of these leaves measured two inches in length, and an inch and five-eighths in width. Self-seeding, the species nevertheless increased very slowly, the plants located for most of the time in the heavy shade of a mature beech tree and its family of fifty or more offshoots from the roots (plate 70).

What has added greatly to the fascination of this study, as mentioned before, was the fact that the mini-wood community continued to evolve and change, new things appearing even as some of the older inmates gradually faded out, or might soon do so. Thus, just as the final draft of this book had been completed, still another species of violet new to the mini-wood appeared, and it is here described, although not pictured.

This was a diminutive and still unidentified species.

From many small heart-shaped leaves in a tiny cluster, a few long-stemmed, bicolored delicate blossoms finally appeared. Measuring a half inch in width, these lateral petals were partly white, dark-veined, and bearded. The other three petals were similarly colored, and the lower one, or landing stage, was thrust boldly far forward, as if extending a special inducement to insect pollinators to traverse its violet-colored white-outlined veins.

As with the fungi described in chapter VII, herbaceous surprises continued to occasionally reward the mini-wood observer—but, unfortunately, less often than was the case among the fungi. Toward the end of the fifteen-year study period, a small, double-lobed rounded green leaf unrolled at the end of a four-inch stem. Close to the butt of a mature sassafras tree, it was first seen on April 15, and recognized at once as the leaf of the bloodroot, *Sanguinaria canadensis,* a member of the poppy family. Named for the color of its juices, this was the species whose horizontal rootstocks we used to exhume as kids, breaking them apart for the exquisite childish pleasure of seeing the "blood" ooze out. This sap was actually orange color rather than red, but no matter; to us it was blood, and it stained our young fingers delightfully.

The single precious leaf was surrounded at once with protective wire. Hopefully then, with the march of time, the author visualized a typical bed of these desirable wildflowers, each pure white eight-petaled blossom and its yellow stamens set off against a double or many-lobed leaf below.

But such was not to be. Although two leaves appeared the following spring, they were puny, signaling of course that, as far as the mini-wood was concerned, the species was doomed.

Who can say how the seed or the tiny new rootstock arrived under the sassafras tree? Perhaps a seed was dropped by a bird in its feces, although the author cannot describe which bird, or state that birds partake of them at all. Again, the seed may have been brought by wind. But whatever the case, the plant had found the mini-wood humus temporarily favorable, and it does thrive elsewhere in acid soil. How long the seed was there—how many months or seasons may have passed before it produced a rhizome and sent up that first leafstalk—are unanswered questions of the sort which so fascinate a naturalist in a study of this kind.

Where properly established, the bloodroot makes large, heavily veined, palmately lobed, rather dark green leaves, lighter on the undersides, and occasionally curled about the flower stems bearing their single, inch to inch-and-a-half blossoms. Of very short duration, one must hit just the right few days in late April or early May to see them in all their beauty, for they may be gone even if one returns within a week.

Sometimes the leaves grow horizontally, and they may remain here and there in the beds long after the flowers have withered away. It seems to be the normal habit of this species to expand into beds of considerable size, once it is established, and it was with deep regret that the author witnessed the unsuccessful efforts of this plant to become a member of this exclusive mini-wood "club".

In a similar case, also during the latter years of this study, a species of *Trillium* endeavored to establish itself in the mini-wood, appearing suddenly from cryptically planted seed or rootstock. At first it seemed to fare better than the bloodroot, its three-leaved stems increasing for two seasons, but it became stationary before dying out in the fourth year.

While the loss of these two possible additions to the mini-wood community was saddening, it nevertheless served to illustrate very clearly how important the exactly proper soil conditions must be for the continued success of many a flowering species.

Leaving early spring behind, we find many additional herbaceous plants of the woodland coming into bloom in May or later, some of them, indeed, not blooming until late summer or early fall. Those which flowered in May constituted a somewhat more shade-tolerant group than the earlier species described, but not sufficiently so to continue growing vigorously very long after blooming.

One, however, whose heart-shaped, deeply notched little leaves persisted in full vigor and color long after its spikes of flowers had withered away was the wild lily-of-the-valley, *Maianthemum canadense* (plate 70).

Blooming from early May well into June, this diminutive species, often called the Canada mayflower, grew in small but closely packed masses at the butts of oaks, red maples, and doubtless other trees, and with cushions of bright green moss in the same habitats, in situations where considerable sunlight often reached the plants.

Bearing many white blossoms on a single spike arising through the notch of the topmost leaf, these flowers were never conspicuous even against the beds of green. Individual flowers were less than a quarter of an inch in width, and when magnified, were seen to form little crosses with their petals beneath the four lengthy stamens.

Most of the plants had arisen from delicate rhizomes which sent up many leaves but comparatively few spikes of blossoms. The little berries which followed the flowers were at first white, turning pink to reddish with darker specklings at midsummer maturity. Cardinals eating them may have aided the species by defecating seeds in whole or viable condition.

Within a few yards of the laboratory, a small bed of these little plants produced many berries year after year. Above this bed, established at the butt of a white oak, a flying squirrel made its summer nest of green oak leaves in a bird box. Here, the author thought, was a possible consumer of the wild lily-of-the-valley's seeds, maybe swallowing the berries, pulp and all. But as many times and as long as the squirrel's box was watched at dusk, the little mammal was never seen to take any of the berries. Upon leaving the box for its evening feeding, it always went *up* the tree, thus leaving the question of the plant's dissemination by a squirrel unanswered.

The thick beds of leaves, however, may indicate that it may be self-planted by seeds, as well as multiplying vegetatively from the rootstocks.

The wild geranium, *Geranium maculatum,* is a wildflower of great beauty which grows equally well within the woodland or on its sunnier borders. It appeared regularly in the mini-wood area in early April, arising from rugged perennial rootstocks. Growing rapidly into plants two feet in height, its well-branched forms came into bloom by May 15 (plate 71).

This geranium produces many buds, which open into flowers with five pink or purplish pink-tinged horizontal petals about an inch in width. In the center of each, a curious beaklike object, the "crane's bill," is the pointed apex of the seed capsule from which, when mature, the seeds seem to be automatically ejected.

Wild geranium leaves are interestingly cleft, lobed, and otherwise quite distinct from those of any of the other herbaceous species found in the mini-wood area.

Another member of the lily family found scattered singly through the mini-wood was Solomon's seal, *Polygonatum biflorum,* whose spring shoots arose late in April from white perennial rootstocks or rhizomes. Following mild winters, some of these stems reached ten or more inches in length by the middle of May, and at this stage were grayish green, blending into a soft bluish gray toward the ground. By the end of the third week in May they had lengthened to eighteen or even twenty-four inches over all, and were often gracefully bowed, perhaps partly due to the weight of the double row of alternate, prominently veined leaves which formed protective rooflike covers over some of the flowers (plate 71).

In a long row on the underside of each stem, paired blossoms dangled at the ends of long single stems. Tightly closed at first, when freshly opened they resembled narrow little bells of an inconspicuous greenish yellow color. In the illustration they are shown closed, after flowering.

This is a shade-tolerant species, often blooming beneath mature beeches and other heavily shading woodland trees. Following the closing and shriveling of these flowers, small black berries formed on some plants, and these were collected by chipmunks, probably for their little seeds.

Leaves sometimes remained on the stems into late August, at which time they yellowed, finally turning brown at the time of falling. Soon thereafter, all aboveground signs of the plants disappeared under the actions of bacteria, fungi, and animal organisms of the soil.

Each April, the new shoots appeared, as they had doubtless done for generations from the same rhizomes in the mini-wood.

Another wildflower that was regularly found within the mini-wood in the identical kind of habitat preferred by the true Solomon's seal, and which before it came into bloom might easily have been mistaken for that plant, was the false Solomon's seal, *Smilacina racemosa,* a handsome perennial herb which many people do not seem to realize is common in certain Connecticut woodlands. Although its two- to three-foot stems are also bowed and its leaves are somewhat similar, having lengthwise veins, those of the latter species are larger and have wavy borders (plate 72).

It will be remembered that the flowers of Solomon's seal are like little paired elongate bells hanging from the underside of the stem, while those of the *false* species are terminal, and appear as a white raceme consisting of many small blossoms.

The rootstock of the false Solomon's seal grows horizontally a few inches below the surface. It is white, with long white rootlets. An eight-inch length of a still longer example bore the telltale butts of eleven previous seasons' flowering stems, as well as the white, already formed bud for the next spring, when it was dug up on November 10 (plate 71).

In the mini-wood, plants in bloom were observed from the middle of May into early June. Blossoms were followed by clusters of round berries which increased in size quite rapidly. Green at first, they turned basically yellowish green, but were so thoroughly sprinkled with red spots that some berries appeared to be solid red or purplish red until examined with a magnifying glass (plate 72).

These fruits, each of which contained two half-round white seeds, seemed little relished by birds and small mammals, so most of them fell to the ground untouched except perhaps by deer mice.

Between May 15 and 30, in southern New England, we come to that all-too-brief and eagerly awaited time span when plant life all around us has just come into full young foliage and flowering once more.

It is a time when sun-warmed woodlands seem to glow, that period when scents, emanating from young growing things and opening blossoms, fill the air with a complex soul-stirring blend like no other perfume on this earth.

A fleeting period, it is a time to take Nature's tonic in large doses while there *is* time, and to let this elixir flow into your nostrils and communicate with your brain. Go out-of-doors on clear warm nights in mid-May. Shut your eyes and imbibe this scent, this ambience which will so soon be gone again until another spring. Listen to the peeps and cries of the migrating birds winging northward again in legions together far overhead—the cries of Warblers, Thrushes, Waders, Yellowlegs, others unidentified—with gonads swelling, hurrying to the nesting grounds. Now open your eyes to the starry background against which they travel with such precision without need of what *we* call technology!

If May's ambience fails to stir you, if the peeps and calls, cries and trillings from the skies and swamps and ponds, from the foliage and flowers do not move you, if this symphony of sounds and odors declaring that life and its begetting is the thing of the hour fails to prod you, then you have missed something very precious about spring, one that henceforth you should endeavor to experience.

This is the time when jack-in-the-pulpit, *Arisaema atrorubens,* subtly adds to the woodland scents. Within the "pulpit" formed by the spathe and flap above, tiny flowers crowd the base of the erect, fingerlike spadix, an organ seen by many as the preacher at his sermon (plate 73).

The inner walls of this spathe may be plain green or purplish green, and often striped with contrasting color. Whatever the case, small insects are attracted to the jacks in numbers, but the little flowers below are partly shut off by an enlargement of the spadix which leaves room only for a few very small pollinators to squeeze by. While some gain entrance and exit, others find themselves trapped by the smooth walls or unable to find the narrow passage by which they entered, and so die below the spadix.

Jack flowers may be staminate and pistillate on separate plants, or have both stamens and pistil on one plant. Despite the difficulties encountered by pollinators, many of the flowers are regularly fertilized, as evidenced by the clusters of rounded green berries that form before summer and which become bright red toward fall (plate 72). Corm resulting from a seed (see below) is shown in plate 79.

Under the glossy red skin of each berry, the small seeds are embedded in a pulp which does not seem to be relished by woodland birds or deer mice, nor by squirrels, and the jacks reseed themselves in their own immediate vicinities, hence the groups of young plants often found growing close together.

The strong stems of the mature jacks ending in three big leaves arise year after year from robust corms which are the food-storing parts of the plants and the form in which they pass the winter. Some of these mini-wood plants reached thirty inches in height, most of them preferring the wetter parts of the area.

Both the Asiatic dayflower, *Commelina communis,* and the native Virginia dayflower, *C. virginica,* members of the spiderwort family, occurred quite regularly in the mini-wood area (plate 74).

These are shade-tolerant species, spreading rapidly by means of creeping stems which, if pulled up and carelessly tossed aside, will often root from the leaf attachment points.

Usually considered a weed—an unfortunate word too often applied to plants even when they interfere very little with agriculture and may beautify the world around us with their flowers, these dayflowers belong in this category of little appreciated species.

The blossoms are small but very beautiful, the two upper petals being of that deep blue sky color as it is seen intensified through small gaps in a woodland canopy. Below these petals is a noticeably smaller one—white in the Asiatic species, but blue in the native. Thrusting out from these three petals are three short sterile stamens and three much longer ones whose pollen-bearing anthers are turned gracefully upward. Each flower is borne singly at the top of a leaf stem, where it is held between two folds of a bract,

characteristics which add to the interest of the plants.

Arising from seed each spring, the dayflowers of the mini-wood area came into bloom late in May, then continued to make flowers for most of the summer, each flower beginning to shrivel in a single day or a little later after having offered its pollen to the insects.

Despite the brief life of these blossoms, many were fertilized and produced seed which was sought when still soft by Cardinals and Towhees, and they, by squeezing the swollen parts, extracted the seeds just as they and the House and White-throated Sparrows extracted the single hard dry seeds from the great numbers of seed-bearing single wings annually scattered by the tulip-trees.

As the dayflower plants formed no rootstocks, they no doubt seeded themselves, and birds consumed the seeds when they were still so soft that they were probably digested, or maybe destroyed by crushing.

Beds of dayflowers, some of which are so extensive that they might be mistaken for vines, will sometimes be found by midsummer. At the author's place they often grew in heavy shade along the mini-wood's northern border, as well as in much more sunny situations. Self-seeding was not always successful either, and no plants at all sprouted some years where dense beds had developed the year before.

In the mini-wood area, only two more or less poisonous herbaceous plants were found as members of the community. One was the poison ivy, which, being a vine, has been included in chapter II. The other species was hellebore or Indian poke, *Veratrum viride* (plate 74), another member of the lily family.

In a rather open part of the wood which received considerable sunlight even after the trees had come into full leaf, a single specimen of hellebore arose each spring sometime between April 10 and 26. From a thick, erect rootstock it sent up double stems which bore many large, broadly pleated leaves. It reached four feet in height by the middle of May, and in some years fell over in the wet soil of its own weight. It never made flowers, however, which occur in panicles with numerous six-petaled greenish individual blossoms formed like six-pointed stars.

At the base of each blossom's three short styles, the ovary is a three-lobed capsule in which each flattish seed is surrounded by a flat, thin wing. The starlike flowers in combination with the large and elongate pleated leaves serve to distinguish this plant from the skunk cabbage, which, as stated, often may be found in the same habitat with it.

All parts of hellebore (of which there are several species) are said to be poisonous. Horses and cattle believed to have eaten the leaves have been reported affected and made ill, but as far as the author knows, not in this country. Some have called the species illustrated in this book the "itch" plant, but the author having often walked among these plants where they were growing in swampy land with skunk cabbages and inadvertently making occasional contact with them, was never afflicted with the "itch" therefrom. Those who are easily affected by poison ivy, however, had best avoid contact with hellebore "just in case," at the same time keeping in mind that the species often grows side by side with skunk cabbages.

How this single, long-lived, vigorous mini-wood specimen came there in the first place is unknown to the author, but the seeds and wings are flat, and one was probably wind carried to this evidently favorable type of soil in which it continued to grow as a vigorous example of the species, its shoots appearing each April but dying down and completely disappearing by early July.

While it is a conspicuously common wildflower, many are unaware that touch-me-not (or jewelweed), *Impatiens capensis,* has evolved a seed-dispersal device that is truly astonishing, and which accounts for the great abundance of the plants which spread annually over these areas of moist woodland or woodland border soils.

By the first week in May, in the area of the mini-wood, the jewelweed seedlings had already reached the five-leaf stage and were crowding each other in vigorous rivalry for available space. Growing rapidly and making many branches, they reached five feet in height by midsummer, the first flowers appearing early in July and becoming more numerous as the season advanced, and most abundant late in August.

These curious blossoms, sometimes three or four to a raceme, dangle from long slender pedicels and are so balanced that they often hang, so to speak, on a level keel, while others may hang at various angles (plate 75).

They have three petals, the upper one often broader than long, and the two lateral ones below together measuring about a half inch in width. Of the three sepals, two are undistinguished, but the third is extended and ends in a spur. The pedicels to which the flowers are attached extend out from the somewhat thicker plant stems from

one to two inches. The petals vary in color through many shades of orange and orange yellow, and are plentifully spotted with red and reddish brown, the spots reaching well back into the flower throats or sacs. Looking into a flower from the front, the pistil and stamens will be seen as short organs, the latter often hidden under copious amounts of white pollen (plate 75).

Soon after the petals fall, the ovary swells into an elongated capsule, and it is via this part of the species that we will witness the astonishing seed-dispersal method of this plant.

Each capsule contains four seeds. When mature but still green, the capsule becomes greatly swollen with its contents. Suddenly it *dehisces,* that is to say, the five strips, botanically called *valves,* of which it is composed, fly apart, at the same time coiling spirally, and with such force that the seeds are flung in all directions, sometimes for considerable distances. What we witness is a mechanical "explosion" produced by a living organism, and the more remarkable for that reason. The plant's English name—touch-me-not —is thus very appropriate (plate 75).

The author would urge the interested reader to seek some of the plant's "touch-me-not" capsules when ripe and swollen, and by disregarding these "warnings," experience the surprise of the capsules' instant reactions.

Color varieties of the jewelweed are not uncommon. One of these, forma *albiflora,* produces white flowers spotted with red or reddish brown, while another, forma *citrina,* has spotted petals that are lemon yellow.

The common *I. capensis,* which flourished year after year in masses on the borders of the mini-wood, exhibited some variation in petal shading and in the amount of spotting, possibly due to some extent to the soil in which they sprouted, but more probably to hybridization with the very similar *Impatiens pallida,* a few of which grew in the area of this study, and with blossoms basically yellow. Here and there a specimen of this other touch-me-not was found growing close to where the thickest masses of the more heavily spotted species were flourishing.

The name we always used for this abundant wildflower when we were children was "silver-leaf," for somehow, fooling around with everything we came upon, as children do, we discovered that when its leaves were placed under water, their undersides would appear as if dipped in solid silver, an effect apparently due to the presence of a microscopic furry coating on the underleaf surface which held millions of infinitely small pockets of air.

The spotted touch-me-not seemed singularly free from the attacks of caterpillars, but its juicy leaves were eaten to some extent in summer by the large orange-colored slugs then maturing, which are described more fully in chapter XVI.

Two herbaceous species which came into flower early in July were the common but little noticed Virginia knotweed, *Tovara virginiana,* and the less common white avens, *Geum canadense.* A few of the latter became established in the same wet habitat with the skunk cabbages, and later in somewhat drier situations. The two-foot stalks bore small, coarsely toothed leaves above and much larger ones below. Around July 10, several other stems produced small white blossoms with five distinctly separated petals (plate 76). The white avens was one of those plants which just appeared very suddenly several years after this study was begun.

The Virginia knotweed, on the other hand, appeared year after year in the same spot on the woodland border, and had been doing so ever since the author moved to South Norwalk.

Here is a plant whose tough, broadly ovate, untoothed leaves are arranged alternately, some with short stems; other, smaller ones sheathed about the main stem. In August it sends up a tall, slender raceme or flower stalk bearing two dozen or more well-separated little flowers which are greenish, or pinkish white. These are alternately arranged and bend downward, and from them the two styles, each ending in a tiny hook, protrude (plate 76).

Minute flies visit the flowers, sometimes in dancing crowds like motes in a sunbeam, possibly indicating that a scent is emanating, one as pleasant and alluring to these minutiae of the insect world as that of basswood or pepperbush or honeysuckle is to the cognizant human being. Fertilized, each flower develops a small dry one-seeded fruit or *achene* which retains the styles and hooks already mentioned, and which catch in animal fur and on human beings' socks, and are thus distributed.

Sometimes mistaken for a fungus, always exciting interest, the white and fleshy Indian pipe, *Monotropa uniflora,* appeared now and then in the mini-wood (plate 77).

A member of the wintergreen family, the Indian pipe's ecology is particularly interesting. Like many fungi, for which, as stated, it has often been mistaken, it is deep-shade tolerant and, having no chlorophyll, cannot manufacture its own food. Close examination will reveal that its bent-over flower is recognizable as such, but that its leaves have been reduced to mere scales, and its manner of life accordingly complicated and in need of much further study.

If in late July or early August, when the species appears, the observer will excavate the whole plant with all the care of the professional paleontologist exhuming a fragile fossil bone, he may find a delicate white network, like so many fine threads, invading the roots of the Indian pipe and extending into the soil or humus as well. These threads are the hyphae of a fungus through which it would seem the pipe receives some cryptic sort of benefit. There also seem to be interrelationships at times among the fungus threads, the Indian pipe, and certain large trees close to which the latter becomes established. Investigations in Sweden, in the case of another species, the pinesap, belonging to the same genus as the Indian pipe, seemed to indicate that this plant's roots, together with those of its fungus partner, and also rootlets of a close-by species of conifer intermingled, and the pinesap thus received nourishment from the tree!*

This whole involved subject is one of those most worthy of more study to be found in our woodlands. In the mini-wood, the Indian pipe has always appeared in situations underlain with the roots of beech trees, and where the leaf carpet was fairly heavy and wet.

In one case, four tall Indian pipes popped up close to the base of a glacial boulder, on the top of which the author had constructed a natural birdbath and from which, all summer, water trickled down its side and to the plant's exact location (plate 77). Nearby was a tupelo tree, and a little farther away, a mature beech. A Virginia creeper vine had grown upward in the same habitat.

When the Indian pipe flowers are fresh and white, and turned downward, as stated, they will be seen to consist of from four to six petals, each about one inch in length and hiding ten stamens surrounding a pistil with a five-chambered ovary, a capsule in which the seeds develop from the fertilized eggs, as in other flowering plants. Seeds are made in good numbers, and as the flower ages, it turns straight up, and the entire aboveground portion of the plant turns hard and black. As previously stated, the leaflets are represented by thin scales on the tall stems which push

---

*See Carl L. Wilson, Walter F. Loomis, and Taylor A. Steeves, *Botany,* 5th ed. (New York: Holt, Rinehart and Winston, 1971).

Above: A white ash, growing in a habitat that was always wet and often flooded for days at a time, developed very large buttresses and withstood a storm that blew down a mature beech nearby. *Below:* A red maple, growing in a portion of the mini-wood which was moderately wet in spring, grew these buttresses. (Page 12.)

Left: A thirty-five-foot basswood tree on the edge of the mini-wood, photographed on a typical brilliant but breezy day in late May. (Page 10.)

Above: In full bloom July 15, the flower clusters at the ends of long slender supporting stems bob and swing in every slightest breeze. Shown much reduced. (Page 10.)

Basswood buds, shown much reduced. Date: June 24. (Page 10.)

Above: Basswood flowers and bracts, natural size. Date: July 10. *Left:* A flower cluster, much enlarged. Same date. (Page 10.)

Plate 27

A fine example of the red maple which in youth split near the ground into two trunks. At left, it is shown May 4, with the leaves just unfolding; at the right, it is in full leaf and beauty, photographed September 2. (Page 10.)

Below and right: Staminate flowers of the red maple, shown greatly enlarged. Date: April 5. *Bottom right:* The pistillate flowers, about natural size, photographed on the same date, same tree. Staminate and pistillate flowers may occur on separate trees also. (Page 10.)

Left: Normally, by May 2, as seen here, the red maple's winged seeds have been produced and the young leaves are rapidly expanding. Shown about life-size. (Page 10.) *Below:* A branch with young but fully expanded leaves. Date: May 25. About one-fifth natural size.

Left foreground: Two sugar maple saplings growing in dense shade. *Above:* Newly expanded leaves, photographed May 23. (Page 11.)

Right: In the mini-wood, the sugar maples grew as understory trees in the heavy shade of the dominant oaks and beeches. As shown, some produced branches not far above the woodland floor. Leaves are shown about one-fifth life-size. Date: May 28. (Page 11.)

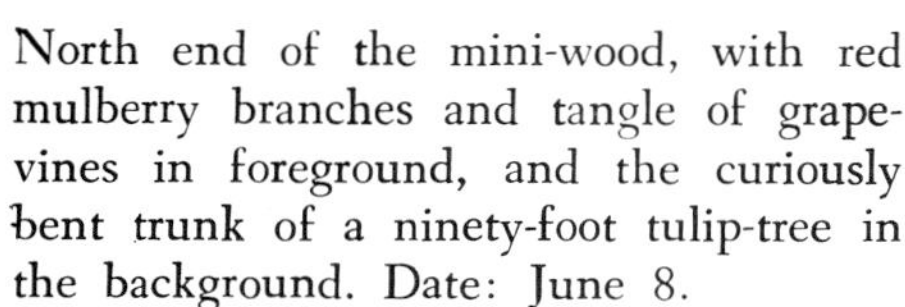

North end of the mini-wood, with red mulberry branches and tangle of grape-vines in foreground, and the curiously bent trunk of a ninety-foot tulip-tree in the background. Date: June 8.

While vast numbers of tulip-tree seeds are shed annually, only a very small percentage result in successful seedlings, most of the seeds disintegrating on the ground. *Left:* Life-size first-year seedling. Date: August 21. *Below:* A second-year tulip-tree growing beside a second-year red maple, at left. About one-fourth life-size. Date: August 4. (Page 11.)

No other native woodland species produces flowers as large and beautiful as those of the tulip-tree. Six pale green petals, each with a large orange blotch at its inner base, form a cuplike corolla for the pistils and whorls of stout stamens. There is nothing else quite like them. Flower below is shown life-size. Date: June 15. (Page 11.)

Above: Sunlight filtering through the springtime mini-wood about the trunk of a mature tulip-tree, May 15. *Below*: Pistils ripen into conelike fruits, each producing dozens of winged seeds (as in inset, upper left). Dates: Fruits, August 30; seeds, November 20. All shown natural size. See also plate 79. (Page 11.)

Left foreground: A young ash showing a sparse branch arrangement. *Right*: A white ash sapling five feet high and several years old, growing in an entrance path to the mini-wood. Date: June 5. (Page 11.)

The white ash that stood firm in wet ground while the mature beech in the same habitat toppled in a windstorm is shown here in the center between darker-foliaged oaks. Date: July 28. (Page 12.)

Typical leaf arrangement of a young white ash. The stalks to which each group of leaflets is attached are shed in the fall, but not until after the leaflets have dropped. Shown greatly reduced. Date: June 30. (Page 11.)

A double-trunked pignut hickory on the mini-wood border, photographed when not yet in full leaf, on May 23. It produced a crop of nuts but seldom. (Page 12.)

Top: Leaf arrangement of the pignut hickory. *Bottom*: Sutures show plainly on these pignuts, shown slightly enlarged. Date: November 11. (Page 12.)

Left: A tall tupelo with most lower branches growing outward, away from the dense foliage and shade of the trees at its back. *Below*: Leaf arrangements and dense little clusters of staminate flower buds, shown natural size. Date: June 5. (Page 12.)

Above: The female or pistillate catkins of the ironwood, shown at left, are very **different** from the male catkins, shown at right. Dates: Female, April 24; male, May 2. Both shown natural size. *Below:* Leaves and fruits, photographed June 13. (Page 13.)

Above and left: Mature ironwood fruit bracts. Each bract bears a small hard nut at its base, as in the first two bracts seen above. Natural size. Date: August 6. (Page 13.)

A wild cherry tree completely re-clothed with new foliage six weeks after being completely denuded by tent caterpillars. Date: August 10. (Pages 13, 136, 137.)

Wild chokecherry. (A) Leaf arrangement and flower racemes. (B) A raceme shown natural size. (C) Opening flower buds, natural size. Date: May 25. (D) Maturing fruit, natural size. Date: July 23. (Page 13.)

Located close to the twigs, opening spice-bush flowers are shown natural size. Date: April 24. (Page 13.)

Above: A rising green patch of vigorous skunk cabbages with a haze of late-blooming spicebushes in the background. Date: May 10. *Below:* Close-up of a spicebush with pollen-rich flowers in full bloom. Date: April 23. (Page 13.)

Scene in the mini-wood. Dark branches in foreground are those of a red mulberry supporting heavily shading vine of the frost grape. In background, the curiously bent trunk of a very tall tulip-tree. Date: June 8, about mulberry-flowering time. (Page 14.)

Right: Leaf forms, and green and ripe mulberries. Dates: Green berries, June 20; ripe (black), June 30. Shown about two-thirds life-size. (Page 14.)

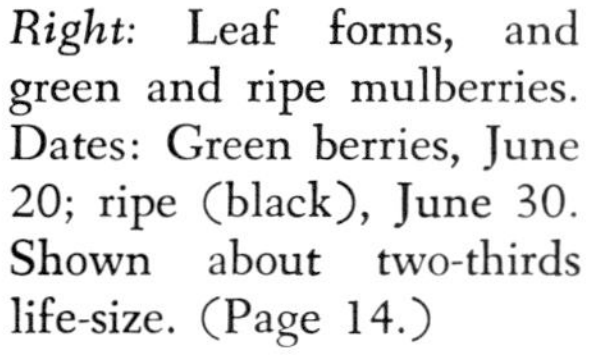

Above: Sweet pepperbush in full bloom. *Right:* Racemes of the pepperbush, whose blossoms give off one of the most alluring and delicate scents in nature. Enlarged about one-fifth. Date: August 6. (Page 14.)

The curious winged objects are milkweed pollinia, impaled at their lower, bean-shaped ends on spines on the legs of wasps which have visited these flowers. Photomicrographs enlarged × 40. Date: July 27. (Page 15.)

Above: Tarsal end of leg of the solitary wasp *Chlorion ichneumonea*, and impaled pollinia. *Left*: Showing how bean-shaped ends of pollinia are impaled on the wasp's spines when it visits milkweed flowers. Both photos × 40. (Page 15.)

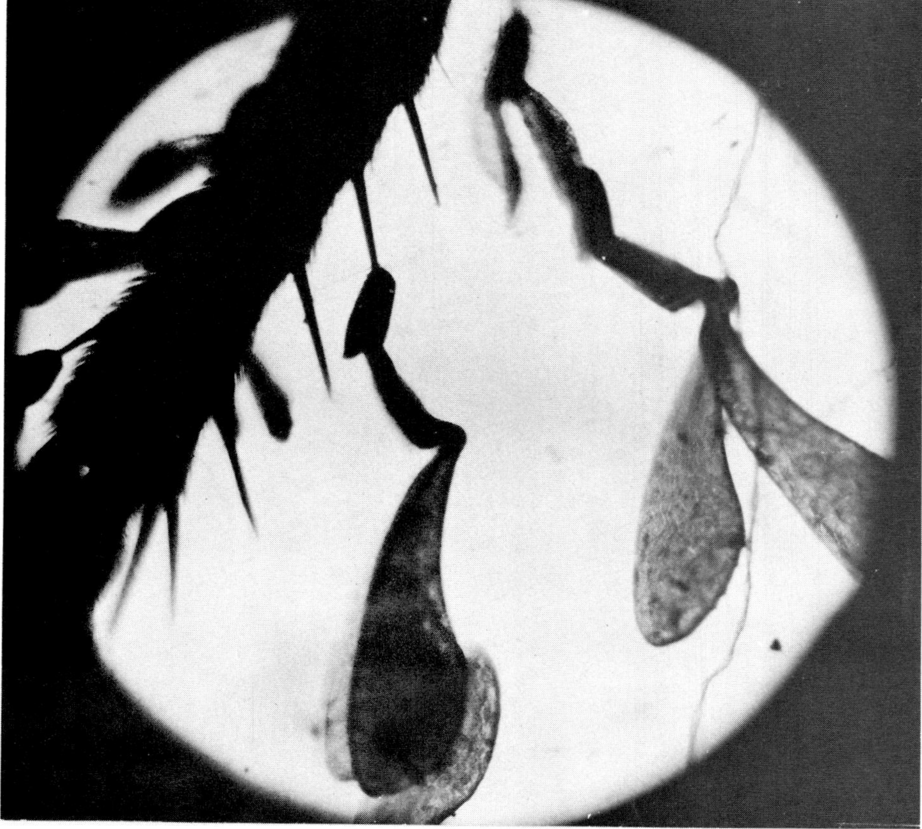

The wasp *Chlorion ichneumonea*, which becomes involved with the pollinia of milkweed, is shown about life-size, and here feeding on pepperbush nectar. Date: August 14. (Page 15.)

Above: Milkweed flowers in readiness to attract insects and to utilize them by impalement of pollinia on the leg spines of these visitors. The flowers often trap and kill insects too weak to pull away. *Left:* Milkweed flowers much enlarged to show details. Date: August 14. (Page 15.)

A cluster of ripe arrowwood berries. Date: August 10. In numerous instances, all of the berries fall off before ripening. Shown natural size. (Page 16.)

The arrowwood's characteristic deeply ribbed leaves, and a cyme made up of numerous small white flowers. *Below:* Close-up of more open cymes, actual size, and (inset) top view of flowering head. Date: June 13. (Page 15.)

Left: Like so many other trees and shrubs, elderberry shrubs in the mini-wood area have prolific fruiting years and off years. Here, even in a prolific year, the shrub continues to bloom even after berries have been produced. Date: July 10. *Right*: Elderberry cyme with maturing berries, July 21, about natural size, and (inset) mature berries, taken same date. (Page 16.)

Highbush blueberry. Blooming branch and woodland background. About half natural size. Date: **May 13.** (Page 16.)

Above: Blossoms and forming berries at the same time. Date: July 23. *Left:* Leaves and arrangements of blossoms. About natural size. Date: May 18. (**Page 16.**)

Left: Winterberry leaves and flower-bud clusters, as they look the last week in June. About half natural size. *Below*: Buds and opened blossoms. Date: July 7. Shown somewhat enlarged. (Page 16.)

Twigs well loaded with the bright red winter-berries which are eaten by many birds, especially Cedar Waxwings. Shown enlarged one-third. Date: November 7. (Page 16.)

PLATE 51

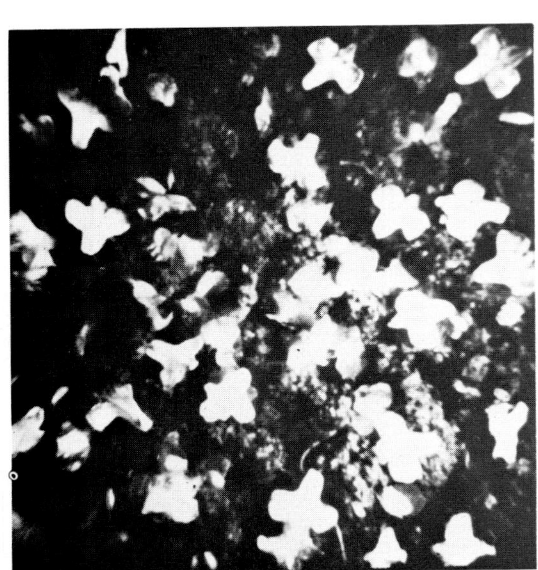

Left: Pollen-bearing catkins of the black birch, four-fifths natural size. Date: May 4. *Below, at arrow:* The bowed slender trunk of a gray birch growing close by a black birch whose branches and twigs fill upper part of picture. Date: March 5. (Page 17.)

Birch seeds resemble tiny birds in flight. On suitable soil they sprout in three to seven days. Enlarged × 2. Date: May 10. (Page 17.)

PLATE 52

Left: Seeds recovered from a birdbath, illustrating how various species of trees and other wild plants become distributed in the excrement of birds. Shown somewhat enlarged. Date: October 1. *Below*: Chokecherry-pit hoard of a deer mouse as it looked after the winter, with contents of pits removed through small gnawed holes. Somewhat enlarged. Date: April 1. (Page 17.)

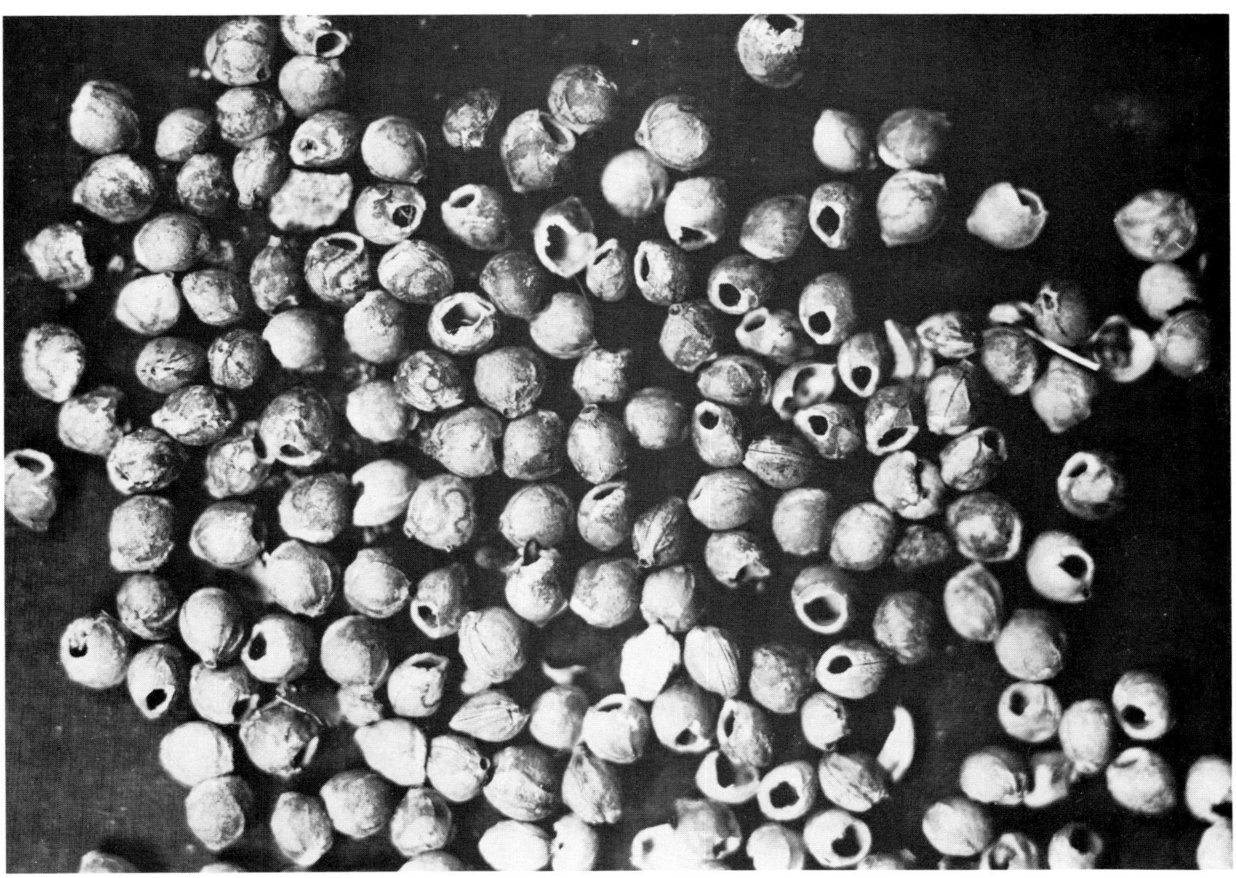

Above: Vista between two black oaks, looking west in the mini-wood in early September. *Below*: A mature beech trunk surrounded by the tree's saplings and offshoots. Date: October 19.

Rocks cut from glacially deposited material in the mini-wood were.used in the author's garage. Squares of basalt at left. (Page 19.)

Right: (1) Hornblende-schist. (2) Gneiss. (3) Granitoid gneiss. (4) Quartz. (5) Hornblende-syenite. (6) Glacially smoothed and polished granite. (7) Basalt. (8) Iron stained granite. (9) Red sand-stone. (Page 19.)

PLATE 55

Above: A Virginia creeper vine, many years old, that long since reached to the top of this white oak. Typical leaf arrangement is shown in center. Date: June 10. (Page 20.)

Top and bottom: "Holdfasts" with disks with which this creeper ascends its hosts. One has grabbed a pebble and carried it upward "by mistake." Shown life-size. (Page 20.)

PLATE 56

up through the dead leaves on the woodland floor.

As an experiment, dead leaves to an unnatural depth were raked onto a known site where Indian pipe flowers had appeared regularly for several years. This was done in the spring, so that this leaf cover had little time to settle. In that July the pipes appeared again, having extended their stems a full *eight to ten inches* longer than normally in order to bring the flowers to the normal height above the dead leaf layer!

It has already been said that the Indian pipe is considered by some to be a saprophyte, but it would seem to the author it is the associated fungus, not the flowering plant, that belongs in this category, since a saprophyte is a plant capable of existing upon dead and decaying organic matter (which wildflowers less often do, although some, like the pitcher plant and sundew, absorb their nutrient from dead insects).

White snakeroot, *Eupatorium rugosum*, a member of the composite or daisy family, bloomed in the mini-wood from the middle of July onward for several weeks, rising three feet or a little higher, and bearing rather large ovate leaves with toothed margins and sharp points. Its inflorescence (a plant's flower arrangement, whether singly on the axis, or in clusters) in the snakeroot consists of rather delicately constructed cymes (two stems arising successively from each new stem), as shown in the bottom photograph in plate 77.

The numerous white flowers have a *fringed* appearance, due to the protruding styles of the pistils. Because of the many stems bearing flowers, this type of inflorescence offers great numbers of individual blossoms to eager numbers of insect visitors, many species of which are useful pollinators. There are bees of various sizes, large and small wasps, flower flies, and certain butterflies, all of which visit the white snakeroot for nectar and pollen, or just nectar alone, as in the case of the butterflies. All of these insects find only tiny offerings of nectar in the individual blossoms, but, collectively, they produce a considerable quantity of this important food for a host of winged arthropods, many of which repay the plants by distributing their pollen from plant to plant, as most everyone knows.

The largest patch of snakeroot in the mini-wood bloomed annually beneath large black oak trees on a slope where the soil was drier than in the rest of the area.

A delicate little metallic blue butterfly which habitually feeds at the snakeroot is *Lycaenopsis argiolus pseudargiolus*, whose common name is the "spring azure." Those which come to the snakeroot are of the second, or summer, brood of the species, and the only form seen at this time of the year.

Another butterfly recorded a few times at these white flowers was the hickory hairstreak, *Strymon caryaevorus*, a mostly pale brown neat little insect with touches of blue, orange, and black on the undersides of the hind wings near the tiny tails. Two chrysalids or pupae were found beneath a board on the ground under a pignut hickory tree within the mini-wood, strong evidence that the butterfly bred therein. (See plate 133, butterflies of the mini-wood.)

Snakeroot is attacked by very minute leaf miners, which, in feeding upon the green matter between the upper and lower epidermises, rout out extensive, oddly twisting paths and irregular spaces, leaving only the translucent leaf skins intact in these areas. This is the work of a maggot (which is later transformed into a tiny fly), but apparently does the plant no serious damage. (See mined blackberry leaf, plate 79.)

While only a few patches of snakeroot grew within the mini-wood proper, the plant is abundant elsewhere in less shady habitats, where it comes into competition for insect pollinators with the similar appearing boneset and also joe-pye weed and other species of the country roadsides and open meadows.

Moving now into September's bloomers, we come to the interesting stoneroot or horse-balm, *Collinsonia canadensis*, a member of the mint family. Usually at its best about the twenty-fifth of the month in the area under observation, occasionally it came into flower by the last week in August; thus, as in the case of many herbaceous species, it was not always predictable.

Arising from knotty brown rhizomes, by May 10, the bright green, beautifully sculptured leaf buds were several inches above ground, the first four opening horizontally. By June 5 the plants had produced a number of oppositely arranged, rather large, coarsely toothed leaves on one or more side branches. By late summer the plants reached three to four feet in height, and the curious flowers on several racemes had appeared, a few opening at a time below the numerous buds, and the whole inflorescence reaching as much as a foot above the highest leaves (plate 78 and rhizome, plate 79).

Individual flowers are hard to describe in simple terms; suffice it to say here that the corolla measured about a quarter of an inch in width, its five pale yellow petals collectively giving the blossom a somewhat bell-like form, but with the lower of the five turned downward and with its border distinctly fringed. The stamens and long pistil project well beyond the corolla, as may be seen in one of the photographs, plate 78.

In years when the weather had remained mild, horse-balm could occasionally be seen still in bloom by October 5, but, normally, the majority of the plants had finished their growing period by that date and were then yellowing rapidly.

A delightfully odorous secretion, not unlike a blend of lemon and lime, emanates from these plants. Pluck and crush buds, flowers, or foliage in the hand, and the odd scent will remain on one's skin for some time. It would seem from the species' common name that, at least in someone's imagination, somewhere, sometime, these plants also charmed horses! Perhaps it was this odorous secretion!

Stoneroot, the other name for the species, is perhaps more appropriate, for its rhizome is a curiously knotty-looking, hard-black rootstock. Growing quite shallowly beneath the surface, it makes a good subject to dig up and examine for the distinct scars, each of which indicates the part from whence a flowering stem arose in a past season. Such an examination will afford the investigator an insight into this particular plant's local history, and tell how many years it has lived in the woodland where found (plate 79).

Plants of the genus *Aster* are very numerous, and as species vary in details, they are very often difficult to identify with certainty, especially for the casual observer. Two species which inhabited the mini-wood, however, were among the more easily recognized kinds and are briefly described here.

The mountain or whorled aster, *Aster acuminatus*, with a somewhat *zigzag* stem, grew from one to three feet in height. The lower leaves were broad, coarsely toothed, sharply pointed, and measured up to six inches in length. Higher up, just beneath the groups of flowers, the leaves were much smaller, as shown in the photograph in plate 80.

The white rays (surrounding the central bright yellow disk) numbered from twelve to eighteen, some of them measuring up to three-quarters of an inch in length, and few of them overlapping (plate 80).

The yellow central disk, when examined with a magnifying glass, can be seen to consist of numerous diminutive flowers; thus, in plants such as this, there are *both* ray and disk flowers, and the seeds in this case are produced only by the latter.

This member of the huge composite family bloomed in the mini-wood at various times, from July onward, and occasionally as late as October.

More numerous in the mini-wood were the white wood asters, *Aster divaricatus*, growing from long-living rhizomes and mostly in the shade of oaks, sassafras, and other large woodland trees. The species reached two to three feet in height, had many side branches, and bore abundant, heart-shaped, toothed leaves.

Their profuse flower buds commenced opening, a few at a time, around August 20 and continued for several weeks on different plants; the corolla measured up to an inch in width and consisted of ten or more narrow, elongate white rays or petals and many blunt-tipped green bracts below (plate 80).

The central disk in this species, like that of the mountain aster, was also yellow. Under a magnifying glass, the tiny individual flowers of which it consists appeared to have five petals, more or less hidden by dense white hairs. Upon parting some of these with a needle, these little flowers with their stamens and pistils could be more plainly seen.

A few days after coming into full bloom, these disk flowers changed from bright yellow to reddish brown, but the long white rays of the corolla remained unchanged for some time.

Asters belong to the composite family whose members have been the most recent of the angiosperms to evolve, and like so many other plants sharing this distinction, they disseminate their seeds by means of fuzzy parachutes to which the seeds are attached singly. Normally, the parachutes with their little burdens detach themselves from the plants and may be carried great distances by the wind. In some cases, however, chutes and their seeds may remain within the dried bracts that stay attached to the dead stems all winter, being held in place for months after other seeds have made good their escapes. Perhaps this is a nice provision by Nature to accomodate small seed-eating birds hard put to find sufficient food in severe winter weather. Tree Sparrows, Goldfinches, Siskins and Redpolls all know this secret of the wood aster.

The foliage of both of these tall asters seemed mostly immune to serious insect attacks, although the tiny leaf-miners, whose galleries between the epidermises of the snakeroot and blackberry leaves are so common, may sometimes also be found in aster leaves. (See plate 79.)

The white wood aster's rhizome was found to be a much-branched, rather slender, shallow-growing object, sometimes zigzag in form, and with large numbers of fine rootlets. Both rhizome proper and rootlets were, surprisingly, *black*, but with many lighter scars indicating from whence stems had arisen in former years. This aster's rootstock may send up a substitute flowering stem if the first one has been injured or broken off. When a winter has been mild, the new underground spring shoot may already be an inch in length as early as January, as if eager for the growing season to arrive.

On some occasions in the mini-wood, pleasant surprises ended in disappointment for the author. Thus, new herbaceous species appeared suddenly, grew fairly well for a time, and then were never seen again. This may happen in any woodland, some such arrivals even reaching the blooming age. Great as the disappointment may be when such a recent arrival gives up and dies out, the event nevertheless becomes of biological interest, illustrating as it does how exacting many species may be in regard to soil requirements—and many other less obvious factors—in order to remain viable upon introduction into a strange habitat.

The most surprising of such newcomers in the mini-wood area was the turtlehead, *Chelone glabra*, a wild member of the snapdragon family, which is locally common enough outside the mini-wood, usually in wet meadows near ponds and swamps. In this case it appeared at the edge of a shallow drainage trench, with the woods on two sides and the author's garden on another. It grew in mid-July, sending up a rugged stalk twenty inches in height bearing ovate to lanceolate, finely toothed leaves. It did not bloom during the first summer, but the following year it seemed on its way to becoming well established, for by September 25 it had made many groups of its oddly bunched white and pinkish flowers.

When fully opened, they were seen to be double lipped, the upper of the petals arching over a shorter, tonguelike lower one. Viewed from the front, the triangular "mouth" thus formed, did somewhat resemble that of a turtle, except for the presence of the stamens and pistil (plate 81).

How this single plant became established remained unknown to the author. Perhaps it was bird planted by some individual bathing or drinking at the trench and leaving a seed in its excrement. No bird was ever observed extracting these seeds, which were spherical, flat, and surrounded with a wing. Whatever the case, its death doubtless indicated that the soil on the edge of the mini-wood was not just right for it in some subtle way, for this is a normally rugged plant arising from a perennial rootstock and preferring wet ground.

Turtlehead is the very special food plant of the intensely local occurring Baltimore butterfly, *Euphydryas phaëton*, a medium-sized species which is black with a row of tawny red spots on the margins of its otherwise white-spotted wings, the white spots being in two rows behind the marginal row, which in some specimens may be quite a bright red.

The species was never observed near this turtlehead plant, or anywhere else about the mini-wood. Where this wildflower grows abundantly near ponds and swamps, the Baltimore butterfly may sometimes be found concentrated in considerable numbers, and as it wanders hardly at all, it would seem that its entire life is normally spent in these limited areas where the turtlehead is available as food for its caterpillars. Being naturally weak fliers, one might say that the butterflies become "captives" within the limited areas suitable to this species of plant. Many fine examples of this insect in the author's collection were obtained on single days and all within a quarter of an acre of swampy land. Such ecological associations are what add greatly to the interest of this kind of study.

Another wildflower that suddenly appeared on the mini-wood border, bloomed for two seasons, disappeared for two more, and then reappeared a few yards distant from its first site, was the cuckooflower, *Cardamine pratensis*, a member of the mustard family.

This was an erect plant with small, rounded basal leaves, and bearing short little branches with linear leaves on the stalk above. Its pure white flowers, each with four horizontal petals, were borne singly or in clusters at the apex of the erect stem (plate 81).

Making an attractive showing, delicate and pleasing to look at, its disappearance was a distinct disappointment, and its sudden reappearance a most pleasing surprise. Would that the bloodroot and trillium described earlier in the chapter as other suden newcomers to the mini-wood might make similar comebacks! But change is in the nature of things, and even in a mini-wood this holds true, as we shall see again and again in these chapters.

*   *   *   *   *

Other wildflowers which appeared and disappeared during this fifteen-year study of the mini-wood area, but which grew a little too far away to be fairly included in the community, were the wood sorrel or "sour grass," the swamp buttercup, Indian hemp (which suddenly set up shop in an abandoned part of a vegetable garden), the beggar's tick, boneset, a thistle, and of course chickweed, dandelions, and pink knotweed or lady's thumb.

CHAPTER IV

THE MINI-WOOD FERNS

No lush or seasonally lush bit of woodland in the northeast would seem a normal botanical community without ferns to beautify the forest floor. Steadfast, calm, dependable year after year, once they are established, rarely eaten by insects and other consumers of woodland foliage, they constitute a particularly satisfactory part of the sylvan scene for the naturalist and all who revere Earth's chlorophyll-endowed organisms.

Ferns excite the imagination of those who know something of their story. Like certain other organisms to be described in later chapters, they are believed to have evolved from *marine* forms over thousands or millions of years, finally becoming adapted to a terrestrial existence. They are, as most of us who are interested in botany know, among the oldest plants on earth. That they were thriving in enormous numbers millions of years ago, anyone may ascertain for himself by studying the quantities of fern fossils to be seen in every museum or offered for sale in every list from houses dealing in scientific study materials. A large percentage of coal is locked-in carbon that was once in living ferns.

Today, there are many hundred living species of these plants, the majority of them tropical; some minute, others *twenty* to *forty feet* in height. Tree ferns such as these, plants of great beauty, shaded the author's headquarters while he was working in the tropical island of Dominica in the West Indies, and are illustrated in detail in the book *Photographer in the Rain-Forests.**

---

*Paul G. Howes, *Photographer in the Rain-Forests* (Noroton, Conn.: Sylvanus Books, 1970).

Common as ferns are in our northern woods, those who do not know their proper names remain in the majority for the reason that these are not *flowering* plants, and identification of them is therefore more difficult. Their method of reproduction is remarkable and very different from that of the flowering, seed-making species, as will be explained presently.

Emerging from the ground in spring, fern fronds are tightly coiled, not unlike a watch spring, or the withdrawn tongue of a butterfly (plate 82).

At first pale green—or almost white, in some instances—and often downy or furry, the *fiddleheads,* as these early shoots are called, quickly push upward, uncoiling as they lengthen, their *pinnae* (lateral branches) opening on each side of the main stipes or stems, and collectively forming the familiar, sometimes almost feathery-looking fronds we know as ferns (plate 82).

Some of the fronds are sterile, others fertile; but, instead of making buds, flowers, and seed, as in herbaceous and seed-plants in general, ferns produce *spores* within capsules known as *sporangia*. Each cluster is called a *sorus* (or "fruit dot"). These form on the undersides of the *pinnules,* the smaller leaflets into which the pinnae are divided and which in some ferns occur on stalklike fronds.

Associated with, and frequently covering, the sorus is a delicate structure called the *indusium*. As this occurs in various shapes on different ferns, it is a useful aid for identifications, in many cases. Fruit dots became visible in mini-wood ferns (according to species) as early as May 25, in June and July, and even later.

In the cinnamon fern the spore receptacles completely

cover the pinnae, which in this species form in compact masses on the stipes so that these look like tall green or brown wands quite unlike the fertile fronds of many of the other most familiar ferns (plate 82).

When the spores are ripe, the capsules split open, releasing them for distribution, the slightest air currents then carrying them far and wide. So minute are these that a million or so brushed together appear merely as a dab of brown or green powder or dust. Spores may be globular- or other-shaped, and some are studded with tubercles. Under the microscope (the only way they may be seen individually by the human eye) spores of the cinnamon fern appeared like ultra-minute green grapes. (The magnified sporangia at spore-releasing stage are shown in plate 83.)

Soil and moisture conditions must be just right for the successful germination of these micro cells. Only a very small percentage of the billions produced do reproduce, for if things were otherwise, there might be little room for anything but ferns in our woodlands.

Unlike a seed plant, which sprouts and very soon becomes a miniature of the parent, a fern spore on suitable ground splits open, and the first cell of the new organism, called the *prothallus,* appears; this represents the next stage in the developing fern. Repeated cell division soon produces a thread or filament which is attached to the soil by still more delicate threads called *rhizoids.* Already the prothallus possesses chlorophyll, and upon further cell division becomes a *mature* prothallus, a heart-shaped flat object less than half an inch in diameter, and bearing the sex organs—*archegonia,* or female organs, and *antheridia,* or male organs—both of which may appear on one prothallus, or separately.

Each archegonium begets a single egg; each antheridium, numerous sperms. As the latter microscopic male cells are *motile,* or self-propelled, fertilization may take place when dew or a raindrop allows the sperm to *swim* to the attracting archegonia in a water film. As in so many other organisms, embryos are created upon the union of a single sperm with an egg cell, and in this case it will eventually grow into a fern, as most people recognize them. (See plate 89, illustrating the prothallus and very young fern.)

The above is, of course, only the barest outline of that which takes place; there is a great deal more to the story of fern reproduction. Although more will be said about it presently, full details would be out of place in a book of this nature, which is designed only to point out how much there is to be learned about what goes on among the organisms in a half-acre or less of woodland.

To whet the reader's appetite further, other spore-bearing organisms, such as the extraordinary slime molds, and many other fungi have been treated in later chapters. For full details on reproduction by spores, however, the interested reader is referred to the list of Recommended Books at the end of this volume.

After observing spore production, the "birth" of the prothallus from a single spore, and the latter's development of the sexual organs, it will be realized that in ferns there are alternating generations. Thus, the comparatively huge fronds represent the *sporophyte* generation, and the prothallus the *gametophyte* generation, a gamete being a protoplasmic body unable to give rise to another organism until fusion with another gamete—in the case of ferns, between an egg cell and a sperm produced by the sex organs of the prothallus. The male organ, as we know, is the antheridium, and the female or egg-producing organ, the archegonium.

Sperms from the former are curiously coiled objects equipped with numerous hairlike *flagella* or "whiplashes" with which they are able to propel themselves to the opening of the archegonium. Only a single sperm penetrates and fertilizes the ovum, much as a human sperm penetrates and fertilizes a human egg, and the cell resulting from any such union is called the *zygote.* From the zygote the *embryo* then develops, and in the case of ferns, the tiny stem and leaflet of the new sporophyte generation. (See plate 89.)

Fertilization may occur only if a film of water is present between the soil and the underside of the prothallus, affording a suitable medium for the sperm's journey.

As we know, some of these fronds of the sporophyte generation will bear *sori* or fruit dots containing the spores of the alternate generation.

Curious as it may seem, when the author was curator of the Bruce Museum, fronds bearing sori were brought to him on several occasions by people inquiring what disease or insect had caused them!

Within the mini-wood area, seven species of ferns and one fern ally were to be found, most of them long-established plants.

While it is quite possible that additional species might very well have done well in the same humus or soil, their microscopic spores had evidently never germinated within the mini-wood. This was a curious fact, as the maidenhair fern, *Adiantum pedatum,* was found growing a very short distance beyond the mini-wood area.

In one astonishing instance, a tiny spleenwort (fern ally) appeared from spores that must have long been dormant in buried glacial clay. Soon after the material was dug up from part of the mini-wood border, the delicate little plant germinated from clay placed in a glass jar, wherein it continued to grow for *over five years!* Its full story is related in chapter V.

It should be mentioned here that no native species of plant of any kind was introduced into the area covered by this fifteen-year study either before or during this period. Thus the mini-wood community, as described and illustrated in this book is just as the author found it with its plants and animals, undisturbed.

Cinnamon ferns, *Osmunda cinnamomea,* grew luxuriantly even in the wettest situations, their sterile fronds reaching four feet in length, and fertile fronds thirty inches (plates 82 and 83). Arising from *aboveground* black cushions of rootlets, the white fiddleheads appeared about the first week in April, then uncoiled rather slowly. By May 15, this fern's graceful and beautiful fronds were usually well expanded, but nevertheless continued to lengthen, sometimes late into June.

This is one of the most satisfactory ferns to transplant into situations about the house or garden. The tough hairlike clumps (rootstocks) may be dug up entire and replanted in the same way that the ferns grew naturally in the woods, with the rounded humpy hairlike portions of the rootstock exposed. The species can stand a lot of sun if the soil remains moist, thriving even if exposed to the sun's rays for many hours each day. For a covering along a foundation or a garden wall there is no more pleasing species.

As some fronds become old and ragged or discolored by midsummer, they may be cut out, and the plants will continue to send up fresh green ones for some time. The ferns require no special care or feeding, a natural acid soil seeming to be their only requirement aside from moisture. They will come up year after year for those who will meet their simple needs.

Fertile fronds of the cinnamon fern are quite different from those of most of the other common woodland species. The uninitiate, searching for easily seen and identified sori, may in this case look for them, wondering why typically fruit-dotted pinnules are not clearly discernible.

It is such things as this (great differences in the manner of growth of analogous parts of otherwise quite similar-looking plants) which baffle the casual observer but which make botanical study all the more interesting to the naturalist.

This species of fern reveals no easily seen fruit dots; instead, their curious fertile fronds are tall and wandlike, with the spore cases densely distributed on the leaflets (plate 83). Green at first, they turn to cinnamon brown upon maturing, and wither somewhat as the spores are released. They die long before the sterile fronds, which remain green until the fall.

Ripe fruiting fronds taken into the laboratory for examination on May 31 soon dried sufficiently for the sporangia to open up in thousands. Under the microscope the individual spherical capsules could be seen splitting, the two halves partly separating as the shells of a clam open part way. In color, the spores were a flat dark green, and in numerous cases, as drying of the capsules continued, they could be observed actually being expelled. The impression the author received upon first viewing myriad unopened sporangia, highly magnified, was as if they were great masses of green grapes lying upon what looked like spun sugar but was really thick beds of silky hair.

Fiddleheads in April are coated with this same silky material, and the hard, tangled black rootstocks from which they arise are dug and sold commercially under the name "odmundine," much used as a medium in which to grow orchids.

Just below the surface of the rich black humus in the woodland grew the very different and numerous rootstocks of the upland lady ferns, *Athyrium Filix-femina* (plate 84).

Most delicately beautiful of the mini-wood species, in the author's opinion, their beds, which might remind one of other kinds of living things which purposely thrive best as gregarious organisms, are of course only found growing thus in large aggregates because of the creeping tendencies of these rhizomes.

Fronds of the upland lady fern may be found with either green- or burgundy-colored stalks; those ferns with stems of the latter color are known as the form *rubellum.* In the mini-wood, all the stalks in some groups were green, while in other groups, stalks of both colors occurred, with some of the wine-colored stems being much paler than others.

In July or later, the species may be recognized by the inverted U-shaped sori or fruit dots on the fertile frond pinnules. In some cases a few sori may be shaped like the letter J, but as time goes on, these also seem to change into an inverted U or horseshoe form. (Plate 84 shows some of these forms quite plainly.)

As in many other species, there was much variation in the smallest leaflets or pinnules. In the lady fern these were found scalloped, fairly smooth, or even toothed. The beds grew luxuriantly throughout the summer.

A very similar variety, the lowland lady fern, *Athyrium F-asplenioides* (but differing in that its stems do not show any of the burgundy coloring of its close upland relation), was also an inhabitant of the mini-wood area, although it was chiefly a more southern plant. Where the lowland and upland lady ferns occurred almost side by side, only an expert would have been able to tell them apart.

While many little insects, arachnids, and occasional isopods may dwell in the litter and humus beneath beds of these ferns, it is a very interesting fact that such green fronds are attacked by very few foliage consumers. While those of the cinnamon ferns were just freshly expanding, however, during the second week in May, some of their tips were found gathered into small compact wads bound together with silk. The culprit was a small larva of a moth, a pale yellow caterpillar with a black head. But before it could do much damage, and before it had reached a half-inch in length, a tiny wasplike insect with long antennae and a longer ovipositor no greater in diameter than a fine hair came to the rescue of the fronds.

It deposited an egg within each caterpillar encountered during its deliberate search. Its grublike offspring which hatched from these eggs thereafter ate the caterpillars, later spinning minute white cocoons within the wadded shelters constructed by the victims for themselves! By the latter part of May, the new parasitic wasps were emerging from these cocoons, ready to carry on their good work.

This minute insect was only one among hundreds of parasitic species of braconids, chalcids, and ichneumon flies, all members of the great order known technically as the Hymenoptera, to which the ants, bees, wasps, and sawflies also belong, and about which much more appears in later chapters.

Many of the parasitic species, often almost microscopic in size, are all-important control insects, and our guardians against the foliage consumers, whether arboreal, herbaceous, or otherwise. If we spray with dangerous pesticides, we may kill many of these natural caterpillar controls at the same time.

The only other fern attacked by caterpillars during the fifteen-year study of the mini-wood was the upland lady species. Twice during this period, the larvae of two species of small moths were discoveerd on the fronds. Feeding at their tips, both species had so neatly gathered and fastened some of the pinnules together with silk that, although globular, they somewhat resembled the spring fiddleheads (plate 85). Within these hollow balls of vegetation, the inhabiting caterpillars had deposited all of their moist green excrement in one ball-like mass at one side of the shelter, thus leaving ample sanitary space for living. Probably these caterpillars fed mostly at night, using the hollows only for daytime shelter, and as a place in which finally to pupate (plate 85).

The outer skin or integument of a pupa or chrysalis is mostly composed of a material called *chitin*, which becomes very brittle once the inmate has emerged from it. This "housing" for the moth or butterfly-to-be is formed as the larva transforms from that stage. On the ventral surface of the chrysalis, the observer will find perfect *molds* of such parts of the coming *imago* (adult form) as its eyes and antennae, its legs and wings, all neatly in place and ready to contain the parts long before they have completely formed from the broken-down raw materials of the caterpillar body. Here is something for the uninitiate to well ponder and puzzle over—an insect mostly disintegrating within a shell, and being rebuilt from the same stuff into a better, more highly organized winged creature. These molds or forms on the pupal surface thus afford a "preview" of the moth or butterfly before it has been constructed!

Pupae of the first-found species constructing these fern balls were enclosed in crudely diamond-shaped meshes forming networks of silk. The actual pupae were of a rich brown color spotted with black, and the molds or forms for the antennae resembled strings of little beads (plate 85).

A close watch kept on the mini-wood beds of lady

ferns each season showed that the second species of cater-pillar, found in numbers, rolled the fern pinnules into shelters very similar to that constructed by the first and larger larva found only once.

When full grown, in late June or early July, these caterpillars of the second species measured around sixteen millimeters (five-eighths of an inch) in length. They were grayish yellow and translucent, and with a rounded black spot on each side of the first two segments, and with rounded brown heads. Their bodies were naked except for a few almost transparent spines protruding laterally (plate 85).

Some of the larvae of this species pupated within the shelters, while others moved out, away from the excretory accumulations when they became too large, finally trans-forming inside a few fern pinnules drawn together with silk. The pupae were rich brown in color and measured up to fourteen millimeters in length. The moths which issued from these pupae during the second week in July, identified as *Macrobotys aeglialis*, were pale golden brown little insects with a wingspread averaging twenty-eight millimeters, or a little over one inch (plate 85).

Long and carefully as the lady fern beds were watched, neither the larvae or the moths of the larger species were ever again found.

That little or nothing goes to waste in the woodland was here evidenced once more by the presence of a little fly whose larvae fed on the masses of green excrement (chewed leaf tissue) left within the pinnule shelters by the caterpillars.

Here and there, often on the gentle slope of the mini-wood, and usually in situations with numerous glacially strewn rocks nearby, the Christmas fern, *Polystichum acro-stichoides,* seemed to thrive best, sending up its tough new fronds above the old ones each May, and from soil which sometimes became quite dry.

An evergreen species, keeping its color even when deeply covered by snow, it possesses that cryptic antifreeze property with which so many other northern plants are endowed (plate 86).

Its sterile fronds in the mini-wood grew to twenty or more inches in length, arising in more or less of a circle from a robust rootstock which sometimes partly protruded above ground. The pinnae (branches from the stalk) are fairly long, and quite narrow in this species, and they are not divided into *numerous* pinnules, as in the cinnamon and lady ferns. The pinnae are sharply pointed, and their margins bear soft bristles. Another peculiarity to be ob-served close in, by the main stalks, are the upper margins of the pinnules, where, protruding like blunt little "thumbs," each is surmounted with a single bristle.

The fertile fronds arose in the mini-wood in mid-July and soon grew taller than the leathery sterile fronds which, as the season advanced, often lay almost flat on the dead-leaf carpet. Toward the tips of these fertile fronds, where the pinnae become smaller, the fruit dots, which are round, appeared in two rows along either side of the midrib (plate 86).

While it is always a satisfaction to find a plant that stays green all winter, it is this character in the Christmas fern that endangers it, and it is often picked for wreathes and thrown away as soon as the yuletide festivities are over.

Experimenting with various kinds of natural vege-tation for use in habitat dioramas at the Bruce Museum in Greenwich, the author found that the fronds of the tough Christmas fern, if carefully dried in a plant press and then sprayed in the proper natural color with an oil medium, would hold their natural forms indefinitely. By inserting and cementing the stalks (in circular arrangements) in blocks of balsa wood, and then burying the blocks in the dead-leaf carpets in the dioramas, very natural effects were achieved. Only a few ferns were so sacrificed from the mini-wood, but all were thus preserved indefinitely in museum exhibits.

Another fern which sometimes remained green even after ice had formed on woodland pools and the first snows had fallen was the leatherleaf woodfern, *Dryopteris marginalis,* whose tough fronds when fully expanded measured from twenty-four to somewhat over thirty inches in length. In July it may be identified with fair certainty by the two rows of round fruit dots, one row along each outer margin of each pinnule, on the fertile fronds (plate 87). The pinnules of this fern are also diagnostic when considered with the fruit dots, they being bluntly rounded, and each pinna tapering to a sharp point.

In the mini-wood, these "leather" ferns occupied rockier situations than most of the Christmas ferns, right in among rocks at times, which provided accumulations of the highly acid soil between them and which often remained wet for considerable periods. Like the Christmas fern's, the fronds of this species grew in more or less of a circle around the sometimes partly exposed rootstock.

The leather fern had no insect enemies in the mini-wood, but some such surprising relationship may very well yet be found in this species, as was the case with the lady fern.

Another fine fern which could hardly be mistaken for any other in New England—a very common inhabitant of roadside waste places, wet meadows, and along woodland borders, where brambles and other vegetation form small thickets—is the sensitive fern *Onoclea sensibilis,* whose characteristics are very clearly shown in the photographs in plate 88.

In the mini-wood area, the fern was not found *within* the woodland proper, but commonly on its borders and in thickets of brambles and weeds, where it received much sunlight, and where the soil was somewhat less acid. If wanted, this species may be successfully transplanted to soil that is neutral as an ornamental plant for the house.

Its interestingly scalloped pinnae are very different looking from those of the other large ferns, and its fertile fronds are short and stalklike, with many almost erect short branches bearing closely packed rows of beadlike spore cases. These capsules remain on the dead brown stalks well into the winter, and although apparently dried up, some retain large spores even until the following spring (plate 89).

Abundant and healthy, but sensitive to the first good frosts, this fern is one of the author's favorite species. Its sterile, almost triangular, fronds, growing from one to nearly three feet in length at times, here arose from comparatively slender rootstocks which crept shallowly beneath the surface and formed large beds by early summer.

We come now to a small species of plant: fernlike, but not closely related to the other species from the mini-wood found and described. According to some botanists, it is not a true fern, although, along with several other known species, it is classified as a member of the family Ophioglossaceae, one of the succulent ferns.

The plant in question was the cut-leaved grape fern, *Botrychium dissectum,* a species whose sterile fronds in the mini-wood specimens were only a few inches high. While delicate looking, these proved to be quite hardy little plants, not evergreen, but remaining green until late in the fall, when they turned bronzy in color, not shriveling very much even in really cold weather.*

As with the true ferns, the new fronds arise perennially in this family, according to the descriptions in other books. In July the mini-wood example sent up two stalks at the same time, one an unfolding sterile frond, the other a somewhat taller stalk bearing many short branches so arranged as to make this fertile frond resemble a miniature tree. Each of its branches was clustered with yellow spore capsules containing great numbers of yellowish spores.

It is from these closely clustered capsules that the species takes its common name, and also its generic name, *Botrychium* (Gk. *botrychos,* a stalk or a bunch of grapes).

The grape fern's rootstock consisted of many fine ramifying branches, like some tree branches, but only two or three inches below the surface. Here and there were more bunchy roots, from which had developed tiny new frond buds by late summer, being thus in readiness for the following spring at this early date.

Its habitat preference seemed to be the borders rather than the mini-wood proper, and in the author's experience, it proved to be a very slow grower at all stages. During the fifteen-year study period it proved extremely rare, only two plants having been found, and neither lived long. Both were growing in acid media well shaded by large trees, and close to an area well grown over with moss on a surface soil that tested pH 5 on the 0-14 scale on which 7 represents neutrality.

The remarkable story concerning two other delicately formed but long-lived little plants, one of which the author believed to be a fern of the genus *Woodsia* and the other a hairlike moss not definitely identified up to the completion of the book, is related in full detail in chapter V, and accompanied by several illustrations. Both of these plants appeared and grew successfully for long periods on lumps of glacially deposited clay taken from a foot below the surface.

While one of these was a fern, rightfully belonging in the present chapter, it has been described in the chapter with the moss in order to bring together all of the sur-

---

*Not illustrated only because the negatives were destroyed by mistake, and no other living specimen of this plant had been found up to date of printing.

prising facts about this plant life which arose from long-buried material on the mini-wood border.

As previously mentioned, maidenhair ferns (genus *Adiantum*), doubtless the most beautiful of all Northern American species, were not found growing within the mini-wood. Possibly the soil there was too acid, and these ferns are known to thrive in limestone areas. It is interesting, however, that within a minute's walk of the mini-wood per se and its prolific community, maidenhair ferns were found growing, thus possibly indicating a limestone deposit.

CHAPTER V

CONCERNING
THE MINI-WOOD MOSSES

Within the mini-wood proper, few species of mosses were found, and while the soil there was adequately acid, the mosses seemed to prefer locations receiving considerable sunlight and were therefore more often met with on the woodland borders.

Probably most people who are not especially observant of nature believe that they know all mosses on sight, the general popular conception for their identification as such being that they are simply bright green cushionlike growths never bearing flowers. This is a good definition as far as it goes, but to those who really study their natural surroundings, it is indeed far from adequate.

No doubt the mosses most often noticed *are* the bright green cushionlike species seen growing from chinks in walls and sidewalks, and in acid soil about the butts of oaks and maples and other large trees. But mosses occur in many other forms; some are so thin and flattened, so hairlike, or have such erect tall branches that they would not be recognized by many as mosses at all.

The surest way to recognize any true moss as such is to wait for the fruiting stage, when slender stalks or *setae* bearing tiny capsules (containing the spores) arise above the main leafy bulk of the moss plant. These capsules are fitted with neat little caps or lids, which at a later stage are thrown open to release the spores. Occurring in various sizes, shapes, and heights above the main body of the moss, these spore receptacles are very important in identification of the species. (See plate 94.)

Mosses and liverworts* are organisms in which, as with the ferns, a gametophyte generation alternates with a sporophyte generation. Botanically, the mosses and liverworts were formerly referred to as Bryophyta, a division of non-flowering plants commonly spoken of as bryophytes, but more recently the mosses have been classed as Atracheata—green plants lacking the specialized water-conducting cells of the wildflowers, shrubs, trees, and other larger plants.

Being possessed of green cells, however, we know at a glance that the mosses are *photosynthetic*—capable of manufacturing their own food as the higher green plants do.

Although a moss starts as a microscopic spore, there is no prothallus arising from it, as in the ferns. Instead, when the moss spore, swollen with moisture, ruptures, a threadlike object called the *protonema* and consisting of green cells, creeps forth. Later, from buds on the protonema, the first leafy stems of the new moss arise. As with the fern prothallus, the protonema is anchored to the soil by rootlike threads called rhizoids.

When a moss matures, it produces sex organs—male organs or antheridia and female organs or archegonia—on separate leafy stems or both together on a single stem of the plant, and in some cases on a scalelike object or thallus. In the base of the vase-shaped archegonium is the egg cell which, when mature, exudes a secretion from an opening or mouth at the top, and which exerts a powerful attraction for the male germ cells. These, as we have seen before in the ferns, are self-propelled microscopic coiled bodies able to "swim" long enough in a film of dew or rainwater to reach the archegonia. The antheridia may produce a number of sperm cells each, but only a *single*

____

*Liverworts are mosslike little plants, usually flat, or up to about an inch in height, and often ribbonlike in form. None were found in the mini-wood during this study.

sperm enters each archegonium. Swimming down its neck, the sperm cell eventually penetrates the egg at its base, thereby fertilizing it, and the union of the two gametes forms the zygote, or first cell of the new organism.

As we know, the new cell, by repetitive division into two, four, eight, sixteen, thirty-two cells, and so on, produces the new stem and spore case (capsule and seta). At length pushing its way up the archegonium, the seta and capsule break through into the open, and, if a true moss, carry along the tip of the archegonium as the cap of the spore receptacle.

These few notes about a very intricate and beautiful process will serve to explain to those unfamiliar with moss biology the presence of almost hairlike stems surmounted by tiny capsules which the reader may see at times rising above the familiar, more often plain body of a moss (plate 94).

Spore capsules may be found that are erect, bent over at various angles, or even hanging bell-like but firmly attached to the setae.

The spores of a moss, which are liberated from these capsules soon after the setae raise them above the main leafy body of the plant, float away on the merest zephyr or air current. Only a very few out of the thousands or millions thus set free by each moss ever find suitable ground and germinate.

Although many common mosses do occur as just neat-looking rounded cushions of green, their individual leaflets, when taken from even the tiniest separate stem, become objects of great beauty when highly magnified. They will be found arranged radially, each whole stem when thus viewed seemingly a complete little plant in itself. As the leaflets may be in most places only one cell thick, their structure may be studied to great advantage with a good stereo microscope, transmitted illumination, and a little patience.

And now for the story of the two little plants which arose from lumps of glacial clay, mentioned in the preceding chapter on ferns.

As described in chapter II, huge fragments of ice, separated from the main mass of the retreating glacier, left us numerous deposits of clay, and hardpan (clay mixed with stones), all of which were washed out of vents in the ice in streams of meltwater.

Here and there within the mini-wood, and along its borders in one spot, almost pure clay of a puttylike con-

sistency occurred, mostly about a foot below the surface, and hardpan strata also underlay parts of the wood itself.

Spring and summer rains, after washing the dust from the mini-wood canopy and its understories, were partly retained by the spongelike properties of the dead-leaf carpet and the humus below, while the remainder, penetrating further, met and followed the hardpan strata out of the woods, leaving pure clay deposits undisturbed.

Gardeners dislike clay for obvious reasons, but let us not forget that some, perhaps taken from this very locality, may have supplied the late Eastern Woodland Indians with the prized medium with which to fashion their pottery and other baked artifacts. In the present case, it gave the author of this book material from which some astonishing little plants soon revealed their presence and their viability.

April 10, with the ground safely thawed out for another season, a lump of this sticky, glacially deposited clay was dug from a layer twelve inches below the surface on the western edge of the mini-wood. A fist-size lump was placed at once in a thoroughly sterilized glass jar and its cover screwed on, but not so tightly as to exclude all air. To prevent the mass from drying out, a little distilled water was occasionally added with a sterilized medicine dropper, but nothing else was added.

Left thus on a windowsill with a western exposure in the author's laboratory, the jar was closely watched, and on the fifty-sixth day, minute rounded greenish forms were just visible on the tacky clay. Examination (through the glass) with a hand lens also revealed a delicate green weblike blanket through which these objects seemed to have risen.

From these odd beginnings, tiny shoots followed which in another three weeks were recognizable as young moss stems. Many others now appeared, and as the weeks went into several months, a threadlike mass of the stems had grown over the upper surface of the clay, and another mass had grown onto the glass, where its extremely minute leaflets could be examined without disturbing its progress.

Definitely a moss, with each delicate stem having radially arranged leaflets, it grew very slowly, but as much as two inches in length as time went on. The stems were of two kinds, the great majority being extremely slender, almost hairlike in diameter, and bearing elongate, pointed leaflets, while a far lesser number were considerably shorter, with ovate leaflets forming little rosettes at their tips. The midribs of the leaflets extended beyond their tips and terminated in needle-sharp micro spines which were only clearly visible when highly magnified. Possibly these rosettes of ovate leaflets held the moss's sex organs, but none was ever observed by the author, nor were any setae with spore capsules ever discerned among the green stems (plate 90).

Most individual green cells were almost perfectly hexagonal in form, and as the leaflets were in most places only one cell thick, they afforded wonderful *living* objects to study under the microscope, the substance of the *chloroplasts* (chlorophyll) appearing an extraordinarily brilliant green under transmitted light.

This moss, still unidentified, has baffled several specialists, because no fruiting bodies have ever been found, although the plant at the time of completion of this book was already *over five years of age,* and still growing within the *same* glass jar, to which nothing but distilled water had ever been added!

Rightfully, what follows now belongs in the previous chapter on the mini-wood ferns, but as stated elsewhere, classification has not been followed in this book except in a very general way. It was thought best, to make their stories more interesting, if both plants begotten from the glacial clay were described here together.

Finding it hard to believe that plant life would again arise from this dense material taken from well within a deposit which had doubtless lain thus buried for centuries, and with deeply embedded boulders close by helping to verify its origin, it was decided to start fresh and experiment with greater precaution than before.

Employing a thoroughly sterilized trowel with which only lumps from the interior of the deposit were selected, the material was then transferred with sterilized forceps to new glass jars, which with their screw caps had been boiled for two hours.

Having been convinced up to this time that the appearance of the first moss simply represented a single freak occurrence, it was with real pleasure and astonishment that the author again witnessed the appearance of this same plant on one of the fresh lumps of clay, and then watched it gradually cover part of the surface with the same delicate threadlike growth.

Still more astonishing was the appearance of an entirely different kind of growth in one of the glass jars. On this clay, dug April 4, along with the other lumps mentioned above, and handled with the same great care, things began to happen just two months after the material's removal from the ground.

The jar in question had been standing on a windowsill where it received short periods of sunlight in the afternoons. Here, upon the otherwise bare clay surface, tiny domelike objects arose, singly, and in little groups. These were of the color known as sap green, and by June 4 were present in numbers. Although nothing like fern prothallia, there must have been something in the jar akin to them (perhaps covered up by them), for by June 12 slender stems bearing fragile translucent leaflets had arisen well above the green "domes," possibly algal growths unrelated to these newcomers except by their "choice" of the present habitat. The leaflets of the new plants were three lobed, each lobe growing to about a quarter of an inch in width, and the three together forming fan-shaped leaves (plate 91).

Delicate, slightly darker veins were visible in each lobe, and with the aid of a microscope, ultra-minute hairs or setae, probably sensitive to factors beyond human conception, could also be seen.

When six months of age (October 4) these first stems had been succeeded by others, and some were then over an inch in length. At fourteen months (the following June 4) the plant had developed stems up to three inches in length, their beautiful little fan-shaped leaflets confirming the author's suspicions—and the astonishing truth—that from this glacial clay had now arisen a delicate fern which seemed to be a species of *Adiantum,* although no maidenhair had ever been seen within or on the mini-wood borders. But no, it was not this fine species; closer examination indicated it might be a species of *Woodsia,* but as the plants had revealed no fruit dots up to the writing of this book, its proper technical name is still in doubt, and the author trusts that some botanist may identify it from the photographs in plate 91.

In any case, the plants which came forth from this clay raised interesting questions.

First, one might ask, what nutrients did these organisms find in this densely packed material which sustained their growth, minus additives other than distilled water? It would be interesting to know also which nutrients therein were used, and in what proportions, by each species, the source (let us not forget) being pulverized rock, or fragments of mountains if you like, leached by the waters of the melting glacier, and deposited in what is now the author's own "backyard."

From whence, we wonder, came the spores from which the plants were born? Could it possibly be that some of them came in this meltwater, and lay dormant but viable throughout the centuries, embedded in this mildly acid clay testing pH 6?

That spores might retain the spark of life for the thousands of years since the glacier retreated is a more intriguing speculation than the drab thought that they had been present in the air and captured by a one-in-a-million shot while the author was transferring the clay to the sterilized jars. (Here one must not lose sight of the fact that so early in April such spores would not be likely to be in the air.)

Even less stimulating to the imagination is the mundane possibility that the spores were carried down into the clay by comparatively recent rainwater.

A diligent search over the years failed to reveal either the species of moss or the Woodsia(?) growing elsewhere in or near the mini-wood. Their appearance on the clay remains one of the most interesting mysteries of this fifteen-year study.

While not a great many species of mosses may occur in one's locality in the northeastern United States, so that recognizing them would therefore not seem too much of a problem, the novice will find that these organisms are often difficult to identify properly. Mostly depending for this purpose on the form of the spore-case or fruiting body, and still finer details such as the number (or absence) of rows of minute teeth on the capsule's rim, the interested investigator may also find the plants exasperating while waiting for them to produce these identifying features. A powerful hand lens—or, better, a compound microscope— should be available if the reader really wishes to know mosses intimately; and for guidance, a standard botany book, or perhaps one of the other books listed at the end of this unpretentious study of the mini-wood, is recommended.

Within the mini-wood proper, mosses were scarce. One species had difficulty remaining green, growing as it did upon the often dry ground around the butts of the largest black oak trees. Reluctant to fruit, it was not identified.

Two very interesting species, preferring very wet habitats, were neither of them at all common. One, which spread its attractive, rather flat, and slowly expanding mats upon a deeply eroded yellowish brown sandstone rock, which it was helping in its small way to disintegrate, was

technically identified as *Ditrichum ambiguum*. Found on this rock which was half buried in the wet dead leaves on the woodland floor, it did not send up its identifying fruiting bodies until December 15 (plate 92).

A novice would probably least expect to find a moss fruiting at such a late date, and while in this case the fruiting may have been due to unseasonably mild early winter weather, it nevertheless illustrated how carefully some mosses must be watched for their identifying fruiting bodies. While the stalks bearing the spore capsules were reddish in this species, they measured only around five millimeters (slightly over three-sixteenths of an inch) in height, and were inconspicuous even against the bright green leafy mass of the plant (plates 92 and 93).

Preferring even wetter situations than the moss just described was the species *Leptodictyum riparium*, a plant with a delicate-looking but densely matted manner of growth, first found growing just above the water level on a stick protruding from a pool. Later, another example was observed whose spore had germinated on the wet portion of a dead branch that had fallen within the woods and was sticking out of a nearly saturated area on its floor.

The first specimen was found with fruiting bodies on November 17. These were red stalked and even shorter than those of *D. ambiguum*, and are clearly shown two and a half times life-size in plate 93.

Professor Robert L. Hulbary, chairman of the Department of Botany, University of Iowa, who examined and identified these mosses, kindly informed the author that *D. ambiguum* was a rather unusual specimen of this species, usually being found farther west, but reported by Grout in New England under the name *D. tortuloides*.

In regard to these semi-aquatic mosses, in these particular instances found growing on wood just above the surface of the water of a pool and on the continually wet end of a branch protruding from a soaking area in the mini-wood, Professor Hulbary writes: "I believe it is, *Leptodictyum riparium* (Hedw.) Warnst., though it is very closely similar to *L. laxirete* (Card. Thér.) Broth. The members of this genus are quite variable and difficult to distinguish. The latter species was formerly treated as a variety of *L. riparium*."

His comments have been included here for the purpose of illustrating to the uninitiate in the study of mosses that

Above: The flowers, enlarged to show the interesting boat-shaped form of the petals. Date: July 27. (Page 21.)

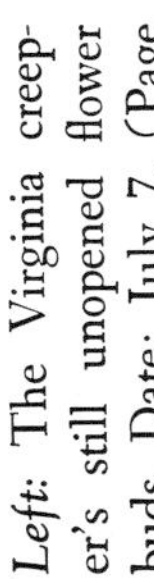

Left: The Virginia creeper's still unopened flower buds. Date: July 7. (Page 21.)

Right: Other, fully opened blossoms on the same date, and a few fruits beginning to form. About life-size.

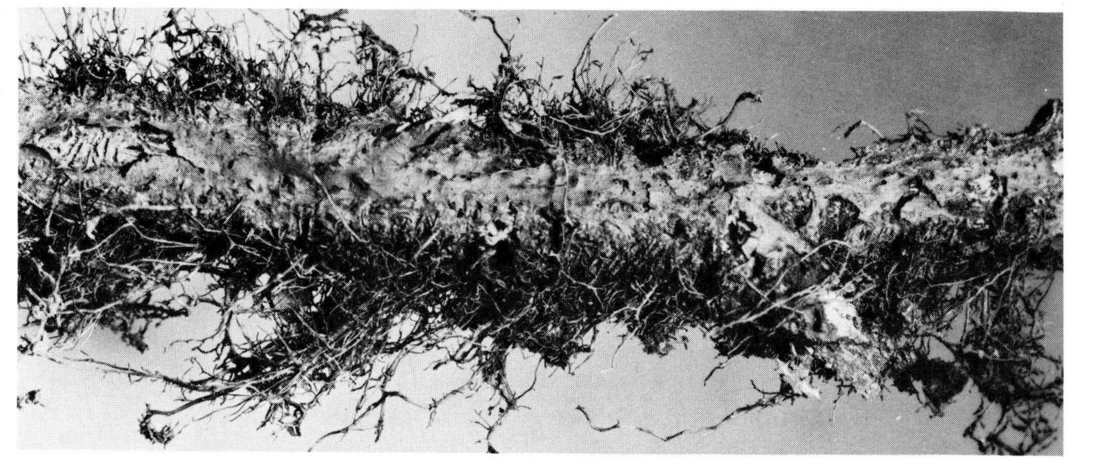

Above: A poison ivy stem, showing the masses of fine, adventitious roots with which the vine clings on the host tree. Shown natural size. *Below:* Blackberry stems (at 2, 3, 4) are armed with *prickles*, not spines. Barberry stems (5, 6, 7) are armed with *spines*, not prickles. (Page 22.)

Above: Virginia creeper vines, showing distinctly as dark, oblique masses of foliage high on the branches of white oaks, have crept away from the main horizontal tree trunks bearing the rest of the vines. The vine at the base of these oaks is shown in plate 56.

Above: Greenbrier growing at the base of a white oak in the mini-wood, shown in full fresh leaf, June 10. *Below:* Close-up of typical leaves, showing the long tendrils, for support only, with which the vine climbs on other species. (Page 22.)

The frost wild grape of the mini-wood. *Left*: Typical leaves, showing lighter undersides, in this case half life-size. Date: July 6. (Page 22.)

Above: An old vine which reached from the woodland floor to the top of a dead pin oak. *Left*: The seldom-noticed, sweet-scented flowers of the frost grape, enlarged one-third. Date: July 3. (Page 23.)

Above: Part of a frost grape stem. Note that tendrils for climbing, which are modified leaves, often arise opposite normal leaves. A cluster of very young grapes is also shown as it looked July 20. *Below, left:* Note that some leaves are lobed. *Below, right:* Frost grapes, natural size, August 21. (Page 23.)

Right: A Japanese honeysuckle tangle, all but hiding its host tree. *Below:* Close-up of vine and leaves. Date: November 14 (and still green). Page 23.

Below and upper left: Tubular flowers, photographed June 15, contain easily obtained drops of delectable nectar. Natural size. (Page 23.)

A wild yam vine which became established on the mini-wood border is here seen in full bloom as it encircled a wire net frame. Date: July 15. (Page 24.)

This bush form of honeysuckle, *Lonicera Morrowi,* was another plant that became naturally established in the mini-wood area, probably being bird planted. Shown about life-size. Date: May 28. (Page 24.)

PLATE 63

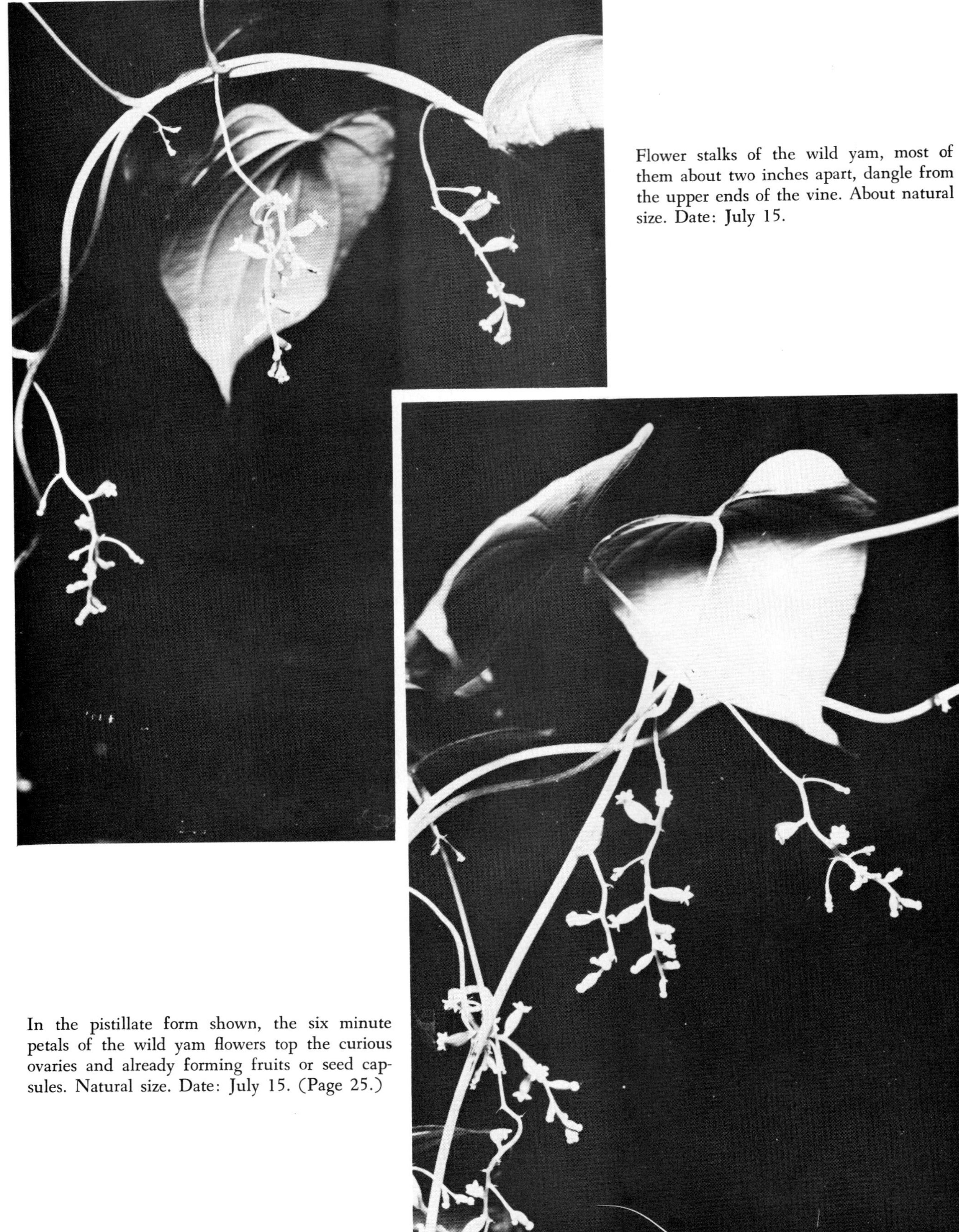

Flower stalks of the wild yam, most of them about two inches apart, dangle from the upper ends of the vine. About natural size. Date: July 15.

In the pistillate form shown, the six minute petals of the wild yam flowers top the curious ovaries and already forming fruits or seed capsules. Natural size. Date: July 15. (Page 25.)

The curious three-winged fruits or seed capsules of the wild yam, shown here about natural size, remained on the vine until late in November. Date: August 4. (Page 25.)

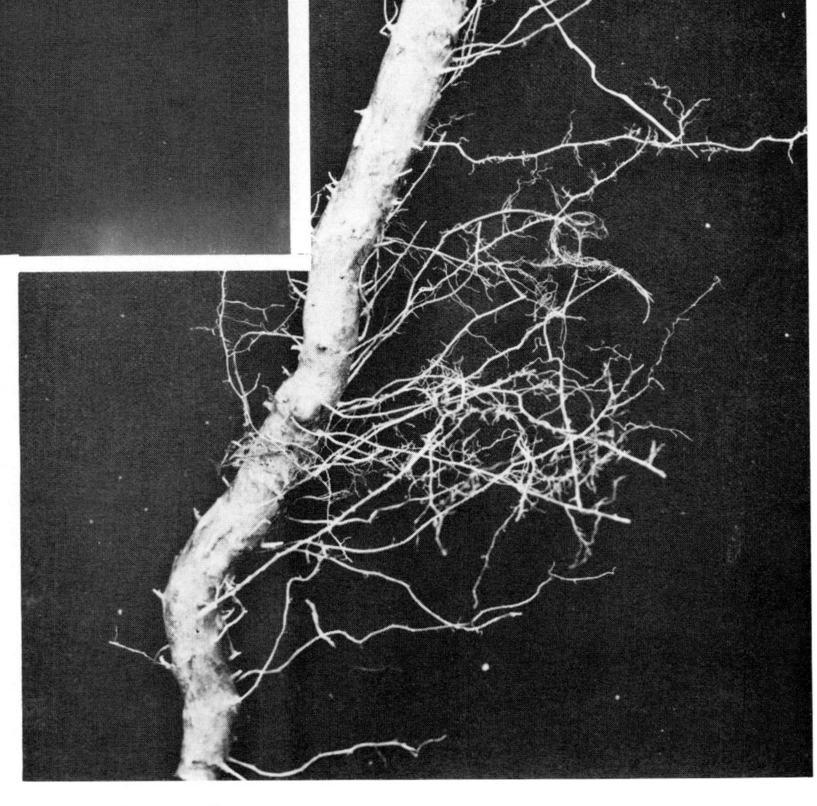

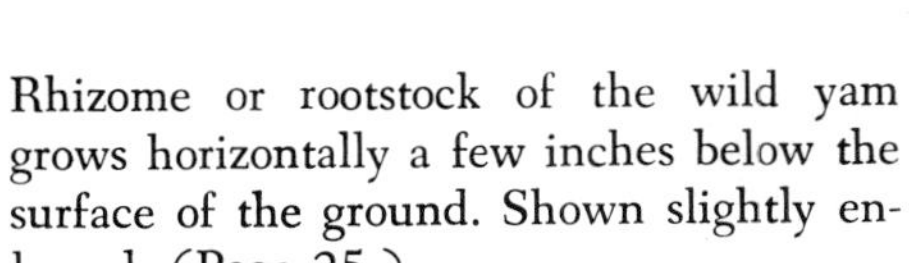

Rhizome or rootstock of the wild yam grows horizontally a few inches below the surface of the ground. Shown slightly enlarged. (Page 25.)

PLATE 65

Left: A typical branch of a common wild blackberry in full flower. Date: May 30. Shown about two-thirds natural size. (Page 25.) *Below:* Ripening fruits of the common blackberry which follow the flower racemes. Natural size. Date: July 21. Inset shows small ripe "black caps." Date: August 14. Many fruits ripen earlier than this date.

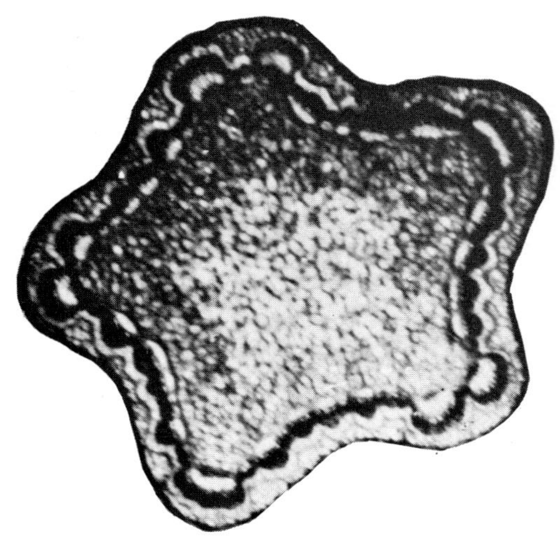

Above: Paper-thin transparent cross section of a common blackberry stem, showing some of the structures, enlarged seven times. Epidermis cells are followed by several tissue layers which form the bark; then by the cambium, which is responsible for the stem diameter; then by another layer, called the xylem. The remaining interior is filled with pith. (Page 25.)

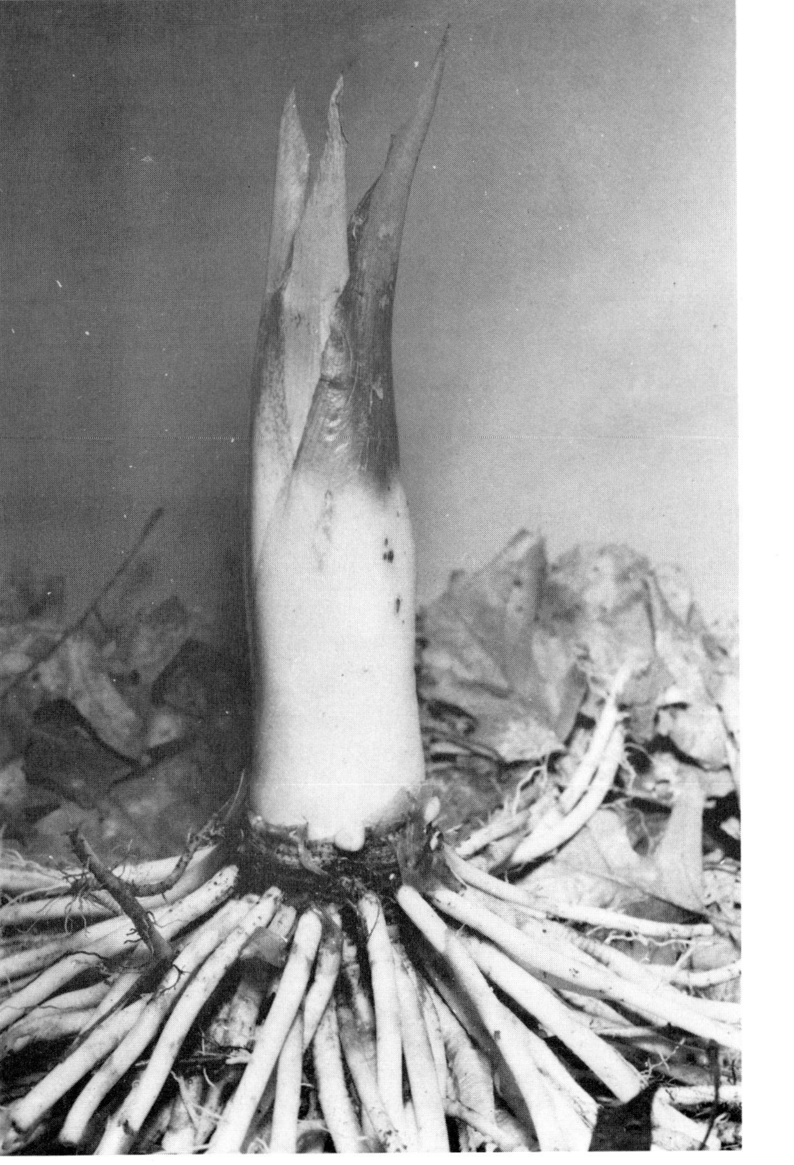

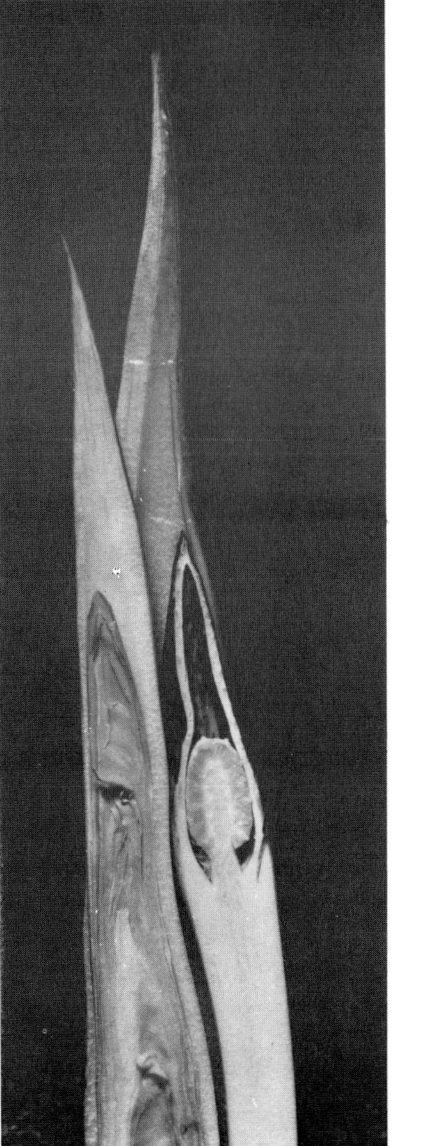

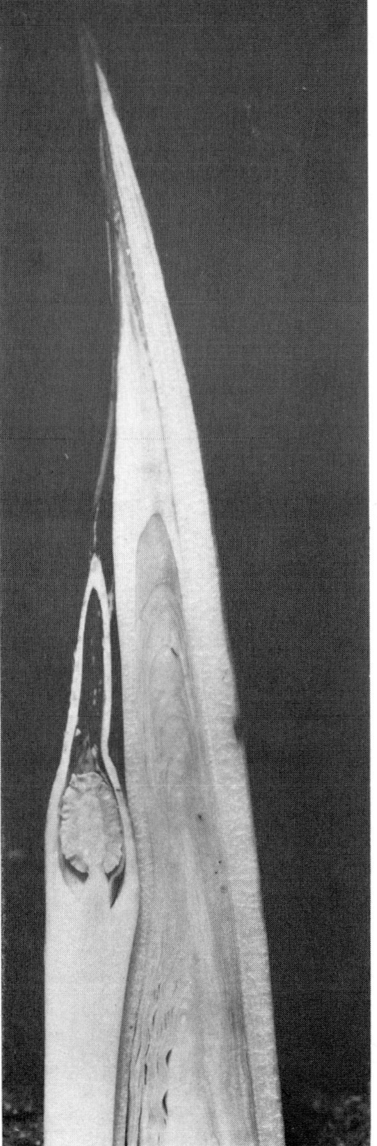

It will surprise some to know that fully formed shoots of the skunk cabbage, shown natural size at left, are already well grown beneath the leaf carpet by late fall in readiness for spring. Specimen with its rugged root system dug October 21. Shoots at right have been sectioned to show flowering parts already formed by January 10. (Page 28.)

PLATE 67

Skunk cabbage shoots and flowers in the mini-wood, March 23. In second specimen from left, spathe has been partly cut away to show spadix and pollen-covered flowers. About half life-size. (Page 28.)

Young leafage following the flowering, photographed May 25, and shown about one-fourth natural size. By June 15, some of the plants reach three feet in height and bear massive leaves, but by September 15, not a sign of them remains aboveground. (Page 28.)

Plate 68

Tiny bulbs of the trout lily, shown life-size. *Below:* Common blue violets in full bloom, May 1. (Page 30.)

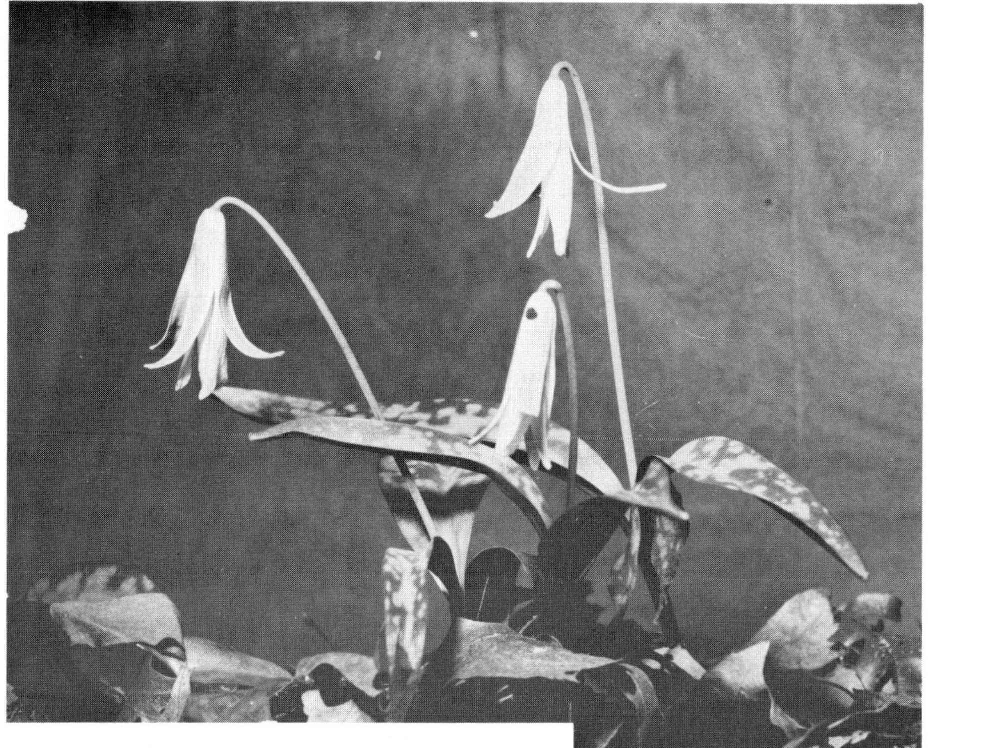

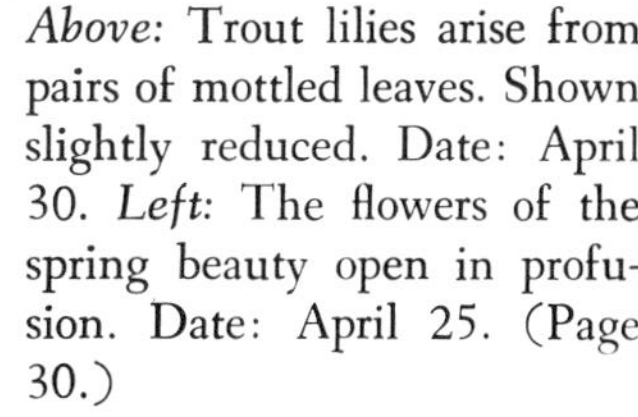

Above: Trout lilies arise from pairs of mottled leaves. Shown slightly reduced. Date: April 30. *Left:* The flowers of the spring beauty open in profusion. Date: April 25. (Page 30.)

Common blue violets. The tall flowers in the container are of the deep violet type. Two of the white type can be seen beneath them, at the rim of the container. Enlarged × 2.

Above: Wild lily-of-the-valley (or Canada may-flower) often grows in small but closely packed masses at the butts of oaks and other trees. Somewhat enlarged. Date: June 21. (Page 32.)

Right: Viola conspersa, a small species of violet which grew in the shade of a mature beech tree in the mini-wood. Shown considerably enlarged. Date: May 17. (Page 31.)

Above: When fully opened, paired flowers of the Solomon's seal resemble narrow little greenish bells. Shown closed after blooming. Date: June 16. (Page 32.)

Above: Wild geranium with its oddly cleft foliage is shown in bloom among other mini-wood plants. Date: May 23. (Page 32.) *Right:* The rhizome of false Solomon's seal, showing butts of former stems and next year's bud (at end). Date: November 10. (Page 33.) All pictures in plate half life-size.

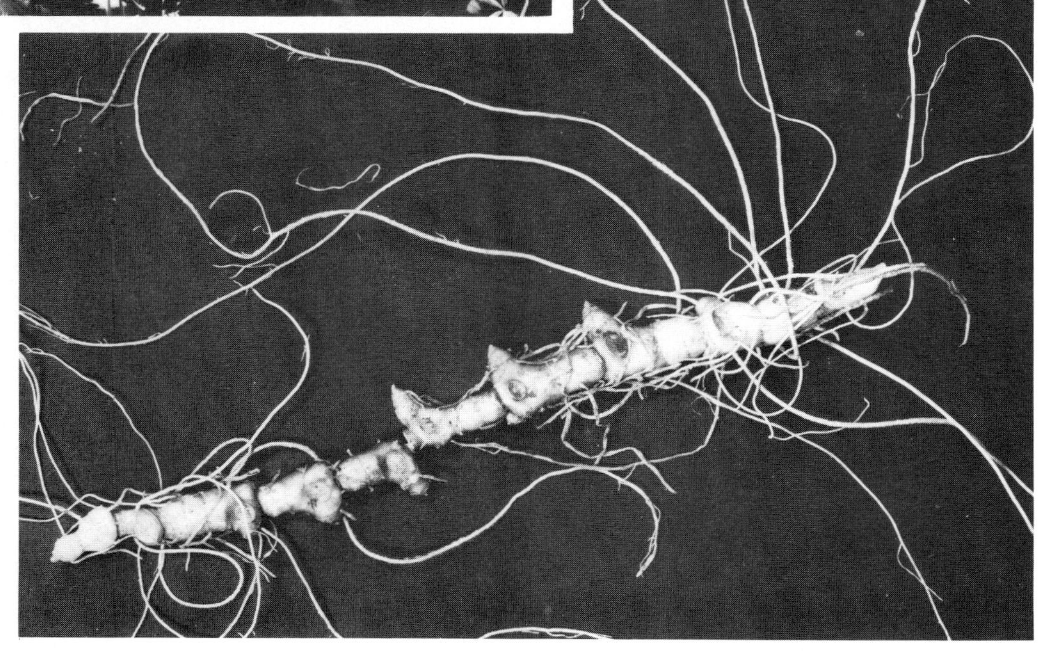

Berry cluster of jack-in-the-pulpit, green at first, turns bright red before fall. Shown about life-size. Date: July 21. (Page 34.)

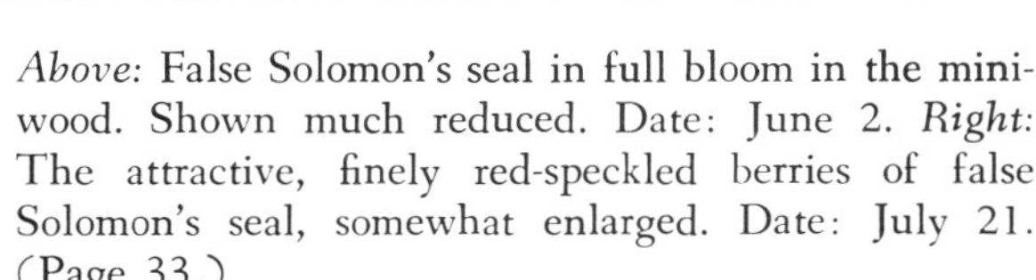

Above: False Solomon's seal in full bloom in the mini-wood. Shown much reduced. Date: June 2. *Right*: The attractive, finely red-speckled berries of false Solomon's seal, somewhat enlarged. Date: July 21. (Page 33.)

Jack-in-the-pulpit. A fine specimen, thirty inches in height, is shown growing in wet ground at the edge of the mini-wood. Date: May 21. (Page 33.)

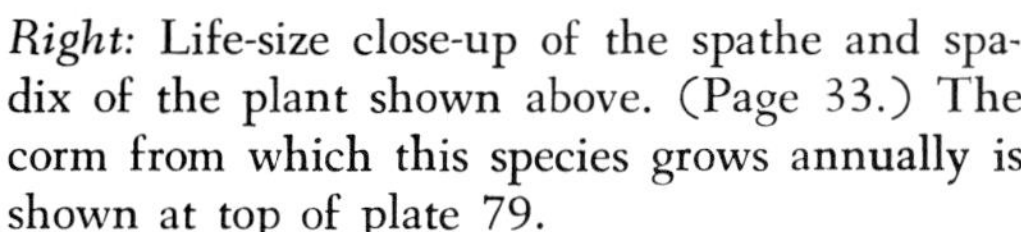

Right: Life-size close-up of the spathe and spadix of the plant shown above. (Page 33.) The corm from which this species grows annually is shown at top of plate 79.

Below: Hellebore, or Indian poke, as it arose from the mini-wood soil. Date: April 19. (Page 34.)

Right: The same perennial plant shown at top left, as it looked by April 26 of another year. (Page 34.)

Left: The Asiatic dayflower, shown about natural size. *Below*: Blossoms of the Virginia dayflower, much enlarged. Dates: July 29 and July 20.

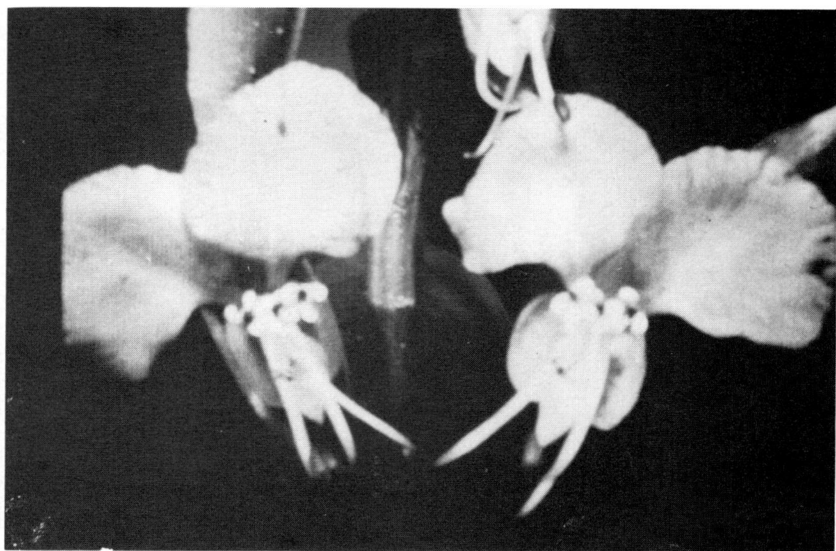

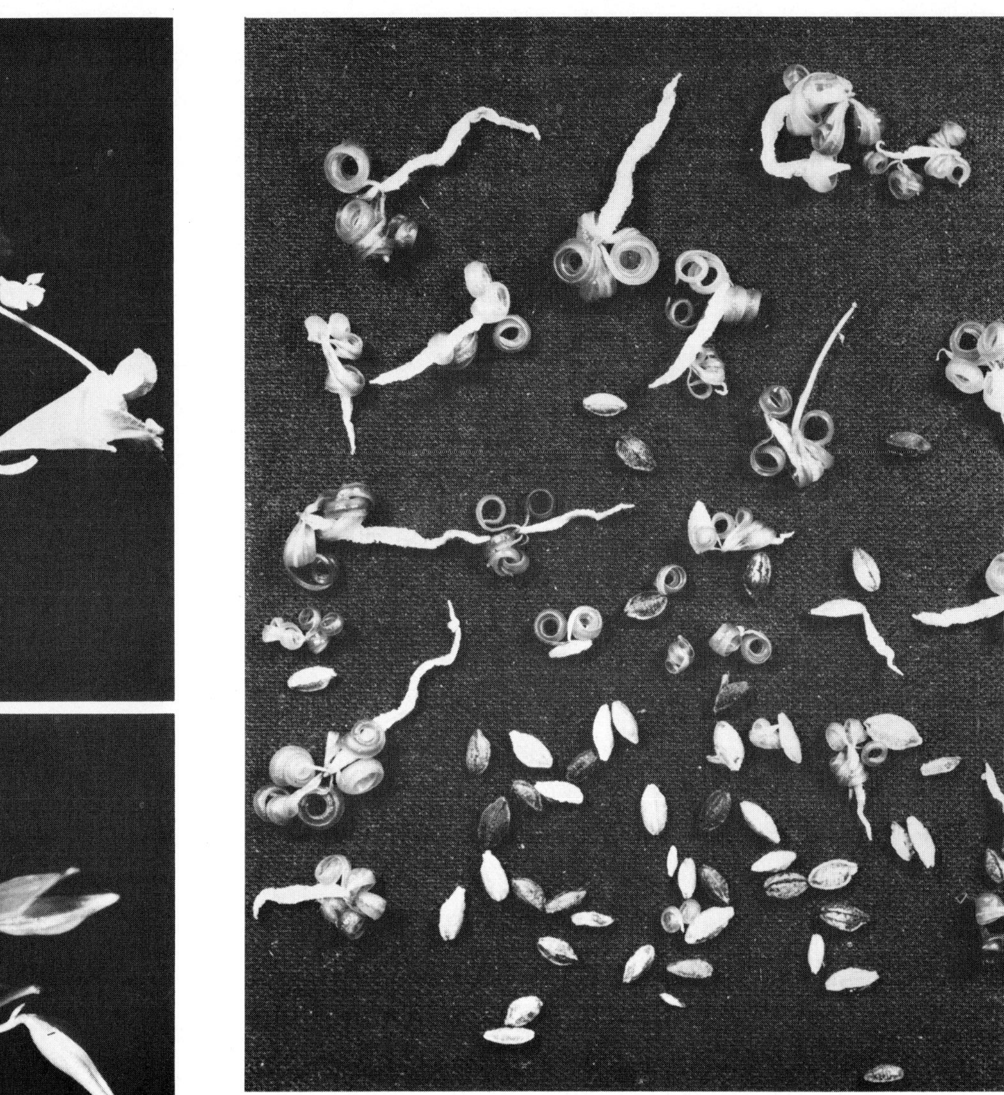

Top: A touch-me-not blossom hangs balanced at the end of a long pedicel. *Above:* Front view into this flower, much enlarged. Date: August 1. (Page 35.)

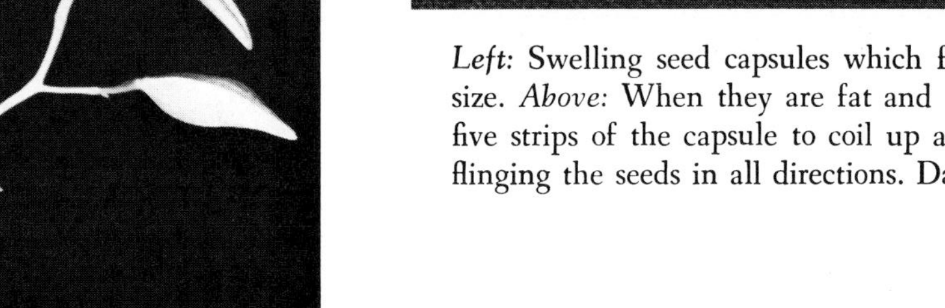

Left: Swelling seed capsules which follow flowering are shown natural size. *Above:* When they are fat and ripe, the slightest touch causes the five strips of the capsule to coil up and fly apart, as shown above, thus flinging the seeds in all directions. Date: August 30. (Page 35.)

Left: A white avens plant. *Above:* A single blossom, shown slightly enlarged. July 11. (Page 36.)

Virginia knotweed in bloom. With the aid of a magnifying glass, some of the tiny flowers may be seen above the more numerous already forming single seeds with their retained and protruding hooked styles. Shown about one-third natural size. Date: August 24. (Page 36.)

Above: Indian pipe flowers and a young Virginia creeper vine which became established beneath overflow from a natural rock birdbath. Date: August 2. (Page 36.)

Above: Indian pipes growing under a mature beech tree, shown about two-thirds natural size. Date: July 27. *Left:* Inflorescence of the white snakeroot. Reduced about one-half. Date: August 17. (Page 37.)

Right: Horse-balm or stoneroot. Note that in fully opened blossoms, the stamens and pistil protrude well beyond the corolla. Date: September 19. (Page 37.)

Left: The horse-balm plant as it appeared before flowering period and already twenty-four inches in height. Date: June 5. *Right:* Typical inflorescence of horse-balm, shown about one-third life-size. Date: October 6. (Page 37.)

Plate 78

Left: White rhizome of a young Solomon's seal and a jack-in-the-pulpit corm, compared. Shown natural size. *Below:* A rugged, old, almost black rhizome of horse-balm and jack corm, compared. "Craters" in rhizome indicate number of flower shoots of former years and age of plant. One-foot rule shows comparative sizes. (Page 37.)

Above: Track left between epidermises of blackberry leaf by feeding leaf-miner larva of a tiny fly. Larvae of tiny moths also leave such tracks. *Right:* Single-winged seeds (natural size) of the tulip-tree may be shed in enormous numbers by a single specimen. (Page 11.)

Above: Slightly enlarged view of the flowers of the mountain or whorled aster. Date: August 24. *Below:* A mountain or whorled aster in full bloom, September 26. (Page 38.)

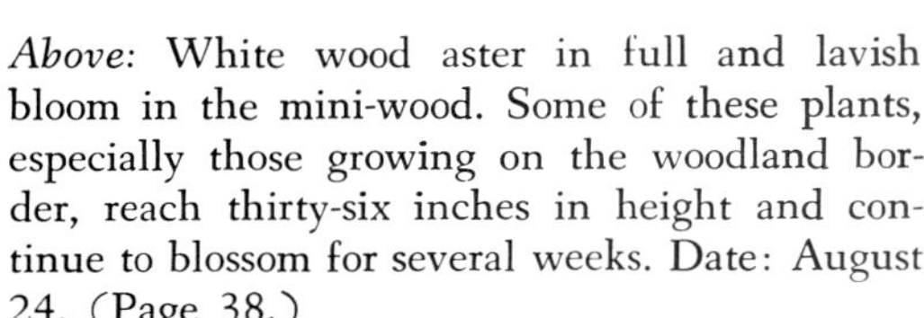

Above: White wood aster in full and lavish bloom in the mini-wood. Some of these plants, especially those growing on the woodland border, reach thirty-six inches in height and continue to blossom for several weeks. Date: August 24. (Page 38.)

these plants are not easy to identify with certainty, and that the kindly aid of an authority is a must for a novice.

An interesting case of two moss species growing intimately together, found on the mini-wood border where the habitat was rather heavy with sandy clay, is illustrated in plate 94.

What a novice would more than likely have mistaken for two stages in the same species was actually the tall setae of a species of *Pohlia* bearing immature spore capsules and rising in crowded groups above the short, massed setae and turbinate (top-shaped) capsules of another moss, namely, *Physcomitrium pyriforme* (Redw.) de Mot.

This plant association became more confusing as the days passed. Found together, as shown in plate 94A on April 15, the partner with short setae and turbinate capsules depicted in 94B had by May 3 also grown tall as in 94C, or could it have been that a third species was here represented? Once again, what had occurred well illustrated the difficulties which may arise for the student of mosses.

A cushionlike moss, found in the area growing at the base of a northern red oak tree, came into fruit by April 5, the dark red setae bearing bright green spore capsules making this species, *Amblystegium varium* (Hedw.) Lindb., most attractive to look at. Fruiting period was very short, but the stalks grew quite tall very quickly and distributed their myriad spores from the capsules within a week. The moss cushion proper was thick and soft, as shown in plate 94D, about one-third larger than life-size.

These few species included in this chapter do not cover the whole moss flora of the area. There were some others growing about the butts of large trees, and rather heavy growths on the bases of poles standing in wet situations— feathery growths which were probably variations of the moss *Leptodictyum riparium*, but not certainly identified. Still others will doubtless appear from time to time, as various other organisms of the mini-wood and its borders have continued to bring fresh surprises to the author through the many years that have passed.

CHAPTER VI

THE MINI-WOOD FUNGI

Knowing a woodland intimately should be one of the greatest pleasures experienced by owner or visiting investigator. Its ever-changing population of fungi, their remarkable forms and manner of growth, and their uses as food for many kinds of woodland animals, both large and small, offer rich examples for absorbing study.

One never knows what ghostly-looking or highly colored species, or what delicately tinted or grossly massive or almost microscopic form any day's visit to the woodland may reveal, be that area a thousand acres of wilderness pine land, mixed deciduous forest, or a half-acre mini-wood, like that investigated for this book.

In the realm of the fungi, astonishing things can happen overnight, and if there should have been a good day's or night's rain, a real soaking of the woodland floor and its litter just before one's visit, then chances are, as if by magic, that fruiting bodies (which we popularly lump under the one word "toadstools") may have arisen since one's last visit, none of them species which were there before. Maybe there will be only one new kind, or maybe a half dozen, or maybe a whole army of very tiny ones studding the saturated dead leaves of this lowly woodland habitat (plate 95).

Fungi arise from spores so small that even were they suddenly to be discharged in the millions, they would appear to the unaided human eye as mere puffs of dust.

The spores are discharged by mushrooms and toadstools from the radiating fleshy laminations or thin walls on the underside of the caps called *gills,* or in a great many other species from pores opening on the underside of the caps.

In the puffball group, with which most everyone living in the country is familiar, the spores are discharged in little clouds after the white or cream-colored spore-bearing tissue becomes dark and dry. They issue from a roundish hole or other rupture in the outer sheath of these stalkless or near-stalkless fungi.

Coral fungi, which are many branched and occur in many colors and do resemble certain corals, bear their spores in layers on the various branches. A species called the hedgehog fungus discharges its spores from projecting fleshy "teeth," and there are still other variations of the process, but in this book we are concerned only with fungi-bearing gills or, in a few cases, pores, with puffballs and also slime molds, the latter being something else again and puzzling organisms described in detail later in the chapter.

On each side of each lamination in the gill fungi, or projecting from each tube in the pore fungi, and lining microscopic cavities in the puffballs, there are organs called *basidia* which, by procedures too involved to be included here, produce and discharge the spores. True sexual union seems to be involved; for technical details, the interested reader is referred again to the botanical works listed at the end of this book.

Once released, the spores fall or are rejected, then remain on the ground near the parent or may be carried short or even very great distances by air currents, which explains why fungus species that have never been observed before in one locality (even in one as limited as the mini-wood) may suddenly appear there.

There are now said to be about forty thousand known

species of fungi, a fact which will bring realization that there are no doubt many, maybe even *as* many others, still to be discovered, and realization too that the interested investigator might indeed find an undescribed species in, so to speak, his own backyard.

Without a doubt, fungus spores of dozens of species blew over and through the mini-wood canopy, through its lower arboreal stories, over its herbaceous plants, and over the woodland floor at all times during suitable weather conditions. Even on the calmest days and nights, billions of fungus spores are airborne to pastures new. Indeed, they are everywhere: out-of-doors, and within our houses. Hence the green and black and pink molds which spring up so rapidly on bread; hence the green spots in Roquefort cheese; hence those white threadlike ramifications that creep over rotting logs and woodwork; hence penicillin; hence the thousands of kinds of mushrooms, toadstools, puffballs, and bracket fungi; hence the enigmatic Myxomycetes or slime molds, which are also classed as fungi.

As with other spore-bearing plants, spores of the fungi must alight upon congenial media and under proper conditions of moisture and temperature in order to germinate. To best visualize these microscopic units en masse, the interested observer should try placing a fungus cap upon black or lighter-colored paper, inverting a bowl over the cap and leaving this undisturbed for a few hours. If air currents have been excluded, a more or less perfect print of the gills or pores will result. Composed of millions of individual spores, the print will be either white or, as is likely, some interesting color. It should be mentioned here that spore color is often important in identifications.

Upon germination, the outgrowing *mycelium,* which becomes a network of interwoven living white threads or *hyphae* too delicate to be called roots, constitutes the vegetative portion of the fungus plant. This remains underground, or within or on rotting wood, or under loosened bark or debris. Mycelium will often be found in interesting patterns, sometimes even growing on the underside of a single wet dead leaf (plate 96).

From the usually invisible hyphae, the fruiting bodies (popularly referred to as *toadstools* or *mushrooms*) arise, mostly well above the substrate in these familiar examples, which include the puffballs. In less obvious and sometimes micro forms, the fruits arise from tiny sporangia, as in the familiar bread molds.

Fungi may be either scavengers or parasites. Lacking chlorophyll, the food-producing substance in *green* plants, they must either feed on dead organic matter or directly, upon living things. In some cases, in the parasitic forms, the mycelium may be microscopic. After spore generation, it penetrates the host victim, which may be minute itself or perhaps as large as a fly or a caterpillar. Such mycelia readily kills flies, ants, and, in the tropics, even grasshoppers, the hyphae (threads) as they extend outward anchoring the dead host's body to a leaf or twig or other object; and then the mycelium, as in macro species of fungi, sends up the fruiting bodies, in cases such as these often in the form of curious straight or wavy stalks.

Fungi and bacteria of course perform many important functions beneficial to other forms of woodland organisms, and together constitute the principal agents of decay. Through their ability to decompose other organic substances, they play an essential role in the production of humus, whose nutrients mean life to many animals of the soil, and perhaps somewhat less directly to the forest vegetation itself. Macro fungi, when fresh, or when themselves decaying, furnish food for many animal species: for enormous numbers of fungus gnats, as well as for beetles, slugs, and snails, certain worms, and such larger creatures as box and wood turtles, deer mice and squirrels, not forgetting man.

It is strange, when one thinks about it, how so many of these lowly plants which are unable to produce food themselves are, nevertheless, able saprophytes capable of hastening, or at least aiding, in the decomposition of the toughest dead leaves, whole limbs, and trunks of dead trees, while many of the parasitic forms are able to kill arthropods and sustain themselves on the victims' bodies.

Fungi often rot woodwork in our houses if undiscovered leaks have softened the tissues (plate 97). Fungi, as we well know, destroy our foodstuffs if these are not guarded. A fungus blight did away with our great American chestnut trees, and others of course are the culprits where many other trees and plants are concerned. Heart-rot fungus creeps into cracks in the trunks of such fine living forest trees as the beeches and aging specimens of sassafras, weakening and sometimes causing their death by windstorms, as once happened in the mini-wood and as is described in the first chapter.

We might go on and on, listing the things we do *not* like about fungi, forgetting for the time being their good points and their importance in the ecosystems of the forests. They are here, all around us, the good and the bad, so henceforth, in this study, we will consider them only as extremely interesting organisms of the mini-wood community, with the same rights to existence and to whatever they may find there that is essential to their nourishment and viability as is enjoyed by the other abundant organisms.

Molds and micro gill fungi were found frequently in examining the litter of the woodland floor and the humus just below, where these interesting forms sustained themselves on negligible amounts of dead organic matter (plate 98).

More easily seen, and perhaps the most remarkable of all the funguslike organisms, were the slime molds, technically known as the Myxomycetes. They are characterized by a merging or assimilating *mobile* mass of protoplasm known as the *plasmodium*, a slimy-to-the-touch, *creeping*, and spreading "object," and which is multinucleate but without cell walls. This often brightly colored plasmodium moves overland in the manner of a giant amoeba, obtaining its food, consisting of bacteria, fungus spores, maybe mycelium, and sometimes seemingly non-nutritive minute fragments of debris, by surrounding and overrunning them by extensions of "itself," thereby capturing the items as an amoeba engulfs food particles with its *pseudopodia* (temporary projections of protoplasm in any required direction, no two of which are ever alike). Plasmodia are commonly found (but seldom recognized) on rotting logs lying on the ground, and very often on those stacked in unprotected woodpiles (plate 99).

Here we come to the baffling truth that, while reproduction at this stage in these molds is very animallike, the type of reproduction which soon follows is plantlike, hence the reason why the true status of these organisms and their proper classification is still a debated matter. What happens is about as follows.

While reproduction is ordinarily asexual, by fission, at certain times individual tiny slime molds *congregate*. Drawn from all directions, they come together in large numbers, and while they are microscopic separately, collectively they form the plainly visible masses of various shapes and sizes which beget the plantlike spore-producing or fruiting bodies. As a plasmodium, the thing creeps along the ground, or under loose bark on rotting logs or stumps, or through refuse, sometimes climbing other plants; on finally coming to rest, it sends up groups of little spore-bearing or fruiting bodes as plumes, stalks, or other forms.

When the mature spores are finally released, they are carried as other spores are, for short or great distances on the slightest air currents. The comparatively few which land and find suitable conditions for germination must still find and fuse with other spores in order to beget zygotes which may then multiply by fission to again produce the creeping plasmodia. The slime mold life history is thus one of the most remarkable in biology. These organisms are often here, right in people's "backyards," but they are seldom recognized for what they are (plates 99 and 100).

As already stated, their spore-bearing bodies take interesting forms, sometimes appearing as globular little pink or brown or almost black objects resembling tiny puffballs, in other cases as tiny spherical bodies held aloft on hair-like stalks, or sometimes they appear as tiny things like toadstools; but in the author's opinion, the most striking of these spore cases (called *sporangia*) were the plumelike, richly buff-colored sporangia of the species *Arcyria nutans*. At first a lighter, almost yellow shade, they turned buff upon maturing. When dry, as shown in the photograph, they averaged five millimeters (about three-sixteenths of an inch) in height, and somewhat longer when first arising on slender stems from minute cuplike objects (plate 99).

A common slime mold whose fruiting bodies are entirely different in appearance from those of *A. nutans* is *Lycogala epidendrum*, which occurs in small groups or in great concentrations as subglobose "domes" on piled logs and other rotting wood. These may be found in many sizes at the same time, often so crowded together that some may become irregularly shaped from pressure. In color, they may be pinkish, yellowish brown, or dark brown to almost black, and usually more or less dotted with tiny scalelike warts. Easily mistaken for small puffballs, the fruiting bodies in this form are called *aethalia*. With age they become very hard, but are sometimes used as food by tiny nematode worms (plate 99), which will be described in another chapter in this book.

Logs in an unprotected woodpile are favored media for this and other Myxomycetes. Indeed, the old woodpile purposely left uncovered and allowed to disintegrate naturally soon becomes an irresistible lure to many kinds of organisms, both animal and vegetable, especially if the logs are within or on the woodland border.

Slime molds often appear on the ends of such logs in which the tissues have decayed. An unidentified species, which once covered the entire end of a stout log within twenty-four hours, produced a coating which closely resembled finely scrambled eggs, both in texture and color.

Still another, particularly interesting, and very different-looking species, identified by the author as *Brefeldia maxima* appeared as suddenly on the end of a log in mid-November. Measuring forty-six millimeters (one and thirteen-sixteenths inches) in diameter, and varying in thickness from eight to ten millimeters, it first looked like shaving cream freshly released from a pressure can. The mass was slightly sticky, and a dissecting needle used to probe this had long threads resembling liquid spider's silk adhering to its metal tip when it was withdrawn.

Around the edges of this mass, extensions which were probably sporangia residues formed what are known botanically as *hypothallia.* In another twenty-four hours a thin and brittle skin had formed over the mass, a cortex that surprised the author by turning a dull but distinctly silver color. As the interior of the organism gradually dried, this cortex ruptured in several places, the widening fissures revealing a dark brown mass consisting of billions of the spores of the species.

Slime molds do the most astonishing things! And now ponder this!

A short meter from an aging woodpile, a small heap of mixed humus and topsoil had been protected with a low fence of wire wickets. July 8, a pure white strange-looking object had *climbed* several inches up a wire and settled its base at a cross wire.

Measuring seventy-five millimeters (about three inches) in height, and forty-five millimeters (somewhat under two inches) in diameter near the base of the cone-shaped form it had taken, this queer plasmodium *seemed* to be a freak example of the slime mold *Fuligo megaspora* and altogether a baffling organism from the mini-wood area under investigation (plate 100).

Pure white when freshly situated on the cross wires, this moist creamy mass dried out gradually. In ninety-six hours the brittle cortex that had formed began to rupture in several places, revealing within it a dense dark purplish brown mass of spores, no doubt spores in the billions.

Slime molds are known to climb stems of other plants during their strange journeys in "search" of suitable conditions, but this example of a species ascending an iron wire was indeed an anomaly. Somewhere, the author read that slime molds require calcium for their metabolism. Could it be that iron (possibly here acquired from the oxydized wires as rust) is also among the slime molds' elemental requirements?

*   *   *   *   *

Any naturally fallen branch, cut log, or whole dying or dead tree trunk may suddenly reveal an outgrowth of what is probably the commonest species of bracket fungus, *Polyporus versicolor* (plate 100).

In such cases the mycelium of the fungus would have been growing for some time, invisible among the softened, decaying tissues of the wood. Then, almost suddenly, the young brackets which are the fruiting bodies of the organism appear on the exterior of the food material. Tiny at first, they gradually grow into much larger brackets, which are soft initially and always more or less brightly banded in gray and white and brown. On their white undersurfaces, myriad pores are outlets from which the spores are released.

As mentioned before, an examination of rich humus will sometimes reveal very minute toadstools or fruiting bodies which have arisen isolated and singly from tiny bits of decaying debris, and which illustrate again how very little goes to waste in a woodland and how very little nutrient may be required to bring these minute objects of great beauty to maturity.

Also exemplifying such modest food requirements are the somewhat larger fungi of the genus *Marasmius,* very often overlooked in the woods. Truly diminutive when compared with such familiar edible mushrooms of the genus *Agaricus* such as those we purchase at the market, the little groups of *Marasmius* are even more beautiful, and conspicuous when seen against the freshly soaked debris of their woodland carpet habitat, from the merest fragments of which they often grow (plates 95 and 101).

The woodland stroller will sometimes come upon large numbers of *Marasmius* diminutives dotting a recently wetted leaf carpet, particularly after a heavy rain. Growing singly, or often in little groups, the fragile fruiting bodies will be seen supported on stems about a millimeter (or about a thirty-second of an inch) in diameter, and up to forty millimeters in length. Beneath the white caps the gills are few and uncrowded. The largest caps found measured eight millimeters in width when fully expanded.

In a great many cases these little toadstools arose directly from the *midribs* of whole, or even small fragments, of dead leaves, and as often from delicate decaying twigs hidden under the leaves, and mostly on such debris fallen from oak trees. The black oak seemed to be the particular host species in the mini-wood, although a few specimens grew from fragments of beech leaves, but most of those found under one big beech were on oak leaf fragments blown there the previous year.

Occasionally, the vegetative portion of the fungus, like

delicate white threads, could be seen on the undersides of the supporting objects, and here it may be well to state again, for those still unfamiliar with the toadstool life-cycle, that all of the stalks or stems, and the caps of various sizes and colors which are so familiar, are the fruiting or spore-bearing bodies of these plants *only,* the vegetative portions from which these arise—the mycelium—being out of sight in the substrate, whether it be soil or decaying wood or another medium. To put it another way, it is, one might say, as if among higher plants an apple tree grew mostly underground, and sent only its fruits, containing its seeds, above the surface.

Overnight a bare woodland carpet of dead leaves may sometimes come alive with *Marasmius* fruiting bodies when such a habitat during the previous day or night has had a soaking rain. It is a delightful sight to come upon, but alas, of short duration, for these tiny toadstools wither away almost at once after the returning sun dries off their precarious fragmentary supports. One feels regret that a fruiting life of such delicate beauty must be so temporary.

We may wonder what there is that is sufficiently nourishing within these tough, often dried-out and brittle, acid-in-reaction leaf fragments and bits of twigs fallen from the oak trees, yet it is there, adequate in supply to bring the *Marasmius* fungi to maturity. We wonder: Do these little plants thrive on cellulose and some of the mineral elements first gathered from the soil by the trees, or is there some undiscovered symbiont organism here concerned? It is pertinent to remember that oak leaves also contain tannin, a compound which undoubtedly helps in rendering them resistant to easy decomposition.

Widely contrasting sizes in allied organisms is always an interesting subject, the more so when examples are found utilizing an identical habitat.

Within the mini-wood, for instance, close to where numbers of these tiny *Marasmius* toadstools showed themselves annually in late June or early July, comparatively huge fungi appeared unpredictably. In one astonishing case, in the eleventh year of this study, several pure white fruiting bodies with variously shaped caps arose only a few feet from where the *Marasmius* species had fruited earlier.

Fat-stemmed with opaque juice, shedding pink spores, the author believes they were a species of *Clitopilus* (and with the aid of a specialist, thought possibly to be *C. orcella*; plate 101).

The first to appear were specimens with white caps averaging a little over six inches in width. They came up near a woodpile beneath a large black oak on September 26, and were followed a day later by a truly giant specimen with a cap so irregularly formed that it could only be described as distorted, and which measured twelve and a quarter inches at its widest part!

Three years then passed before the species reappeared, this time under another black oak and only a few yards from where the first giant one had arisen. They were so pure white that they were conspicuous and instantly spotted from a distance; the date, October 1, and they definitely had not been there the day before!

It is such interesting surprises, offered by such splendid organisms, that continue to make a long woodland study so worthwhile, even if that woodland be but half an acre in area.

Strange to say, neither *Marasmius* nor *Clitopilus* (in those observed) were attacked by insects to any extent when fresh, as so many other kinds of fungi are at once. A few tiny black beetles were found burrowing into the big toadstools, but nothing was found eating *Marasmius,* which quickly dried and shriveled, undamaged by these numerous kinds of fungus consumers.

It is a curious fact that many large fungi dry out quite easily, while other species, after being picked, rot so completely and so quickly under the influence of bacteria and liquifying insect larvae.

When brought in from the woods and allowed to dry, *Clitopilus* specimens slowly withered but did not rot, and emitted an increasingly powerful and lasting mushroom odor with a peppery quality. When placed in a closed jar lacking fresh air, the toadstools quickly decomposed in their own juices, with the process no doubt hastened by bacteria. Exposed again in this condition they attracted numerous fungus gnats, insects of very small size which deposited eggs at once, and as a result the rotting mass became quickly infested with the gregarious feeding larvae of the flies. The many known species of these gnats (family Mycetophilidae) are among the numerous consumers of fungi, yet while these toadstools remained growing in the woods, they were attacked only by the small beetles already referred to, which appeared to do very little damage.

The surprising succession of odd and interesting fungi goes on and on in most woodlands. New ones become established, older arrivals repeat themselves, or die out not to reappear.

One of the most astonishing to appear in the author's half-acre was the hen-of-the-woods, *Polypilus frondosus,* a species consisting of a bewilderingly confused pile of more or less fused caps and stems collectively forming fleshy masses sometimes of great size (plates 95 and 102).

Standing within the half-acre was the double trunk of a long-dead white oak. Rising like a narrow elongated letter V, solid on the exterior, it was naked except for a couple of lianalike twisting stems rising to support the epiphytic foliage of Virginia creepers high above.

With confident expectations, the author watched this tree each summer and fall throughout the period of intensive study covered in these chapters, but it was not until the *ninth* summer that a fungus of any description was observed on or about the butt of this standing old tree.

But then came sultry July 15, when, strangely enough, after a number of clear hot days and nights, there suddenly appeared a massive fungus in the exact area I had been most closely watching. Growing from the wood of the dead tree at ground level, its somber colorings blended so nicely with its surroundings that its camouflage was almost perfect from a short distance away. Here was a ponderous thing called the hen-of-the-woods, a find indeed, a species considered a real table treat by some who know their fungi accurately, for when young and tender, the "hen" is said to be among the best. To the casual stroller, coming suddenly upon this thing in the woods, however, it might easily appear as one more, possibly even deadly, member of a fungus family to be avoided.

This plant—that is to say, its fruiting portion outside of the tree—consisted of a crowded aggregation of heavy, oddly stemmed fungi, white underneath, and contrasting with the camouflaged upper surfaces and edges of the tiers. Its actual appearance is clearly shown in the close-up photographs (plate 102, and the fruiting bodies shown in plate 95).

This first specimen to appear measured fifteen inches in height and was supported by a tapering main stem nine inches in width. It weighed exactly ten pounds.

On August 2, another group appeared at the butt of this same tree. It weighed seven pounds. But this did not end the story, for a little over a year later, two still larger clumps had fully expanded by August 1, again after a period of several dry days that were this time interrupted by heavy rains July 27 and 28. These masses measured twenty-two and twenty-six inches in width, and weighed eight and eleven pounds respectively! (Plate 95.)

This seemed to represent the plant's supreme effort and zenith year, for no fruiting bodies were produced afterward for three years, when a partly dry example was found at the tree on October 21 which probably had been at its best a week earlier (strangely enough, this time as a fall rather than a summer example).

Again the fungus skipped a year, but on November 11 of the following year it again put forth a group of fruiting bodies measuring seven and a half by nine inches, which weighed only one and a half pounds (plate 102). With this specimen the short, erratic history of the "hen" came to a close in the mini-wood.

In retrospect, we wonder how long its mycelium, once established in the tree from a floating microscopic spore, had been ramifying through the woody tissues before it commenced to produce these bulky fruiting masses. We wonder also how many additional years it might continue to "eat" the old tree before putting forth more of its astonishing spore-bearing aggregates.

What mycelium does in wood is help in the rotting process. In the case of the tiny *Marasmius* toadstools which arose from mere bits of leaves and twigs, no rotting of such media was visible, yet the mycelium threads must exude enzymes capable of digesting (for expansion and the nourishment of its fruiting bodies) such difficult leaf and twig materials as lignin and cellulose. In the case of the hen-of-the-woods, its mycelium seemed even more remarkable to the author, feeding as it must have on the even tougher materials of a dead hardwood tree.

At one side of this tree's base, just behind where the first hen-of-the-woods appeared, there was a cavity just large enough to admit the author's hand. Inside, one could feel a large amount of soft humuslike material which later proved to be almost black and moist, and which extended into the center and to an undetermined distance upward into the trunk. Enough of this stuff was removed to twice fill the funnel apparatus for collecting small forms of animal life, as described on page 88 of this book.

Recovered *alive* from the two samples (making about a quart of this organic, humuslike material) were 25 small brown isopods, 20 white enchytraeid worms, 30 microscopic glossy brown (eight-legged) mites, one larger oval glossy mite, two pink spiderlike mites, and one pale yellow dark-spotted spider. Numerous other insects were also recovered, including 260 miscroscopic springtails (Collembola), brown ants of two species representing at least two colonies, and two predaceous rove beetles, doubtless living on some of these other animals which will be given more space in later chapters.

These notes have been added here to illustrate interest-

Left and above: The cuckooflower, an attractive, white-blossomed member of the mustard family, made erratic appearances on the mini-wood borders. It is shown about half life-size. Date: May 13. (Page 39.)

Right: Turtlehead. The plant shows foliage and clustered blossoms. *Above:* The "mouth" of a flower, with tonguelike lower petal. About half life-size. Date: September 26. (Page 39.)

Left: Typical uncoiling fronds of the cinnamon fern, photographed April 24, natural size. Most ferns appear from the ground in a similar manner, uncoiling gradually. (Page 41.)

Above: Tall spore-bearing fronds of the cinnamon fern. *Left:* Sterile fronds of the cinnamon fern, not yet full grown, and three fertile fronds (in opening at left of photograph). Date: Both May 31. (Page 42.)

Uncoiling cinnamon fern fronds or fiddleheads, at various stages. Natural size. Date: May 8. (Page 46.)

Above: Close-up of fertile fronds of the cinnamon fern shown in plate 82 and at time (May 31) of releasing millions of dark green spores. *Right:* Sporangia actually scattering same (white dots much enlarged).

Upland lady ferns in the mini-wood were represented by both green-stemmed plants and those of the form *rubellum,* with pink- and burgundy-colored stems. Date: July 1. (Page 43.)

Left: Fertile frond of upland lady fern, underside showing *sori,* or fruit dots. Date: August 9. (Page 44.)

Above: Pinnules (small leaflets) at the frond tip, gathered into a hollow-ball "hideout" by the larva of a small moth, to be used between feedings. Natural size. Date: June 30. (Page 44.)

Top: Hollow-ball shelters at tips of lady ferns, formed by larvae of the moth *Macrobotys aeglialis*, shown somewhat enlarged. Date: June 20. *Above*: A *Macrobotys aeglialis*, also shown somewhat enlarged, that emerged July 20. *Upper left*: Fern ball (enlarged), opened to show larva and excrement neatly deposited at upper left of shelter. Date: June 15. *Lower left*: Much enlarged pupa of first-found species of moth whose larvae also construct fern balls. Date: July 17. (Page 44.)

PLATE 85

Left: Distinctive sterile fronds of the Christmas fern as they appear in the spring. The species often stays green all winter. Date: May 30. (Page 45.)

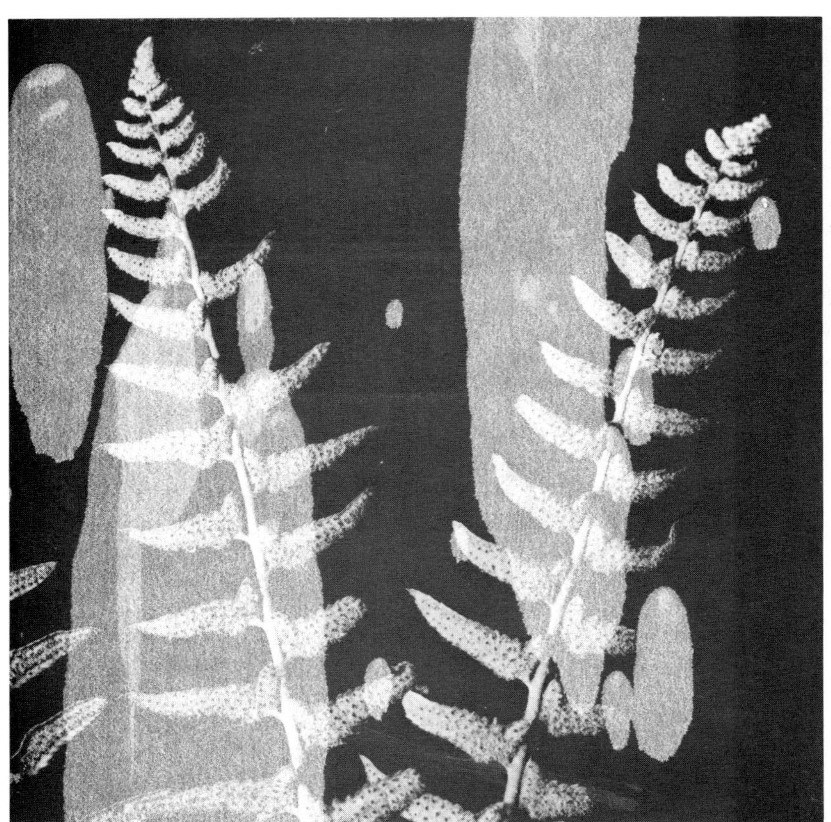

Left: Underside of a fertile frond of the Christmas fern, showing the characteristic fruit dots and the thumblike projection on some pinnules. About half life-size. Date: July 21. (Page 45.) *Above:* Unfolding sterile fronds of the leatherleaf wood fern. Date: May 22. (Page 45.)

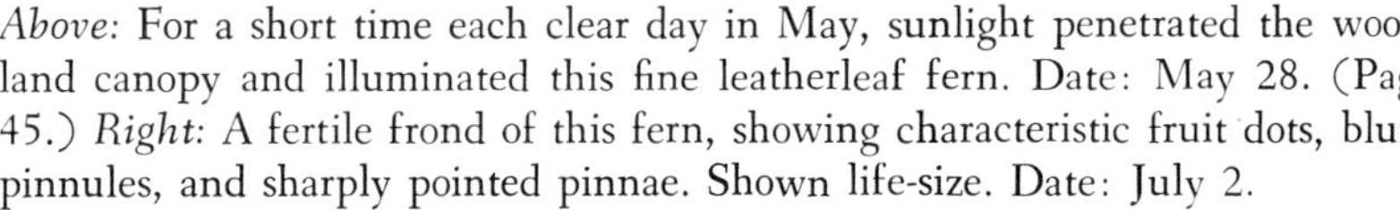

Above: For a short time each clear day in May, sunlight penetrated the woodland canopy and illuminated this fine leatherleaf fern. Date: May 28. (Page 45.) *Right:* A fertile frond of this fern, showing characteristic fruit dots, blunt pinnules, and sharply pointed pinnae. Shown life-size. Date: July 2.

Upper left: The sterile fronds of a young sensitive fern, photographed natural size, August 30. *Upper right:* Fronds arise from the rootstocks, one of which is shown at right, from May onward. (Page 46.) *Bottom:* A luxuriant bed of these ferns in full shade. Date: July 14.

PLATE 88

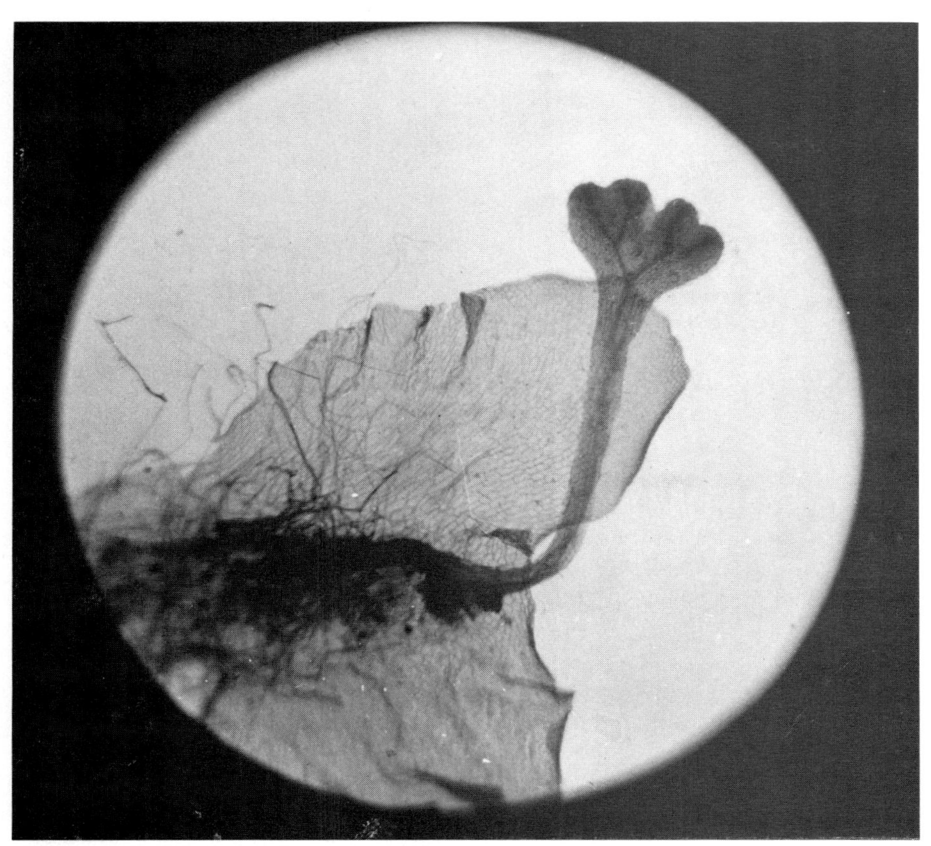

Left: Photomicrograph of a fern prothallus, sixteen times life-size. Showing are the rhizoids, stalk, and first leaflike frond of the sporophyte generation. Prothallus individual cells may be seen with a magnifying glass. (Page 42.) *Below*: Fertile stalks of the sensitive fern. Date: December 1. Life-size. (Page 46.)

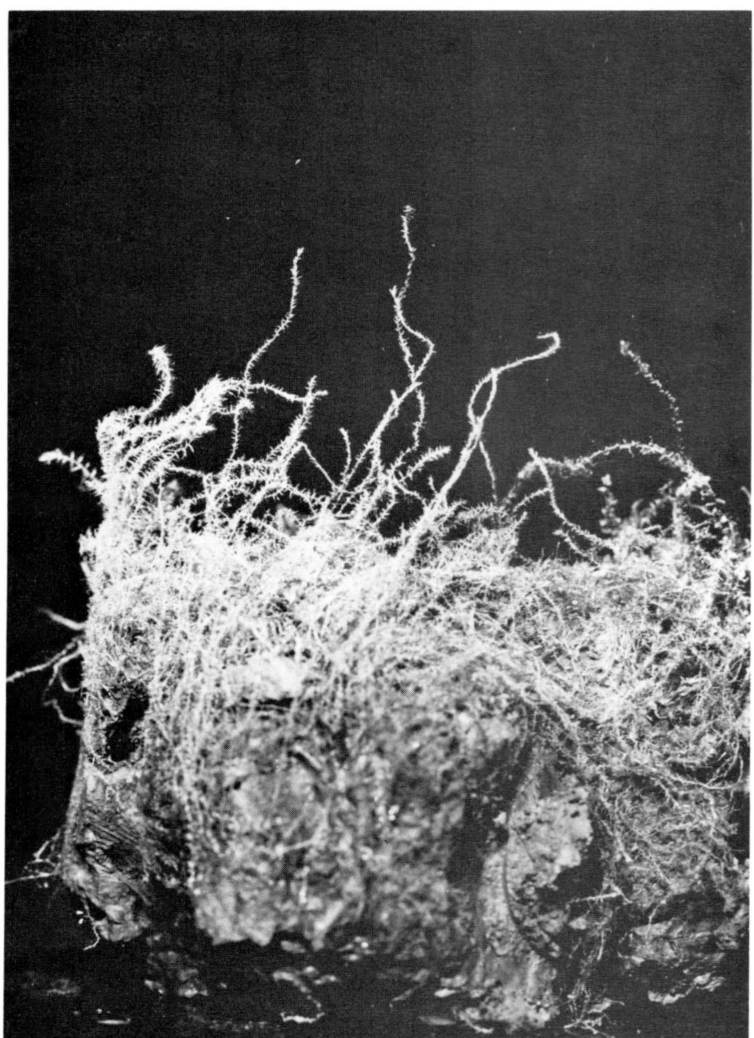

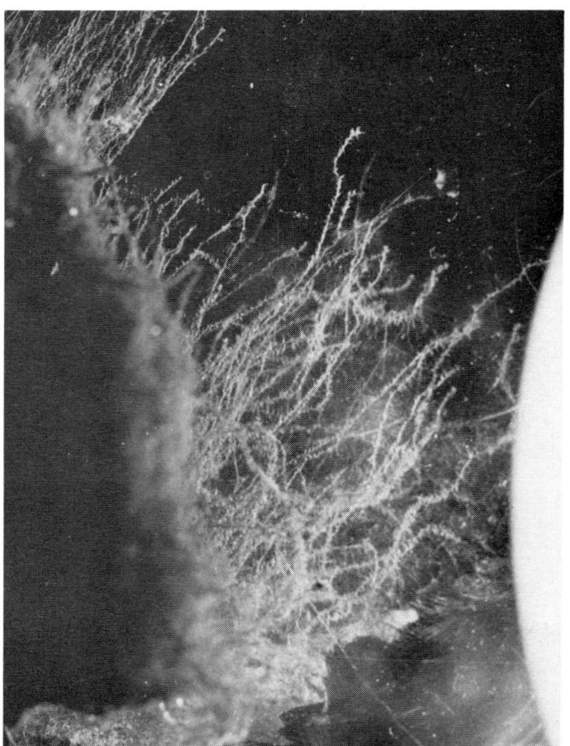

The unidentified moss grown from glacially deposited clay dug on the mini-wood border. *Left:* The moss at seven months, twenty-two days after sprouting. *Above:* At five years, nine months, twelve days of age! *Below:* The four-year-old moss covering clay. All shown × 2. (Page 50.)

This remarkable occurrence has baffled specialists, and the moss has remained unidentified because it has not produced setae or fruit stalks and spore capsules. Its full, strange story begins on page 50.

The fern which arose from lumps of glacially deposited clay. *Above:* Four months after appearing. *Top right:* Six months old. *Left, below:* Ten months old. *Right, below:* At fourteen months. All shown $\times$ 1½. (**Page** 51.)

The delicately formed stalks with their spore capsules first appeared during a mild spell in mid-December. The moss is *Ditrichum ambiguum,* and was later found fruiting as late as January 17. *Below:* The leafy part of the plant, shown in detail. Both pictures enlarged somewhat more than twice life-size. (Page 52.)

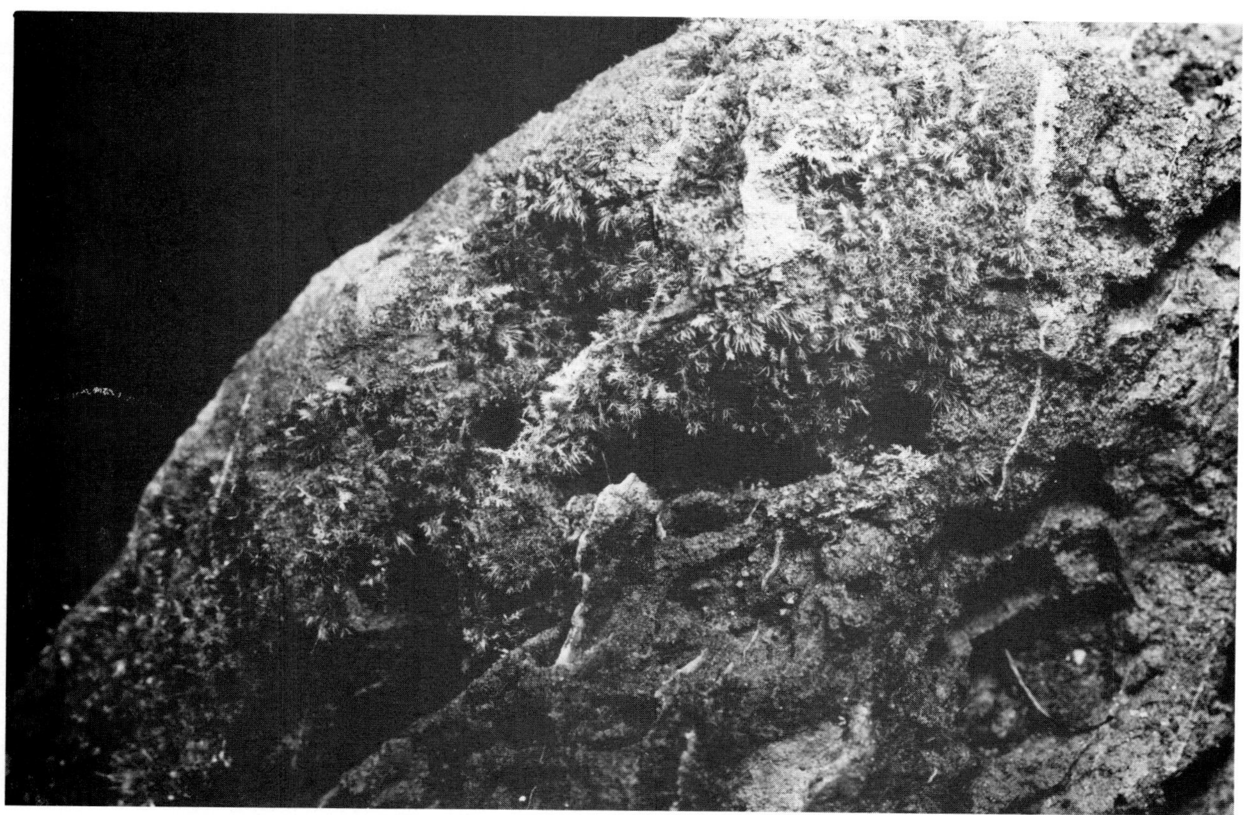

Shown much enlarged, the edge of the sandstone rock overgrown with the moss *Ditrichum ambiguum* illustrated in plate 92. Seen thus, it resembles an elfin forest covering a mountainside. Date: August 28. (Page 52.)

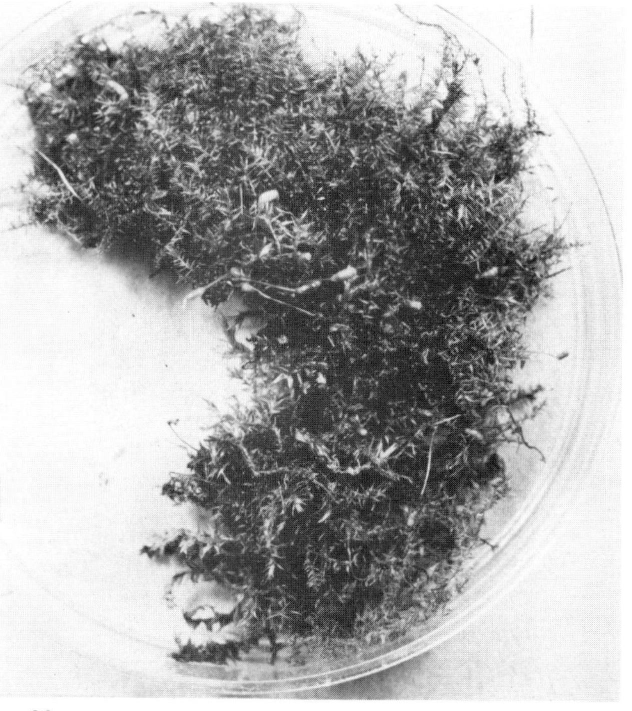

Above: Growing on wet wood, this delicate moss, *Leptodictyum riparium,* was found fruiting as shown, November 17. Whole plant at right. Both shown more than twice life-size. (Page 52.)

PLATE 93

(D) Fruiting stage of the moss *Amblystegium varium*, with dark reddish setae and bright green capsules, which grew at the base of a northern red oak tree. Date: April 5. Shown enlarged about one-third. (Page 53.)

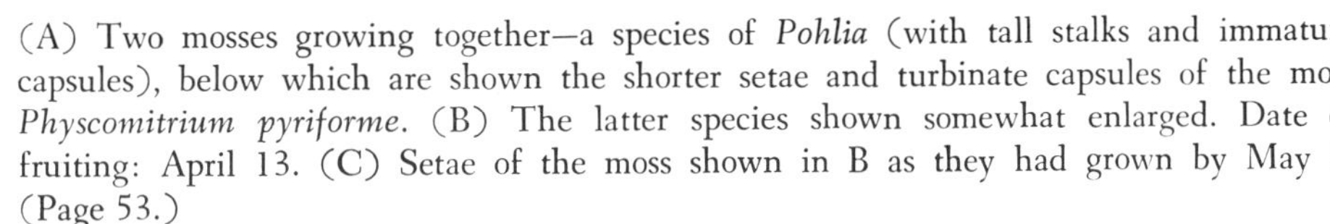

(A) Two mosses growing together—a species of *Pohlia* (with tall stalks and immature capsules), below which are shown the shorter setae and turbinate capsules of the moss *Physcomitrium pyriforme*. (B) The latter species shown somewhat enlarged. Date of fruiting: April 13. (C) Setae of the moss shown in B as they had grown by May 3. (Page 53.)

Contrasts in size of woodland fungi. *Above:* Hen-of-the-woods fungus which appeared very suddenly. Weight of left specimen, eleven pounds; right example, eight pounds! Note yardstick in foreground. Date: August 1. (Page 59.) *Below:* Flashlight photograph of tiny *Marasmius* fungi growing on fragments of twigs and dead leaves on woodland floor. Life-size. Date: June 23. (Page 58.)

Unidentified minute cold-weather fungus found in humus with oak- and beech-leaf overcarpet. Enlarged 2½ times actual size. Date: November 26. (Page 57.)

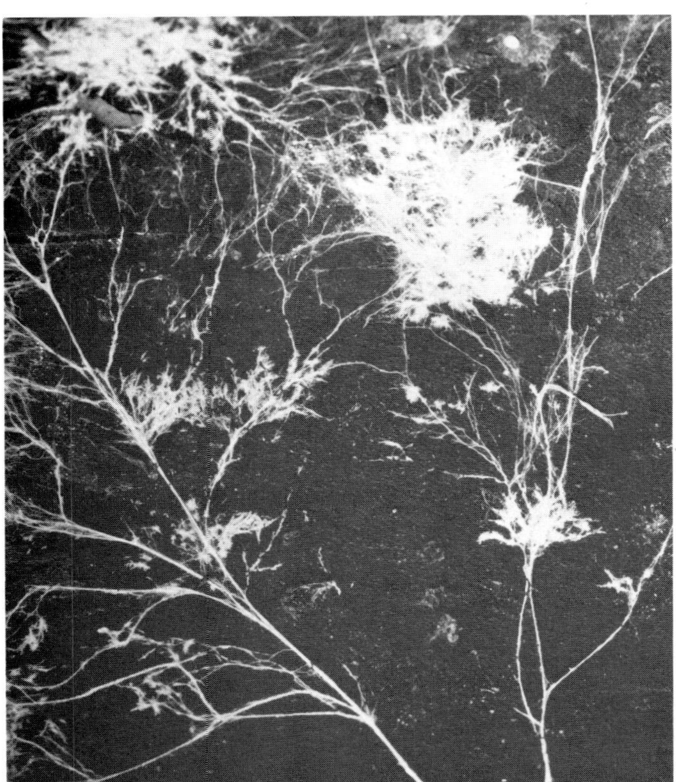

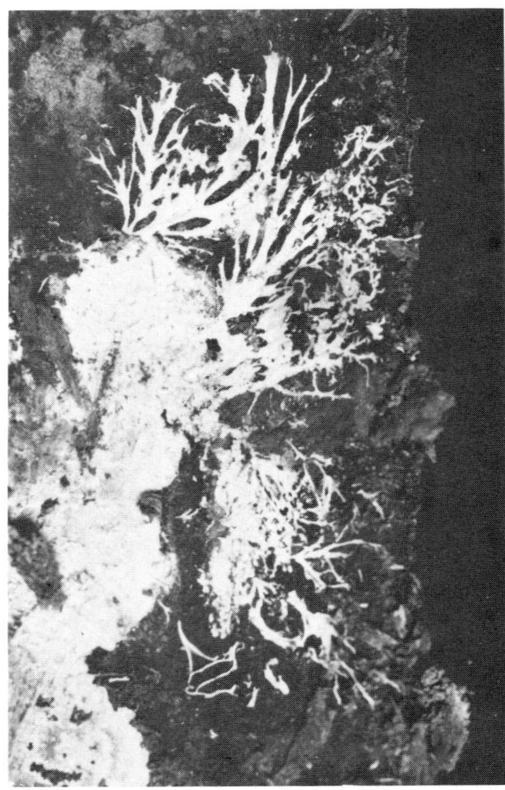

The white, threadlike growths which are the vegetative hidden portions of fungus plants and from which the fruiting bodies or toadstools arise are called "mycelia." *Top:* Two examples of mycelium on the underside of an oak leaf. *Left, below:* An extensive mycelium from the underside of rotting bark. *Right:* Mycelium from underside of a rotting log. All shown somewhat enlarged. (Page 56.)

PLATE 96

Spores landing by chance on a decaying doorsill of the author's toolshed brought forth these vigorous clusters of fungi. Of two species, they appeared as shown, just eight days apart and within a foot of each other. The group above, with yellowish caps, yellowish salmon-colored gills, and white stalks, was apparently a variant of the species *Naematoloma fasciculare*. Date: September 5.

The toadstools in this group were similar to the well-known orange-yellow species *Clitocybe illudens*, but had all-white caps and stalks. At night they were slightly luminescent, probably due to the presence of bacteria. Date: September 13. Both groups shown about two-thirds life-size.

PLATE 97

Several species of micro gill fungi inhabited the mini-wood humus and litter. This one shown three times life-size. (Page 57.)

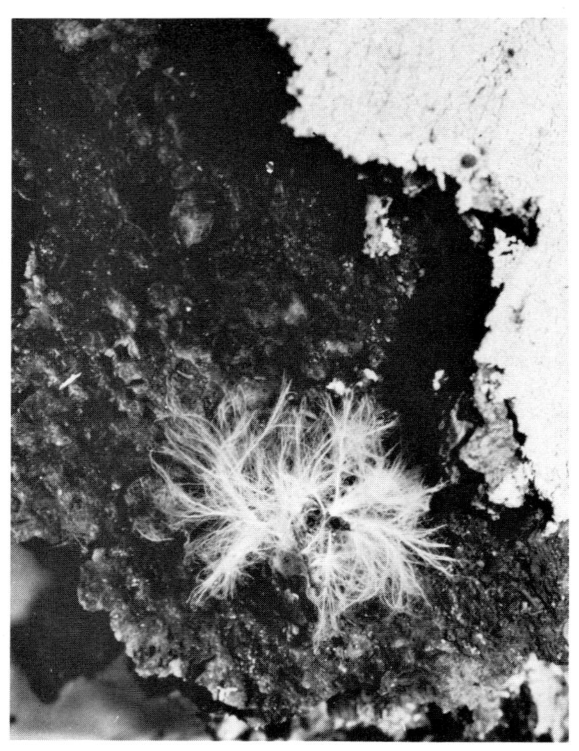

Above: Feathery white mold, which appeared on the end of a rotting log, waved its myriad branches in the slightest breeze. Shown much enlarged. Date: August 25. *Left:* A slime mold covers the end of a rotting log. Characterized by a *mobile* mass of protoplasm, the organism *creeps* over decaying wood or over the woodland floor in search of it. Remarkable fruiting stages come later. Date: November 8. (Page 58.) See also plates 99 and 100.

Left: A slime mold in the spore-bearing stage looks nothing like the creeping plasmodium from which these dozens of richly colored plumes arose. This species is *Arcyria nutans.* Date: June 28. Shown slightly enlarged. (Page 57.) *Right:* Slime mold *Lycogala epidendrum* in the spore-bearing stage. In the plasmodium stage it is coral red. Shown one-half natural size. Date: November 13. (Page 57.)

This slime mold appeared on the butt of a log November 19. In twenty-four hours it had turned a dull silver color with sufficient glow to show well in the photograph. Its name: *Brefeldia maxima.* Shown enlarged $\times$ 1½. (Page 58.)

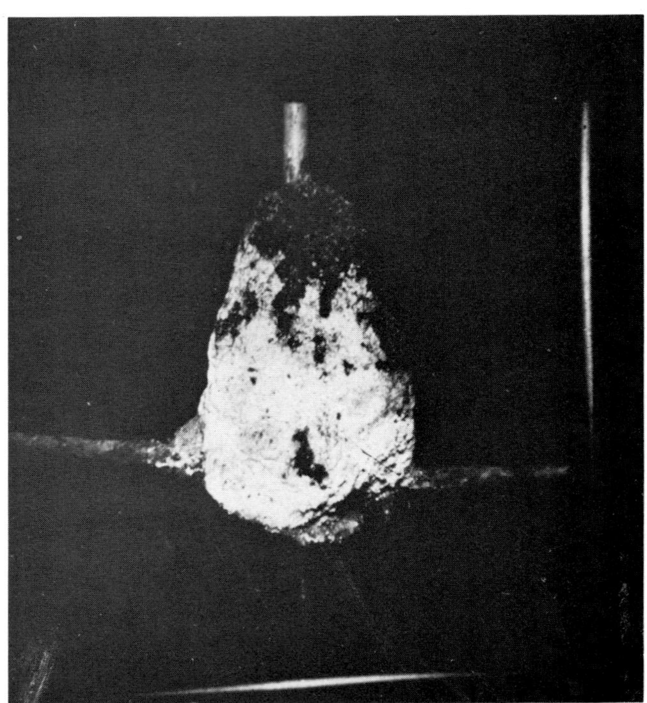

Left: Slime mold enigma. Climbing up the iron wire of a fence, this specimen in the plasmodium stage formed this cone-shaped white mass. Date: July 8. *Right:* Same mass rupturing and revealing brown spore mass. Date: July 12. Both reduced one-third. (Page 58.)

Polyporus versicolor, the commonest bracket fungus, feeding upon and aiding in the disintegration of a log on the mini-wood floor. When young and soft, the brackets are banded in gray, white, and brown. Date: October 21. (Page 58.)

PLATE 100

Top left: Diminutive fungi of the genus *Marasmius* on a decaying oak twig; another on a fragment of dead leaf. Natural size. Date: July 1. (Page 58.) *Bottom left:* The two *Clitopilus orcella* illustrated had caps over six inches wide and pink spores. Date: September 26. Found only during eleventh year of this study. *Above:* Species of *Agaricus*, probably *A. sylvicola*, of an odd shape. Appeared suddenly after a twenty-four-hour rain. Date: September 24. Shown natural size. (Page 58.)

A hen-of-the-woods fungus which weighed only one and a half pounds, shown about one-half life-size. Date: November 11. Others in the mini-wood weighed eight and eleven pounds. See also plate 95. (Page 60.)

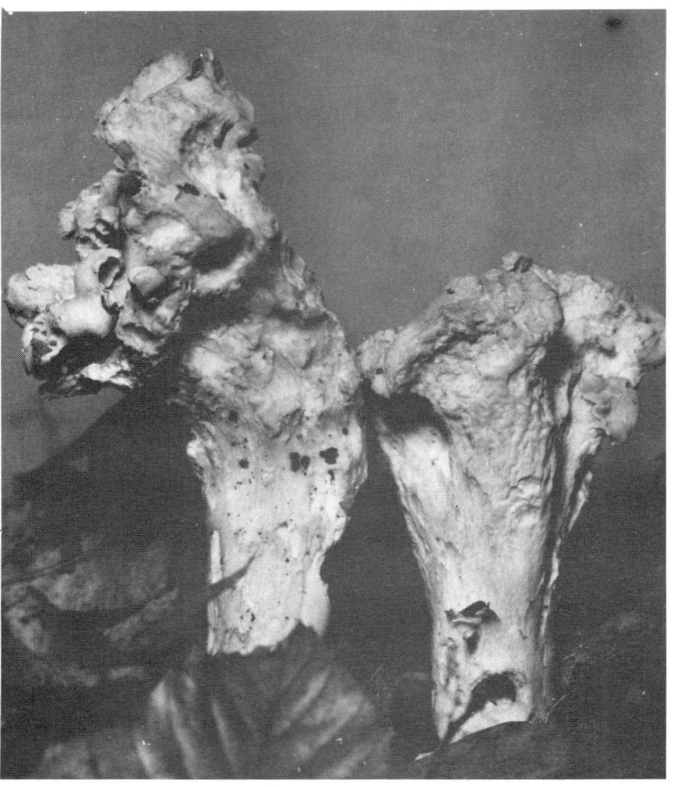

Above: Large clumps of *Naematoloma* toadstools such as this were reduced in two days to brown or black liquid by fungus gnat larvae. Date: September 17. (Page 61.) *Right:* A fungus anomaly, so oddly formed that positive identification was impossible. (Page 62.)

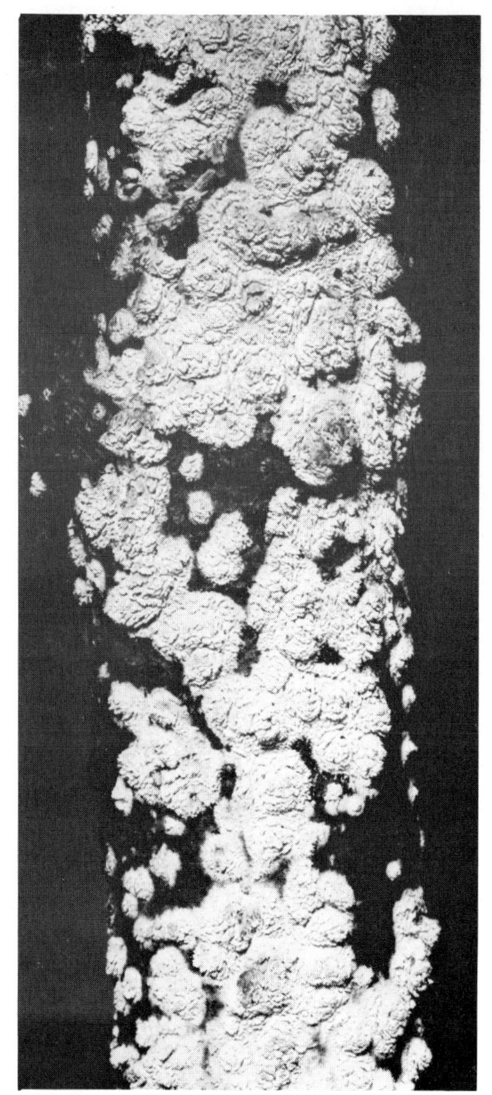

Fallen dead branches often bring curious fungi into view that otherwise might not be seen. This species, resembling miniature cauliflowers, is a good example. Shown one-third life-size. Date: November 20. (Page 63.)

Above: This species of puffball fungus, *Lycoperdon perlatum,* was a late fall or early winter arrival in the mini-wood area. Photographed November 7, when ground temperature was 34°. Shown half natural size. (Page 63.) *Below:* Another cold-weather species fruiting on dead oak limbs was the jellylike Jew's-ear, *Auricula Judae,* shown half life-size. Date: January 12. (Page 64.)

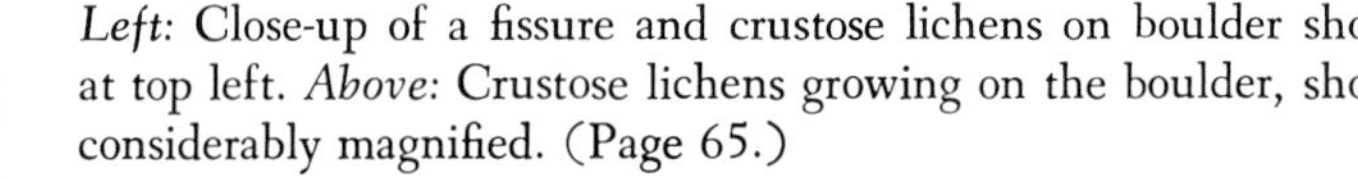

Left: A boulder deposited thousands of years ago by the melting glacier is being slowly eroded by the elements and a rich growth of crustose lichens. Enough primitive soil has accumulated in crevices in the rock to support a few moss plants also. (Page 66.)

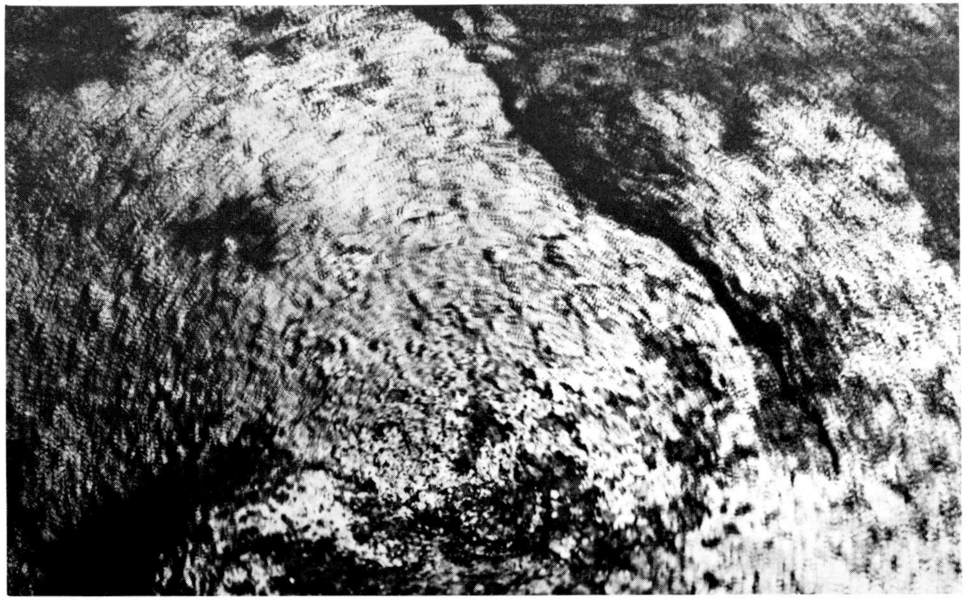

Left: Close-up of a fissure and crustose lichens on boulder shown at top left. *Above:* Crustose lichens growing on the boulder, shown considerably magnified. (Page 65.)

A rock-splitting process is under way. *Left:* Crustose lichens covering the face of a boulder on the mini-wood border have helped to prepare soil in the crevice. Moss and young ferns have already lodged there and are part of the amalgam. From a red maple seed fallen by chance into the crevice, a sturdy sapling has arisen that may one day split this rock in two. (Page 66.) *Right:* Foliose lichens on the bark of a northern red oak. Now examine the close-up magnified view in plate 106. (Page 65.)

PLATE 105

Below: Foliose lichens on the butt of the oak shown in plate 105, magnified five times life-size. Note the tiered, branching lobes. These lichens become pale jade green when wet. (Page 65.)

Foliose lichens on the bark of a red maple tree, enlarged five times life-size. (See page 65 for an explanation of why these lichens change color when wet.)

ing interrelationships, and also to reveal how a fungus such as the "hen," in helping to rot and reduce dead wood, may benefit many other organisms of the woodland community.

Represented in the fauna of this old tree were both "waste"-consuming and predaceous animals representing widely diverse orders. Countless bacteria and probably numerous micro fungi were also present.

Supposing that a considerable portion of the tree's interior had been filled with humuslike matter: keeping in mind the numbers of organisms found in a single quart of it, one may well imagine the total population which this dead tree was sustaining.

Here was a little world in itself, its population isolated, as on an island, its various organisms intimately interrelated, but its dispersal predictable, for one day not too distant, this tree would fall to the ground.

Strange as it may seem, however, there would be many survivors of such an event, for many of the tree's living inmates being so small and so light, they would be merely shaken up by the fall, and would doubtless carry on in the humus of the woodland thereafter.

Unlike most fungus fruiting bodies which are attacked by many kinds of animal consumers as soon as they emerge, hen-of-the-woods was extraordinarily immune. In no example were these big fleshy masses attacked by fungus gnats or fungus beetles, nor by slugs or snails which are regular feeders on many other fungus species.

"Hens" were therefore easy to preserve and study in all cases in which they were encountered by the author, although such specimens may be attacked under different circumstances elsewhere. Large fresh masses of the fungus brought indoors dried out intact, and retained a very powerful mushroom odor for many months.

It might be of interest to mention here that the mycelia of some fungus species occur only in connection with the roots of certain trees and other seed plants. They form what are believed to be *symbiotic* (mutually agreeable or beneficial) associations, each plant receiving something useful from the other. Such interrelationships between fungi and seed plants are technically known as *mycorrhizas*. It is interesting that specific fungi seem to require specific seed plants in many such associations.

Of interest at this juncture for comparison with the hen-of-the-woods (relative to insect attackers and consumers) were large clumps of a species of *Naematoloma* toadstools which were seemingly involved only with partly ex-

posed roots of red maple trees on the mini-wood borders, and a few times on the exposed root of a large red maple that had doubtless originally been within more extensive woods.

With pale orange-brown caps and somewhat lighter stalks, masses of these toadstools appeared on one or two particular exposed roots, and were fully expanded within twenty-four hours.

Clumps of various sizes, up to some fourteen inches in diameter and seven inches in height in the longest-stemmed specimens, were observed between August 13 and September 17 in eight consecutive years before the species petered out almost altogether. (Plate 102.)

Beautiful when in their prime, even these largest clusters were reduced within two days thereafter to horrid liquefied masses, brown to black in color, and literally rippling under the ravenous onslaught of tens of thousands of wriggling feeding larvae of various species of fungus gnats that were aided in the demolition process by countless bacteria.

As soon as the fruiting bodies of the fungus appeared on the root exposures, they were visited by the egg-laying gnats, mostly tiny flies of the family Mycetophilidae, that seemed always to be ready for them and whose larvae hatched from these myriad eggs within a day. These larvae were dull white maggots, one species with tiny projecting spines along the body. Many were tapered in form, and such numbers were present that they kept bumping and awkwardly wriggling over and under one another. None measured over five millimeters (about three-sixteenths of an inch) in length when full grown, many of them being only three millimeters.

As the fungus mass was reduced almost to zero, only a black, thin pancake remaining, and this soon dried altogether, hundreds of the larvae that had not reached full growth in time to dig into the ground to pupate died of starvation or dessication. Thus did these tiny flies help to control their own numbers to some extent, while the large numbers which did successfully pupate in the ground where the fungus was demolished soon issued as winged adults to start their life cycle again in other fungi.

Why some fungi such as these are so quickly attacked and consumed, while other species may be hardly touched, is a mystery, especially in such a case as the hen-of-the-woods, which is not poisonous even to human beings.

Microscopic examination of disintegrating *Naematoloma* clusters revealed other consumers. White springtails were feeding on minute portions of this decaying material;

microscopic mites were also present, some of them requiring magnification up to fifty diameters in order that details of their transparent bodies might be studied. Doubtless larger mites and insects larger than springtails also fed on these fungi, enjoying the fetid banquet offered; more details about many of these animals are given in later chapters.

Among the mammals, gray squirrels, which frequently nibble at other fungi, would have nothing to do with *Naematoloma*, which is an interesting fact.

Deviation from the normal form is not infrequently met with in both plants and animals, often for no discernible reason. Baffling things happen. Take, for instance, the occurrence of the strange slime mold which climbed and fruited upon a wire fence, as described in this chapter. And among the many normal fungi which we popularly refer to as toadstools, the mini-wood produced one anomaly so extreme that even a specialist in mycology found it all but impossible to identify with certainty.

The discovery of this last-mentioned toadstool served only to heighten the author's interest and enthusiasm for the study of fungi, proving as it did that something entirely new to one was a possibility on any day's investigation, even in a half-acre mini-wood.

Such an anomaly, a truly striking deviation in a species, occurred with great suddenness during this study, but only once, and on August 13 in the twelfth year.

Would that we might know from whence they came and from what normal species these spores floated in, germinated, and formed mycelium in the mini-wood capable of producing such odd forms in the fruiting stage (plate 102)!

Ten of these curiosities greeted the author's eyes on that August morning. With numerous nubbinlike crinkled and confused orange-yellow *pilei* (caps), supported on broad stems of a lighter shade and streaked with perpendicular indented lines, these ten objects had arisen during a twenty-four-hour period from organically rich humus covered by a blanket of dead oak leaves. In little groups, or singly, all were within a six-foot radius. They measured from seven to ten centimeters (up to four inches) in height, and their actual forms may best be understood by examining the photograph in plate 102. A specimen not illustrated had arisen in the form of a long-stemmed Y.

Professor Alexander H. Smith of the University of Michigan, curator of its herbarium, kindly advised the author that these anamolous fungi could possibly be a

species of *Cantherellus* (near what was called *Craterellus cantherellus*) because of the smooth *hymenium* (spore-bearing layer in fruiting bodies), but in the case of these specimens, the numerous small pilei were not typical of this species, and the identification thus remains uncertain.

The above information (not quoted word for word) has been included here to illustrate the caution with which a specialist approaches an identification. It should be carefully noted by the student reader, for this is the attitude to be assumed by the amateur as well as by the professional scientist.

Removed from the mini-wood, specimens placed in covered terraria were soon swarming with the larvae of fungus gnats, whose eggs had already been laid within them. Intact at the outset, each specimen was slowly reduced to a liquid mass by thousands of the translucent larvae looking yellow from the liquid showing through their skins. Constantly in motion, they waved their black heads upon choosing spots to feed anew. By August 25, some individuals had finished feeding, had grown to fourteen or up to eighteen millimeters in length, and were collectively starting to spin a network of silken threads over and around the rotted food mass. Very delicate flimsy silk, it was at first mistaken for mold, but by August 28 the larvae were pupating here and there within this network after shrinking to six millimeters in length. Soon the coming adult's eyes could be seen as two black dots, and within another week the fully developed fungus gnats were emerging—ocher colored, six millimeters in length, and with three black dots on each side of the thorax and nine on either side of the abdomen. By September 7, similarly colored gnats, probably of another sex, issued from the network, these having ten black markings on the thorax, and three black bands and six black spots on the abdomen.

Smaller and less numerous gray larvae also appeared in the rotted fungi. By August 23, these were spinning delicately thin silk cocoons here and there about the borders of the diminishing food mass. From these cocoons, typical, rather long-legged gnats of the family Mycetophillidae, with conspicuous reddish brown heads, emerged five days after pupation.

A few much larger larvae of crane flies were also observed feeding in the liquefied fungus mass as it continued to shrink and finally began to dry out under the onslaught of all these dipterous insects. The mass now resembled carbonized toast in its container in the laboratory, but had it been left in its natural on-the-ground situation, this black residue would no doubt have soon been dissolved and admixed with the other materials being recycled in the soil.

In this story of the fungus-fungus-gnat relationship, we have an interesting demonstration of how these insect species, after taking toll of the fungus in great numbers, reduce it to such a state in the process that it no longer serves to nourish hoards of the larvae that are still too young to pupate, and therefore die out around and under the black and brittle residue. The fungus gnats thus help to control their own numbers, but nature has so arranged matters that enough of the insects reach maturity; and before the fungus cluster has been too destructively eaten into, there has been time for the plant to release its spores in vast numbers. The insects have been fed, offspring of the fungus are assured, and the food chain remains with its links undamaged.

Appearing occasionally within the mini-wood, but more often in somewhat less shady habitats along its borders, were fungi of the family Lycoperdaceae, popularly and familiarly known as "puffballs."

These fungi have neither gills nor tubes (pores) from which the spores are discharged. They are solid fleshy objects when freshly arisen from the mycelium, whose soft white sporogenous substance, called the *gleba,* is surrounded by a continuous outer skin or cortex known as the *peridium.* As the gleba dries to a powder consisting of billions of spores, the peridium ruptures and the microscopic mature spores are then released in little puffs like smoke whenever the fruiting body is jarred, or squeezed by curious human beings. Seen thus en masse, these tiny cells may be brown, dark purple, black, olive, yellowish olive, or clay colored.

Puffballs occur in many greatly diverse sizes, from tiny objects, or the more often seen moderately sized individuals, up to single larger specimens the size of an orange or a basketball, or even a giant-sized species technically known as *Calvatia gigantea,* which measured three feet in diameter and weighed forty-seven pounds!*

In the mini-wood area, small specimens appeared at various times during warm weather, but more often the most interesting and neatest groups came up in cold weather in early November.

Most frequently found were the white puffballs *Lycoperdon perlatum.* With a mycelium of fine white threads

---

*Nina L. Marshall, *The Mushroom Book* (Garden City, N.Y.: Doubleday, Page & Company, 1914).

growing shallowly, and feeding on humus well worked over by earthworms, the fruiting bodies first appeared on these threads as minute white nubbins, but at maturity, on the surface of the ground, measured up to thirty-five millimeters (about an inch and three-eighths) in height (plate 103). Although appearing bare to the naked eye, even low magnification revealed tiny white spinelike protrusions from the surface of the peridium. As the puffballs aged, these dropped off, leaving microscopic indentations or spots where their bases had been. Shown in the photograph are specimens which matured by November 7, when the ground temperature in the early morning was 34° F, and 48° at noon. Another group, found on the same spot the following year, had arisen by November 4 after heavy frosts during previous nights.

Another species about the same size, with a rougher skin, the pear-shaped puffball, *Lycoperdon pyriforme* (some larger in diameter than the former) was observed several times, and on one occasion it was found fruiting on a rotten log. White at first, brown as it aged, its peridium was flecked with small scalelike objects. The gleba was yellowish green and olive green, and in shape the species was somewhat like the elongate brown Bosc pears.

Still larger puffballs, measuring up to three inches in diameter and squatty in form and covered with warts on an ocher or brownish peridium, occurred where a large log had completely rotted out in the mini-wood. Its name: *Scleroderma aurantium.*

In the world of the fungi, many interesting things happen high aloft in the woodland and other trees. Not only are the bracket fungi, such as the ubiquitous *Polyporus versicolor,* mentioned earlier, at work on dead and dying wood, but many other, stranger forms fruit high above the ground as well.

Indeed, we might never know what does grow up there, were it not for stormy weather. As fall and winter winds rock and bend the naked limbs and branches and sway the tree trunks, a natural pruning process takes place which brings much already fungus-weakened and rotting wood to the ground.

Some curious species seem to be at their best in winter, as in the case of an oddity on a white oak limb found lying on the snow in late November. Thickly coating the wood were large numbers of saprophytic fungi (unidentified at this writing) but which for lack of a better common name we might call the "cauliflower" fungus, for even without a magnifying glass, the crowded fruiting bodies did resemble this delectable vegetable in miniature (plate 103).

On another occasion, this time on December 12, after a very windy night, and with the thermometer registering 24°F the next morning, a freshly fallen dead limb was lifted from the snow beneath the largest black oak in the mini-wood. On what had been its upper side when aloft were several growths of a curious fungus whose many variously shaped lobes of a fleshy consistency, almost like jelly, were in no part even stiffened by the cold weather, although moistly clammy to the touch.

On January 12 of another year of this study, this dark brown jellylike species was found again, this time on a fallen white oak branch, and thereafter on several other occasions it was also observed, but always in winter and often with snow on the ground. Here were very interesting cases not only of a winter tolerant, but evidently a winter-preferring species, yet jellylike and wet to the touch, a fungus for some mythological reason known as the Jew's-ear, *Auricula Judae* (plate 103).

Despite its strange consistency and appearance, this fungus belongs to the class Basidiomycetes, or higher fungi, which bear their spores on structures called *basidia*.

Upon bringing examples of this species into the laboratory, they quickly dried up, becoming hard and brittle masses greatly reduced in size, but holding fast to the branches. A few hours soaking, however, restored them completely to their original size and jellylike consistency. This experiment was repeated with different specimens, found on different dates, and always with the same results, the inference being that the same thing probably occurred over considerable periods of alternating wet and dry weather while the fungus groups were still up in the trees.

Strange to say, only a single representative of the genus *Boletus* (family Boletaceae) was found in the mini-wood during the entire fifteen-year intensive-study period. These are fungi in form much like typical toadstools, as popularly visualized, with caps and stems, and occurring in many colors. But in place of the familiar thin platelike gills on the underside of the caps from which the spores are released, the species of *Boletus* have great numbers of side-by-side tubes in which the spores are borne.

These fungi emerge from the ground and rot very quickly. The one specimen found in the mini-wood in August, which was already too far gone for identification, had apparently had a reddish cap and dark brown stem.

And now, upon nearing the end of these botanical chapters, constituting Part One of this book, it is hoped the reader will have gained a clearer idea of how a forest area, be it a vast tract, or a mere mini-wood, is always in a state of change.

Interrelationships in these ecosystems are too often bewilderingly complex, as for instance among several species of insects and their fungus hosts, among themselves, when they are living therein, between these saprophytic plants and *their* hosts, and (not to be forgotten) the possible influence on one or all of these organisms by other animals, especially parasites.

Many interrelationships and other interesting things of course occur where it is next to impossible to study them, as for instance in the woodland canopy, when the trees are in full leaf. One cannot just flit from treetop to treetop, as one might study the organisms of the woodland floor from day to day in intimate contact. Wind-snapped branches and even twigs, as we have seen, may sometimes bring us a little knowledge about these regions, but as to what actually goes on in most such areas, especially in summer, we still have a very great deal to learn.

All we may do is to continue to scratch for facts about these upper levels, but always with the pleasant thought that, year after year, things new to one may be found first hand, and that new acquaintances in the form of plants and animals are distinct possibilities, and that a *new* fact may come the enthusiastic investigator's way at any time, even in a mini-wood.

CHAPTER VII

BRIEF OBSERVATIONS ON THE LICHENS

Lichens, we may say, are compound plants, extraordinarily interesting organisms in which the plant body consists of an alga and a fungus combined in a symbiotic interrelationship.

Easily recognized are the many extremely flat, often crusty, rounded or more oddly outlined blotches in various shades of gray or grayish green growing very commonly on rocks and boulders. These are called the *crustose* lichens, and examples of them are illustrated in plate 104.

Perhaps more colorful, equally common, but somewhat differently formed species are as numerous on tree trunks in New England and elsewhere as are the thin crusty blotches on rocks. Expanding very gradually, like all lichens, these other forms grow in gray or pastel green lobes, branching and rebranching, often overlapping each other, the growths sometimes covering extensive areas of the bark. Within and on the borders of the mini-wood they were found mostly about the butts of large trees. Especially common on borderline oaks, these forms, known as *foliose* lichens, became pleasing, deeper shades of pastel greens when rain soaked, and were most attractive subjects when viewed somewhat magnified (plates 105 and 106).

Lichens occur all over the world in great variety, thriving in many climates and habitats. Being extraordinarily viable and hardy partnerships, they are tolerant of considerable heat and cold.

Some forms, called *fruiticose* lichens, have numerous upgrowing little branches, reminding one of miniature shrubs, while others are more hairlike; still others, known as *squamulose* lichens, look somewhat like rather ragged masses of minute flat noodles. Growing on many large rocks in New England, there are large sheetlike lichens, partly upturned from a single point of attachment; these are called *rock tripe*. The foliose species, on the other hand, are anchored to the substrate by very delicate rootlike projections from the thallus or plant body, called *rhizines*. Very striking pancake-shaped lichens of a bright orange-yellow color were observed by the author on many of the headstones in the oldest Cape Cod cemeteries; this is a species known to thrive on salt air blowing in from the ocean, while the trunks of the Cape pines were veritable "mines" of lichen species.

While most of us know much more about nature now than did our forebears, there are still many who do not realize that these common organisms consist of two very different kinds of plants growing in an intimate partnership which is undoubtedly mutually beneficial. Such a condition, known as *symbiosis,* is of frequent occurrence in nature among other kinds of organisms, some of which will also be met in this book.

The alga partner in a lichen is a microscopic green or bluish green plant, while the other partner is a fungus consisting of a mass of microscopic threads or *hyphae.*

The bright green sheen which we observe on old damp woodwork, on rocks and shingles, consists of great numbers of algal cells called *Protococcus.* Equally tiny algae, of other species in great numbers, constitute a part of every lichen, but these species could not live without the fungus partner, nor the fungus without the alga.

The collective lichen body is called the *thallus.* In this

composite structure, the green algal cells manufacture sugar as other green plants do by the process of photosynthesis. This food is in part utilized by the fungus for nourishment. What the fungus has to offer in return is probably water, which the lichen readily absorbs, and, one might imagine, possibly protection in thickening the thallus, but what else it may supply is left to the reader to ponder.

There are a great many technical details to be learned about lichens which cannot be covered even superficially in this book. Suffice it to say here that, given a microscope and a properly prepared thin cross section of, say, a foliose lichen, the organism will be seen to be stratified, and in such a way that the two different plants of the partnership may be clearly observed. An upper cellular layer or stratum called a *cortex* is underlain by a somewhat thicker stratum consisting of algal cells, which, in turn, is underlaid with a much thicker stratum made up of the threadlike fungal hyphae known as the *medulla*. At the bottom or lower surface of the thallus there is another cortex which of course is also cellular.

On the surface of the thallus, certain tiny cuplike or disklike objects are reproductive organs called *apothecia* which produce spores in their intricate ways. We know also that lichens reproduce vegetatively; fragments broken from the main plant body (if consisting of both the algal cells and fungus hyphae) may grow into full-sized lichens if they fall on suitable media.

Occasionally a large rock or boulder will be observed which, despite its apparent solidarity and great strength, has, nevertheless, been split by a tree sprouted from a seed which by chance lodged years before in a debris-filled crevice.

On a minor, or one might say almost microscopic scale, crustose lichens, in expanding and contracting, by the pull of their rhizines, and by the deposition of acids, very gradually but surely split off minute rock fragments from the host upon which they have located. Aiding in the process are acids which descend in rain, and the alternate freezing and thawing of such water as may work its way into the smallest fissures.

Lichens, as we know, have always been the pioneer plants in soil production. The glacial boulder shown in plate 104, staunch as it might appear, was being slowly eroded by lichens and the elements. Close examination revealed surprisingly large amounts of loosened mineral fragments around and beneath its coating of lichens, material ready to be washed off by the next heavy rain. The same conditions prevailed at the rock shown in plate 105.

Mixed with these fragments in the rock cracks and fissures was dead organic matter, both animal and vegetable, the former from tiny arthropod feces and their own nitrogenous dead bodies. Here also were dust and other matter brought down in the rain, all of these things being mixed into a primitive and continually building soil eventually capable of supporting mosses. These plants themselves gradually build a good black soil beneath them in which the spores of ferns may then germinate, while even the winged seed of a red maple, an acorn, or some other seed may one day lodge and, after sprouting, find sustenance. Thus once again we might have a growing young tree whose roots and trunk could eventually split even a large boulder as a major if not primary event in the macro making of soil, which in an early stage is seen in plate 104.

Commonest lichens in the mini-wood area, growing on nearly every rock to be found there, where for a good part of every year much sunlight reached them, were the unidentified species of whitish, mineral gray, or greenish (when wet) crustose lichens shown in plate 104. Capable of absorbing considerable moisture, it was the algal cells which caused this tint showing through the cortex when wet.

Seen in the dry state, under moderate magnification, these lichens, which to the naked human eye seemed no thicker than blotches painted on the rocks, became coarse-looking crusts resembling earthy landscapes studded with many rocks, and sometimes with cracks or fissures of varying depths running through them in several directions. Other crustose examples, when magnified, were seen separated into irregularly shaped blocks, thus appearing much like dessicated and caked mud in a dried-up pond or lake bed.

How well these lichens were performing and aiding the work of rock erosion, as already alluded to, was clearly revealed, for around their edges were many rock particles broken away and ready to become part of the soil.

Looking at these pioneer plants, growing year after year on the faces of these inhospitable-appearing rocks and boulders left by the glacier where, for the most part, these ubiquitous reminders of the Ice Age in New England lay like icebergs, with the greater part of their bulk buried and invisible, it was interesting to speculate once again whether these grayish white and greenish lichen blotches before one's eyes were not the living descendants of species brought here on these very rocks and boulders riding the ice thousands of years ago.

Immune to great cold and extreme summer heat, propagating from fragments of the original thalli, they may very well have arrived thus, and whatever the truth of the matter, the author likes to believe they did.

Unlike the trees in the pitch pine forests on Cape Cod, for instance, whose trunks often support an astonishing variety of interesting lichens on their rough bark, the trees in and on the mini-wood's borders did not seem to be congenial to many kinds of these organisms.

All of the foliose examples in question were so firmly attached to the bark of their hosts that they could not be removed intact for photographs in detail without cutting the bark itself away with them.

While to the naked eye they appeared from even a short distance as simply amorphous green coatings, often vivid when wet, under the stereo microscope they were seen to consist of great numbers of outward growing sheet-like lobes and branches of various shades of pale green (when dry), and in many instances with the lobes bordered with greenish white. In most places they grew in layers, one above the other, like the pages of a book, but with spaces in between, and none of these had formed neatly rounded thalli, as in many other foliate lichens.

Most lichens are difficult to identify as to the exact species, and chemical tests must be resorted to by the serious lichenologist; thus, the reader who intends to go further into the subject is referred to the botanical works listed at the end of this volume.

Lichens produce unique chemical compounds themselves, among them mild phenolic (fatty) acids. The chemical makeup of each species is so constant that all of a kind will give the same reactions to the tests regardless of where they were collected. Not only may definite identifications thus be made, but species that cannot visually and certainly be separated from other very similar ones may be properly classified by chemical means. At times, microchem-

ical tests must also be employed in cases where two lichen acids give the same reactions in color tests. Lichenologists also resort to crystal tests by extracting the compounds with acetone and allowing them to recrystallize on a microscope slide so they may be identified by comparison with pictures prepared for the purpose. Fluorescent light is also used, as certain lichens fluoresce brightly, while the technical investigator may even enlist the processes of chromatography for positive identifications under certain circumstances.

Tiny arthropods called orabatid mites, members of a large group of eight-legged, somewhat buglike creatures, but not insects (which have six legs) inhabited the foliose lichens illustrated, and were found wandering among the lobes and apparently feeding thereon.

A great many species of mites inhabit the earth, living in all sorts of habitats. In the mini-wood there were legions of them, especially in the humus, where some of them were scavengers and parasites on animal matter, while others were vegetarians.

Mites will be met with in many roles and in many sizes in the following, zoological chapters in this book. Those inhabiting the foliose lichens growing on the bark of oak and red maple trees possessed domelike, almost globular bodies of a rich reddish brown color, as highly polished as a fine carnelian gem. They were sluggish, slow-moving individuals, very different from certain spiderlike mini-wood species which moved with such speed that it was difficult to keep them in view under high magnification.

And now, upon turning the page to the zoological chapters in Part Two of this book, the reader may soon be surprised to find that, while the large number of animal forms found in the little area here under investigation greatly complicate this story, they also serve to add greatly to the interest and fascination of such a continuing little community.

PART TWO

ZOOLOGICAL

CHAPTER VIII

FIRST PHYLUM ANIMALS AND ROTIFERS

At the bottom of the ladder of animal classification (in the phylum Protozoa), zoologists have grouped some fifteen thousand species of mostly microscopic single-celled animals, among which is believed to be some form closely resembling what the most primitive ancestor of all higher animals may have been like.

While the higher animals are made up of various aggregates of cells, forming more and more complex bodies and their organs, and culminating in the Primates, in Protozoa, the protoplasm even in these single-celled organisms performs the life functions carried on by anywhere from a few to fabulous numbers of cells in these higher life forms. In other words, the protoplasm in these single living independent cells carries on such necessary functions as digestion, excretion, reproduction, locomotion, and all other required activities.

Among the vast number of Protozoa, there are some species still baffling to both zoologists and botanists, for it is impossible to say with certainty whether they are animals or plants; and if, perhaps, they bridge the two kingdoms, they may be most like our remotest ancestors. These species are known as *flagellate protozoans*.

Before describing the protozoans encountered in the mini-wood in somewhat more detail, a few more words about the phylum in general may be of interest.

Ubiquitous bits of life, protozoans are to be found almost wherever one might care to look for them. In every pool and pond and swamp, in every roadside ditch or smallest standing puddle, in every field of weeds or hay, there are incalculable numbers of them. In every *bale* of hay—no matter how long dried—there are potential mil-

lions more, themselves desiccated or temporarily encysted, all animation suspended but still viable, only a thorough wetting of their medium being necessary to restore them to active life!

In the dew-moistened grass of your lawn, many may be swimming; in every handful of moist woodland humus, in any temporarily retained rainwater, in any crack or hollow or depression in any rock or boulder, there may be thousands of them, and many multi-celled animals as well.

In the raindrops retained by leaves of lettuce, there are often others, quite harmless if swallowed, which seem to come down from the skies. They may swarm too in those masses of disintegrating fungi liquefied by myriad larvae of fungus gnats and bacteria, as described in chapter VI; they appear in moistened mosses, and from brittle field-dried lichens, when these are put in water and allowed to stand for a day or two.

To demonstrate their persistence and viability, the author once scraped dried caked mud from the underside of a mudguard after coming north from Florida in his automobile, placed this in a jar and covered it with *distilled* water as a precaution against introducing any life forms from tap water; yet protozoa were observed in the fluid in considerable numbers a few days later.

On the leaves of trees, in their fruit and seed, either northern or tropical species, there may be protozoa of one form or another. Thus it will be realized that these often grotesque-looking little entities inhabit thousands of habitats and niches, and that their presence is most often unknown to the majority of human beings; hundreds of active species, and as many in a desiccated or encysted state, are yet viable and capable of emerging from suspended animation to once again eat and excrete, reproduce their kind, and carry on all other normal life functions upon the simple addition of water.

A glance at the drawings in plate 107 gives a clearer idea of some of their diverse forms than wordy descriptions, but the reader should remember always that there are great numbers of other known species, and possibly as many yet to be discovered.

While many of these are still considered to be one-celled animals, zoologists have more recently preferred to think of them as *protists,* or animals consisting of one or small numbers of cells with organizations of varying complexity.

A great many protozoans are equipped with *cilia,* which are encircling or otherwise placed microscopic hairs that act as "banks of oars," so to speak, often moving in unison to propel the animals forward or backward, wave food toward the mouth, and perform other functions.

Unlike the amoeba we studied in school biology courses, that "first" animal capable of extending its protoplasm in any direction to engulf its food, the ciliated protozoans are bounded by a permanent membrane or *pellicle,* and are therefore held to permanent body shapes.

And how do such minute animals nourish themselves? Probably their food mostly consists of protozoa and plant cells smaller than themselves, and in many cases this may be observed.

As to their antiquity, no doubt they antedate the dinosaurs by millions of years, and very likely many of the species differ little today from what they looked like thousands or millions of years ago. No living soul can say from whence they came in the first place, or exactly when their beginnings were, but we do now know that they may be transported over great distances by wind, or in the fur of animals, and on the feet of birds that have waded in swamps or have swum in ponds. In dozens of ways they have been distributed over the whole world.

For a very long time many biologists have referred to the vast populations of these animals (which anyone can raise by simple culture methods), as *infusoria,* because they are so easily obtained for observation by infusions of rotted hay or other vegetable matter in water. A spoonful of chopped dry hay placed in a small beaker or glass jar of distilled water or water from the tap will usually beget them in the course of a few days in fabulous numbers.

Paramecium cordatum, the "slipper animalcule" (plate 107) is one of the most easily cultured forms to be brought thus from suspended animation. Why it is almost always in hay, and how it gets there, no one seems able to say for certain. This species was an abundant inhabitant each spring in a little drainage ditch at the lower border of the mini-wood. A species of *Colpoda,* a bean-shaped protozoan, about which more will be said presently, was also an abundant inhabitant of the ditch and other vegetation, sometimes including the leaves of trees (plates 107, 108).

Reproduction might seem to be no great problem to understand in these animalcules, but to simply state that the processes of fission "take care of this when these organisms split into two, and the resulting daughter cells do likewise, and so on and on until there are thousands or millions of like organisms," is to condense extremely intricate matters almost to the zero point, for in order to divide, these single-celled animals, like multi-celled higher forms, must undergo the processes of *mitosis.* Every cell contains a minute, usually spherical or oval body called the *nucleus,* a center of control which determines reproduction and the traits of the species. In order to understand cell functions, including mitosis and other involved facts relating to fission, the reader must consult up-to-date biology books.

With some protozoans, simple budding as well as binary fission takes place at times, and sexual reproduction may occur and alternate with the asexual method, as in the case of the "slipper animalcule" *Paramecium* and some other protozoans.

Let us now shrink ourselves down to the dimensions of some of these minute creatures in order that we may get a closer look at their often crowded world and habitats.

In bewildering array, they dart about in the culture. Thus, in a hit-and-miss manner, they nevertheless succeed in finding their food items, alive or dead. Through their transparent pellicles we may even see the food digested and the "waste" excreted from some more or less definite area, according to the species observed. Under laboratory conditions, the clear walls (pellicles) of some of these remarkable, sometimes grotesque, animals hide nothing from us, and we may watch all the vital processes with a microscope.

Most of the animals seem to have mere depressions or grooves for mouths, and toward these intakes, currents containing food particles are being wafted by banks of cilia, arranged for the purpose in rows or in incomplete circles.

Related to these infusorians we have already admired in this culture, prepared from a few grams of rotted vegetation, we also have some flagellate protozoans, equally curious creatures, differing from the former in their mode of propelling themselves by having one or more whiplike tails instead of cilia (*Euglena,* plate 107).

The author believes that insufficient attention has been directed to the fact and mystery of the presence of many infusorians in one state or another in so many kinds of *dry* vegetation. From whence came their encysted or dried "bodies" in such numbers in dead grass, and even in *baled* hay, in a state of suspended animation? Why is it that they are not to be seen on the foliage of trees when the leaves are dry, yet may appear when the leaves are wet? Where did those which the author once found on acorns come from? Where did those which inhabit rainwater, temporarily retained in cracks and depressions in rocks and boulders, come from in the beginning?

Let us look at some more of them, animalcules brought back to active life in cultures made from several widely differing materials.

In infusions prepared from handfuls of dead and long sun-dried grass, some from the borders of the mini-wood, some from a roadside at Orleans on Cape Cod, identical species were observed in numerous instances.

Very numerous in these cultures were those ubiquitous animalcules of the genus *Colpoda.* Also appearing sooner or later were many very curious ciliates, smaller than the former, with a bowed profile, and belonging to the genus *Scyphidia.* When viewed through the microscope, they reminded the author of turtles, if such could be so active minus heads and limbs (plate 107). Here and there were elongate animalcules with long gooselike necks and bodies (*Lionotus,* plate 107). These infusorians were ten times longer than the colpodians, adding greatly to the interest of these animated scenes in miniature. Species of slipper animalcules, genus *Paramecium,* appeared in most of the cultures, increasing enormously in numbers by the asexual process of fission within a few days. At times, so many gathered together at the surface of the fluid and against the glass of the containers that, despite their microscopic size individually, they became visible to the naked eye as white bands probably consisting of tens of thousands of these animalcules.

A highly interesting entity in these cultures was a comparatively huge, round-bodied, grotesque organism with a small head and a long down-turned beak that, like most of these other infusorians, rapidly propelled itself with rows of cilia. Named *Trachelius ovum,* it was a startling discovery for the author, what with its complex anatomy in full view through its clear pellicle, and the fact that it had never been seen by him before (plate 107).

In every culture based on rotting vegetation, there were countless numbers of animated specks of life, so small and unidentifiable with the equipment at hand, that henceforth they will be referred to simply as the "gyrating dots." Invisible except under high magnification, at times enormous numbers of them formed oddly shaped figures at the surfaces of the cultures. Commencing as small groups, these would gradually join together, the gyrating objects becoming aggregations of such density (all apparently linked or otherwise adhering to one another) that in places they completely stalled all other infusorian traffic, including the comparatively huge colpodians which occasionally endeavored to break through their lines but were actually held back by these living fences of myriad shimmering organisms so much smaller than themselves (plate 108).

Dramatic swarmings such as these, sights of things so far removed from the everyday world, as we human beings know it, were viewed by the author as clearly as if he himself had been of microscopic size and was down there within the culture, watching. In contemplating sudden enormous populations such as these, it should be kept in mind that they may be begotten at any time from debris lying in your mini-wood or mine, which to all intents and purposes might seem to the average human being to be dead and useless to other organisms in such numbers.

The author has found through long observation and simple experiment that infusorians of this genus *Colpoda* may be obtained for observation from rainwater that has passed through the woodland canopy. What is more astonishing is that these active, bean-shaped but harmless animalcules, as stated earlier, were also found in drops of rain retained on leaves of lettuce, notwithstanding the fact that the plants were of course grown in the open, where the water had reached them from the sky *uninhibited.*

During these investigations, cultures of infusorians were also obtained from such odd sources as wild iris flowers, bits of moss, reindeer lichen, scrapings from weathered shingles, and many other materials. Most remarkable as a demonstration of the viability of protozoans were the active examples of certain species obtained by soaking the huge seeds of a tropical forest tree (*Mora excelsa*) in distilled water for about one week, the seeds having been in dry storage at the museum for *thirty-three years!*

Again and again during the author's simple experiments, it would seem to have been proved (in the case of *Colpoda*) that some of these protozoans were indeed brought down in rain. Possibly first carried aloft, when in a dry state or encysted, by wind in dust from dried-up ponds, puddles, or swamps, they could be widely scattered before being precipitated in raindrops, days or even months later, to populate leaves and other vegetation, and also the cracks and depressions in rocks and boulders where we find them. The theory would of course apply to only a small percentage of these animalcules.

Having many times obtained colpodians from arboreal foliage by the simple distilled water method, the author believes that these animalcules may at times temporarily enter and dwell between the upper and lower epidermises of these leaves.

Thin as most leaves are, it should be remembered that elaborate food factories exist between their upper and lower surfaces. Here, and somewhat resembling frankfurters in form, are located masses of *palisade cells,* packed together perpendicularly. These cells, which may occur in one to three layers toward the upper surface of the leaf, constitute what is known as the *palisade parenchyma.* Below these layers, and reaching to the lower epidermis of the leaf, there are other, irregularly shaped cells constituting the spongy *parenchyma.* Between these cells and in other locations within the leaf there are, of course, air spaces of various sizes. Photosynthesis, the complicated process by which green plants produce sugar within these cells from carbon dioxide and water in the presence of light, and with air spaces still available within the leaf surfaces, has been here recalled to the reader to afford a better understanding of what follows.

The leaf epidermises are composed of single layers of interlocking cells which might be compared to the pieces of a jigsaw puzzle. Large numbers of openings into the leaf, called *stomata,* are situated on both surfaces, the greater number being on the underside. Each stoma is equipped with two crescent-shaped guard cells or "doors," whose openings and closings are controlled by changes in cell turgor or internal water pressure. It is through these openings that oxygen, water or water vapor, and carbon dioxide pass in and out of the leaf complex. To demonstrate this, a wilted lettuce leaf sprinkled with water and placed in a cool place will soon become crisp as the guard cells open and the reviving fluid is drawn in.

Now, if *Colpoda* do indeed come down in the rain at times, and thus populate arboreal leaves, they could easily enter the intercellular spaces through the open guard cells and remain within the leaf when its surfaces are dry. If this actually happens, as seems to be the case in these experiments, it would explain why these animalcules may be obtained by placing dry *green* leaves in distilled water, causing the guard cells to open and allow in and out passage of the colpodians.

And now hear this! To help satisfy insatiable curiosity, the author obtained an all but frozen ear of corn from a supermarket in February, which the clerk declared had come in from Mexico.

Chopped-up portions of the husk and the cornsilk were placed in jars with distilled water in a warm place and allowed to rot. When these cultures began to stink, as all rotting vegetation does, with a vengeance, microscopic examination of drops of the surafce water revealed many infusorians similar to if not exactly like those so commonly found in Connecticut!

In addition to species of *Paramecium* and many *Colpoda* that looked exactly like those obtained from mini-wood debris and foliage, there were oddities named *Scyphidia* which the author, in home cultures, had found in rotted cabbage and hay.

In cultures at home made by infusing reindeer lichens, many of the curious creatures *Lionotus,* with stout midsections and long, drawn-out "necks" at both ends (already mentioned) were occasional inmates. These things would slide back and forth against bits of debris as if deliberately "scratching" themselves, and then suddenly dart off in any direction in the strange corkscrew manner by which they propelled themselves. As interesting in the lichen cultures were many protozoans named *Saprophilus,* which the author calls the "wheat kernel" animalcules (plate 107).

In addition to the one-celled so-called animalcules described above, found in some of these cultures, there were also occasional multi-celled and much more highly organized animals called *wheel-animalcules* or *rotifers,* a large number of species of which constitute the phylum Rotifera.

Not only were they seen in the cultures made by rotting vegetation, but it was found that one species also inhabited the rainwater temporarily held in the cracks and depressions in rocks and boulders in and out of the mini-wood.

Scraping up some of the "dirt" from these cracks and

depressions when it was dry, revealed through microscopic examination only a few eight-legged mites (which seem to occupy all habitats in the mini-wood) and which in this case had rounded polished dark red bodies resembling semi-precious agate.

Upon wetting some of this debris with distilled water, rotifers named *Rotifer citrinus,* arising into active life from cysts at first unnoticed, soon became visible through the microscope. Gradually extending themselves to four or five times the length of the form in which they had remained when dry, their "resurrection" was most interesting to observe (plate 108).

Transparent, possessed of definite, completely visible organs, believed to have even a primitive brain, they progressed by extending their "heads," then drawing up the posterior portions of their bodies, repeating the movements until a suitable anchorage was located.

At each side of the "head," a contractile stalk could be seen; each stalk supported a disklike group of cilia or hairlike extensions. It is from these disks or "wheels" that rotifers take their name, for when the cilia are in full play, and fanning food particles toward the creature's mouth, the illusion of two rapidly turning wheels is remarkably convincing (plate 108).

Some of the many rotifer species also swim about freely rather than clinging safely to debris as *R. citrinus* usually does, and the gyrations of these others may have had something to do with their common name as well.

Food particles wafted to a rotifer's mouth are passed on to a muscular cavity below, which is called the *mastax,* and within which is located a pair of efficient jaws or *trophi* whose purpose is to grind the food. When the ciliated "wheels" above and the trophi below are in full motion, the former begetting the fodder and the latter masticating it with the efficiency of a shredding machine, viewing this through the microscope is like watching some man-made device in a factory, a fascinating sight to observe through the transparent walls of the rotifer's body.

Rotifers are common in many places in standing water and may remain active for long periods, but those to be found in rain-catching rock cracks and depressions must shrink and encyst whenever fair weather dries out their habitats. How long these animalcules have been present, how long they *can* live, alternating between active and encysted existence, and how long the normal life span of *Rotifer citrinus* may be, are questions the author cannot answer.

The astonishing facts remain that they *are* in these crevices in the mini-wood, and this means of course that they must exist elsewhere in similar habitats, and in those places there must also continue to live and propagate the proper organic food items which make up the rotifer diet.

Probably even the few infusoria and rotifer species mentioned in this chapter, an infinitesimal group out of the tens of thousands of species which exist, thousands of which have been named, are unknown to the vast majority of human beings.

CHAPTER IX

TREE-TRUNK TRAFFIC

Small-time animal traffic to various arboreal levels, moving in two directions, is often minor by day but quite heavy at night. The stems and branches of the woodland shrubs, the trunks of the understory trees, and the boles and networks of limbs of the larger species serve as smooth or rough pathways, roadways, avenues, and thruways to the various foliage areas, including the airy meadows of the canopy itself.

Not unlike human beings, many of the lesser woodland organisms become restless as evening approaches, and again, as with so many of our kind, as night becomes a fact, the lesser animals find its screen and shelter compatible with gastronomic and sexual urges, for they too must find their proper food and their female counterparts, they too are pulsating, determined entities, with hearts and blood, and not to be denied the normal satisfactions of life. Blood need not be red to course or bathe small but nonetheless courageous anatomies. The colorless fluid does as well for those lesser animals that possess it as the hemoglobinous red stuff does for us.

Many insect and other arthropod denizens of the miniwood, as well as such apterous (wingless) creatures as terrestrial isopods, which are common creatures, used the trunks and limbs quite regularly, each to seek some specialized habitat or niche, satisfy their sexual urges, and seek the food necessities of their kind. Plant stems, tree trunks, and branches are not unlike road networks. They lead to the good things which shrubs and saplings and full-grown trees and their foliage have to offer the *wingless* small animal organisms of the woodland.

Go into the woods at night when everything is still soaking from a good summer rain- and thunderstorm, or after a few days of showers. Seek out the bole of some century-old tree, a mature sassafras or a majestic beech, for instance. Squat beside it with lantern or flashlight, and observe what is going on in a more or less perpendicular portion of the world.

Creatures will be going up and coming down the tree. Some, like the ants, will be hunting, or traveling aloft posthaste to sip at a rich deposit of honeydew sugar exuded on the leaves by some species of aphid. (Aphids or plant "lice" are those ubiquitous sapsuckers and producers of one of the sweetest sugars in nature, which, strange as it may seem, they eliminate in such vast quantities that the foliage even far beneath their hordes becomes so coated with the syrup that it crystallizes at times, finally turning black from a species of mold which also thrives on sugar.) Honeydew in the woodland is avidly sought by bees and wasps, by two-winged flies, by butterflies at times, and by ants wherever it occurs (plate 109).

Rain brightens the green and gray algae that grow on tree trunks. The commonest green sheen on bark is the alga *Protococcus,* whose microscopic cells multiply by fission; they are spherical singly but resemble overcrowded micro cauliflowers when viewed en masse.

The rains may also have caused other minute plants to "pop" up in the fruiting stage, and whose vegetative portions grow invisibly in the bark's dead tissues, or in debris accumulated in the furrows. Lacking chlorophyll, these miniature fungi are dependent, as the larger species are, on dead wood or rotting debris and vegetation for support of their mycelium and their usually pallid but nonetheless beautiful tiny fruiting bodies.

It will pay one to watch for, and examine in detail, any mini-plant body belonging to the flora of the bark world. When highly magnified, some of these will bring a sudden sense of special pleasure to the observer upon his initial "discovery" of previously even unimagined details of the greatest delicacy and beauty. These minute fruiting bodies also serve to illustrate that sufficient nutrients must occur in ample quantities, sometimes even in the merest accumulations of debris in the bark furrows.

Occasionally, little land snails with discoidal buffy or darker shells (measuring three-sixteenths of an inch or less in width) slowly ascended the wet bark of the big trees at night. When extended, they varied considerably in color, according to how much of the animal's body remained within the shell and, to some extent, upon what the animal had recently eaten (See Plate 145).

These and other terrestrial snails of the mini-wood, as well as the shell-less mollusks which we call slugs, are covered in more detail in chapter XVI.

As the snails were found more often in other habitats than on tree trunks, they may seem somewhat misplaced here, but as a definite part of the tree-trunk traffic, they are very briefly mentioned in the present chapter.

Rains cause a distinct brightening and probably vigorous fresh growth of the alga *Protococcus* as already mentioned, to which at such times many small woodland creatures then resort, possibly because this plant as a food item is then tenderer and perhaps even more "tasty."

Among the regular alga feeders in the mini-wood were two kinds of small slugs (less than an inch in length), probably native species. Most of the larger forms commonly encountered here are introduced species, and as they vary considerably in color at times, are seldom easy for the inexperienced observer to identify with certainty (plate 110; see also plates 145, 146, and 147).

The "natives" which were frequently found browsing at night on the tree-trunk algae were glossy dark gray to almost black dorsally, and a plain light gray on the undersurface or foot, while the raised, saddlelike portion of the body, called the *mantle,* was indistinctly *striated* in these small individuals.

*   *   *   *   *

Those of us with flower or vegetable gardens are more likely to encounter much larger and sometimes very destructive slugs than those in the mini-wood. An introduced species, varying in color from yellowish brown to a much brighter orange-yellow, was once unwittingly introduced into the author's garden in a load of topsoil, with serious results. Two seasons of tedious attention were required to bring it under control.

Hatching from eggs in the spring, when this alien vegetarian once becomes established and a large breeding population has been formed near a garden, these big mollusks may raise hob with such flowers as petunias and, surprisingly, with the rank-smelling marigolds, and with vegetables such as lettuce and tomatoes. Sometimes they eat the blossoms as well as the foliage, and their nightly movements among wildflowers such as the touch-me-not or jewelweed are often recorded in telltale silvery tracks of hardened mucus on which, before it dried, the slugs traveled over the foliage.

In the fall, egg clusters of these slugs were often observed when damp woodland litter and other refuse was turned over. The author believes that most of these eggs winter over, hatching in the spring.

Several species of worms occurred in the mini-wood in large numbers and have been given more space in a later chapter, but one of these, the familiar earthworm, *Lumbricus terrestris,* must also be mentioned in the present chapter as a not-infrequent member of the nocturnal tree-trunk traffic.

Customarily, the earthworms fed on dead leaves and other litter of the woodland floor, ascending from burrows to reach the humus and leaf carpet above for the purpose, and there, in enormous numbers, depositing their excrement in the form of pellets, very different in form from the familiar worm "casts" of clay and other matter which they leave on the surface in more open ground. These pellets are important in the woodland economy, being utilized by many smaller refuse-eating animals of the soil, while the residue becomes a part of the humus.

On wet nights, small earthworms often leave the ground to ascend tree trunks up to two or three feet, where they apparently feed upon algae. Few people realize that earthworms can climb, and the author was surprised to find them well up even on such comparatively smooth bark as that of the beech trees. Only the small earthworms appear to feed thus, none of the "night crawler" sizes having been so observed.

During the tree-trunk study, a strange wormlike animal was also found in a micro "meadow" of *Protococcus,* the commonest alga on the bark of a beech in the mini-wood. Apparently a rarity, only two were found by the author during long nocturnal searching, an effort made worthwhile, however, if only to prove once again that rare oddities are still an exciting possibility in any "backyard" woodlot.

With a more or less cylindrical, unsegmented body, it proved to be a member of a large group of mostly flattened but sometimes round-bodied organisms constituting the phylum Platyhelminthes or flatworms, and in this case a terrestrial turbellarian probably of the genus *Rhynchodemus.* As this animal did not seem to have an English name, for convenience we may refer to it here as the "twin-spot" worm.

Measuring nineteen millimeters (not quite three quarters of an inch) in length, and two millimeters in diameter at the broadest point, it tapered sharply toward either extremity and extended into a narrower "snout" anteriorly. Basically pale brown, minute dorsal pigment dots formed two lines the length of its body, and two black marks on the "head" end may have been eye-spots, but were visible only under high magnification.

These animals moved slowly, with a gliding motion, apparently undisturbed by the rays of the author's electric lantern. One wonders exactly what the role of such woodland animals may be in its overall economy—wonders also where they abide by day, for never were any worms like them in every detail found elsewhere in the mini-wood during the whole period covered in this book. Evidently an epiphytic vegetation feeder, this species was sluglike in its motions, and in its ability to exude a protective slime when disturbed.*

What its enemies might have been (other than the microscopic nematodes [see chapter XII] which, as internal parasites, ate the author's two prize specimens before they could be photographed), was undetermined. Perhaps the common centipede, *Lithobius* (see chapter X) was one of these, notwithstanding the turbellarian's repellent slimy exudations, for this powerful little centipede regularly attacks and avidly devours small earthworms.

Some turbellarians are themselves predators. In these, a thickened cylindrical and muscular portion of the alimentary canal, called the *pharynx,* can be extruded for the capture of their prey. The twin-spot turbellarians, however, were probably habitual feeders on *Protococcus* and other algae growing on moist bark, as observed. Nocturnal creatures, they represented another interesting plant-animal relationship unknown to the author previously.

One real difficulty in keeping all small and minute animal forms in good health in small jars and petri dishes (for easy observation) is mold, which grows when media are confined and too wet. Constant vigil must be kept to prevent lethal mold from taking over and wiping out these little animal colonies. Whatever medium is employed must be kept damp but not soaking, and a few drops of water applied with a medicine dropper now and then will maintain proper conditions.

Curious little terrestrial isopods, commonly but improperly referred to as sow "bugs," pill "bugs," or wood "lice," will be discussed here, as some of them form part of the nocturnal tree-trunk traffic. Mostly, however, they dwell under loose bark on rotting logs, under damp refuse, or in the woodland humus in great numbers, and are also to be considered part of the "army of the soil"; as such, they are treated in later chapters under that title.

These organisms are not "bugs" or "lice" in any sense of those words. They are land-adapted sessile-eyed crustaceans of the order Isopoda, those which inhabited the mini-wood being members of the suborder Oniscoidea and referred to henceforth in this book simply as *oniscoids* (plates 111, 112, and 113).

Anyone examining oniscoids even superficially would doubtless notice almost at once that they are equipped with many more pairs of legs than insects, which have only three pairs. They would also see that, dorsally, these isopods are protected by overlapping plates covering the seven thoracic and six abdominal segments.

Oniscoids carry the heritage of marine creatures in their beings, and this makes them the more interesting as organisms now dwelling in our woodlands and in many other situations close to our homes. Oniscoids indeed might be said to have the sea in their chromosomes, and although they have now become land animals permanently, are, nevertheless, more closely related to such other crustaceans

---

*Epiphytic plants are those which use other plants for support only, taking no nourishment therefrom, as true parasites do. See chapter II for details.

as the lobsters and crabs, the shrimps and beach hoppers, than they are to "bugs" and "lice."

The oniscoids which were so common in the mini-wood and which are abundant elsewhere in suitable habitats are today's representatives of ancestral forms which left the water ages ago and became successfully adapted to terrestrial existence.

Ubiquitous as they have become, most of us pay them little or no attention and know nothing at all about what they are doing in our woodlands; and because they are so seclusive, many of us are unaware even of their existence. Except for drawings and rather overworked old photographs in zoology textbooks, the author does not recall seeing recent detailed pictures of these isopods. It is with pleasure therefore, that the new photographs (plates 111, 112, and 113) are offered in this book.

Observing these woodland isopods in life was indeed a pleasure, especially when keeping in mind the belief that, in age as species, they greatly antedated the glacier which perhaps twenty-five thousand years before arrived and then melted on the very land which they were now inhabiting.

Often gregarious, harmless, and beneficial, their requirements for the abundant life in the mini-wood were merely minute portions of dead vegetation, mostly in the form of rotting wood and humus. In return for such meager fodder and nutriment, they deposited excrement in the form of millions of minute brown or black brick-shaped pellets, "waste" matter useful as a part of the humus amalgam which is the food source of so many other animals and plants.

Dampness in their habitats is a must for oniscoids. They succumb quickly if their gills, placed on modified pleopods or abdominal limbs, become dry. The interesting fact that they possess these organs but themselves are at all times terrestrial organisms clearly links them with ancestral animals of the sea and suggests the ages that must have been required for their evolutionary transition. We may speculate that, at some very remote time, certain marine forms probably ventured out of the water while the littoral remained wet between tides. Finding things safe under such conditions, and with plentiful washed-in food available, they could have remained thus for longer and longer periods, eventually extending their explorations inland as their pleopods were gradually modified with gills (capable of utilizing moist air) like those we see in these terrestrial woodland isopods today. As with many species

of crabs which have thus been adapted to a partly or entire life "ashore," this mode of existence has become common to many isopods in various parts of the world, although the majority of species are still marine, and fresh-water inhabitants.

The largest species in the mini-wood and the oniscoid most often uncovered from damp debris, or when bark was removed from rottings logs, was *Oniscus asellus* (plate 111). Measuring approximately fifteen millimeters (about five-eighths of an inch) in length when mature, its rather long oval body, consisting of thorax and abdomen made up of several somites each, has become dorsally hardened, but on the underside there is no carapace to shield the legs and gills or the eggs which, in the female, are carried in a pouch until the young have hatched.

Examined under the microscope, this isopod was seen to have compound eyes, a group of about twenty-four on either side of the head, and appearing as tiny beadlike objects even under strong magnification.

Why it should be thus provided, living as it does by preference a life in total darkness, is anyone's guess, but probably eyes are another mark of its ancestry which may now be greatly modified and useless; however, a tough, shieldlike projection close to each group (like the "blinders" on a horse's bridle) may, if fulfilling a protective role, indicate that the eyes are still light sensitive.

In color, the dorsal surface of this species varied considerably, from various shades of gray to brownish gray, and was marked with rows of yellowish spots and many white dots in no regular patterns. Underneath, it was plain white or near white, its appendages fully exposed (plates 111 and 113).

These oniscoids were kept in captivity for many months in large gregarious groups, in terraria containing a layer of humus from the mini-wood, and with strips of bark taken from rotting logs scattered loosely on the continually damp surface. The atmosphere within these "cages" was kept sufficiently moist by placing glass sheets over the open tops, but raised slightly to allow for air circulation and to avoid the growth of mold.

The preferred diet of this species being rotting wood, these captive colonies did especially well on the decaying bark supplied them, feeding entirely on the softer undersides of the fragments taken at random from the well-weathered woodpile where the animals themselves had been collected.

A dozen of these oniscoid adults, maintained on a diet of rotting bark in a petri dish measuring four inches in diameter and a half an inch in depth, almost completely covered the glass bottom of the receptacle with their dark-colored fecal "bricks" in a period of seven months, a fact which gives some idea of the rate at which such organisms help to decompose dead wood. Although the pellets averaged only 1.50 mm. in length, 0.75 mm. in width, and 0.50 mm. in thickness (one millimeter being the equivalent of 0.04 inches), the vast numbers deposited year in, year out, by all of the oniscoids in the woods, collectively play an important role in its economy, aiding as they do in the production of humus.

In the male oniscoids, the organs for the transfer of spermatozoa are situated in the pleopods, and as far as the author is aware, on the first and second pairs, wihch are specialized for the purpose. This arrangement alone serves to illustrate another of the oniscoids' differences from insects, including the *true* bugs for which they are so often mistaken. Male insects in general are equipped with intromittent organs for sperm transfer. In the female oniscoid, platelike expansions may be seen on some of the basal segments of the first four pairs of legs. These are receptacles, technically known as *oöstegites,* or sometimes appropriately referred to as *marsupiums.* Within these the eggs are incubated and the young hatched, and sometimes held for a time. Newly released young measure about one millimeter in length, and except for their very pale color, closely resemble the adults.

It is interesting to remember at this juncture that the familiar mammal, the 'possum, which was also an occasional inhabitant of the mini-wood, was thus not alone as a carrier of its young temporarily in a protective pouch. Nature, it would seem, has made use of her good ideas in organisms sometimes far apart anatomically. Indeed, we have a third kind of creature, which also, like the 'possum, sometimes carries "litters" of offspring on its back; this is a large non-web-building, wandering Lycosid spider, and has been observed on several occasions with its crowded young "aboard," scurrying over the mini-wood carpet.

In common with all other animals, oniscoids have their consumer enemies. The commonest small centipede of the mini-wood area, a species of *Lithobius,* was observed in the act of attacking these isopods, dispatching them with its poison fangs, and after imbibing some of the ample oniscoid juices, partly devouring what tissue was left within the drained shells.

As another example, an adult *Oniscus asellus* in one of the author's maintained colonies, suddenly became paralyzed. Unable to move, although still alive, it remained in one spot as if anchored there. Three days later, the victim's platelike somites commenced to fall apart, and when inspected with the microscope, were found to be swarming with extremely minute organisms called nematodes, which, having eaten all suitable materials within the body, were protruding in all directions, waving their head ends to and fro, as if in agitation and bewilderment that nothing more was available!

Nematodes of hundreds of species constitute an enormously successful phylum. Most often small to minute, frequently parasitic, these worms occur all over the world: in plants and animals, in the woodland humus, in water, in mud, and in many other habitats. We wonder what happens to the multitude of such immature parasitic forms when the food gives out. Most of them no doubt just die very quickly, and thus are their numbers kept under natural control, while the comparatively few that reach mature growth remain to lay the eggs for another generation. (See plate 124.)

Perhaps the most curious of these little oniscoids to be found in the mini-wood, and evidently much less common, were occasional individuals appropriately named *Armadillidium vulgare.* Having the interesting and amusing habit of rolling themselves into neat balls for protection when disturbed, they do remind one strongly of the armadillo's similar protective habit.

The seven rather broad, armored thoracic somites, or segments, are so constructed that they neatly overlap as the animal rolls up. The smaller abdominal segments are tucked or pulled in, and anterior plate number one is so excavated that it accommodates the head as the rolling takes place, and thus rounds out the contours of the ball (plate 112).

These rather sluggish isopods measure around sixteen millimeters (about five-eighths of an inch) in length when mature and in walking position. They are almost a uniform dark gray, and sometimes have a slight gloss. In captivity, they have sometimes crawled up a piece of the bark in their receptacle and remained thus in the light for days at a time, whereas the commoner oniscoids remain hidden at all times. Hatched from eggs, the armadillo isopod's life history is essentially like that of these other small terrestrial isopods.

Interesting is the fact that *Armadillidium vulgare* occurs in many other parts of the world too. It has been found (locally distributed) over most of New England, and some years ago, Professor B. W. Kunkel* gave its range as Louisiana, Texas, Mississippi, Kentucky, Washington, D.C., South Carolina, New York, Ohio, Massachusetts, Rhode Island, Bermuda, Algeria, and the Azores!

As abundantly as the larger sow "bugs," *Oniscus asellus* are found in the woodlands beneath rotting logs, loosened bark, and refuse heaps, there is a still more numerous, smaller, brown species which seems to be chiefly an inhabitant of moist humus, while a third, minute species, in color pure white, is also found in the humus in its upper layers. All are consumers of dead vegetation, and again, because of their numbers, important in the woodland economy.

The tiny white oniscoid is so transparent that a black line may be seen traversing the dorsal surface; this is the consumed food showing through the outer integument from the alimentary canal. This seems to be the species *Scyphacella arenicola* or a closely related form.

These tiny things measured from one to three millimeters (many being less than a sixteenth of an inch) in length. Their bodies, elliptical in form, were well equipped with antennae whose terminal joints were extremely delicate, the last one ending in a needle point. They moved slowly, even when disturbed in the small petri dishes in which the author kept colonies successfully for as long as eight months on thin layers of sifted humus with fragments of bark for shelter. As in the case of the larger *O. asellus,* their fecal pellets (visible only with a microscope) accumulated in large numbers, as was ascertained by keeping colonies on fragments of damp bark in mini-petri dishes lined with filter paper, against the white surface of which the pellets in a limited circle could be counted. Less than half a millimeter in length, less than half as wide, and less than a third as thick, in form they were, like those of the larger species, shapsd very much like bricks: straight-sided and almost square-ended.

Some may feel that these are insignificant observations, but not so, for their countable forms made it possible to visualize how enormous must have been the total number and value of these micro "bricks" excreted by all of the woodland oniscoids, from these tiniest ones to the comparatively giant species. The role of these small animals in the processes of humus accumulation was obvious.

As this smallest white species lived very well on bits of bark alone, it would not be surprising if, as in the case of the wood-eating termites, these oniscoids harbored microscopic living organisms (Protozoa) *within* their digestive tracts as *symbionts* (animals or plants forming mutually beneficial partnerships with a different species), and in the present case aiding in the digestion of the cellulose ingested by the bark-eating oniscoids.

The first of these tiny white specimens to be seen by the author appeared on decaying bark being microscopically examined for silver collembolans and minute white millipedes, organisms treated in chapters X and XI. Thus it was demonstrated once again how it often pays to examine every possible habitat minutely, for indeed there seems to be animal life wherever one chooses to look for it.

Within a week truly micro young of these oniscoids appeared on the bark; these must have hatched from eggs carried by one of the mature specimens hiding in a crevice in the old decaying wood.

The third or brown species, already mentioned as inhabiting the mini-wood's richest humus in large numbers, was the species technically known as *Metoponorthus pruinosus,* a name apparently in part referring to a hoarfrostlike appearance of the animal but which has remained a mystery to the author, as none of the numberless individuals of this species examined during these studies seemed to fit this description.

Differing very much from the still larger *O. asellus,* these brown specimens were slimmer in build, with abdominal somites very noticeably narrower than those of the thorax. Over all, they measured from eight to ten millimeters (up to about three eighths of an inch) in length. Often, in employing the funnel-collecting device described in chapter XI, the author obtained up to two or three dozen individuals from a single humus filling of the apparatus.

Preferring total darkness down in the rich medium, this species appeared to require a wetter habitat and a cooler microclimate than its larger relative, *O. asellus.*

It may be that more than one species was here represented, as some were found with a uniform reddish brown dorsal surface, while many others were of lighter shades, mottled with darker colors. In structure, however, they all seemed identical, and may have been only color varieties of the same *M. pruinosus.*

---

*B. W. Kunkel, *The Arthrostraca of Connecticut* (Hartford: State Geological and Natural History Survey Bulletin 26, 1918).

These brown oniscoids moved with great rapidity when uncovered, burrowing out of sight into the humus before one could get a good look at them. They occurred in the *thousands* in the wettest humus.

There are several additional species of these terrestrial isopods which inhabit Connecticut and many other parts of eastern North America, while some identical species are found in other countries. Teachers and their students should find enjoyment in searching out and reporting what they may find out about these isopods as worthy projects in biology.

The species *Philoscia vittata* reaches eight millimeters in length. It is of a brownish color with lighter margins on the segments and with a lighter median line separating the darker ground on each side of the back. *Cylisticus convexus,* measuring about twelve millimeters in length when mature, is light brown, with a row of yellow spots on each side and rows of smaller spots of the same color paralleling the larger ones. *Porcellio spinicornis,* slightly larger than *C. convexus,* has a yellowish gray dorsal surface, with varying dark brown patches arranged in five more or less extended series. *Porcellio scaber* is a ten-millimeter dull-looking oniscoid of worldwide distribution. Its shell is grayish black, sometimes with irregular black spots. *Porcellio rathkei,* about the same length, has dark gray to black segments bearing white spots. Finally we have *Actoniscus ellipticus,* slaty gray, only four millimeters in length, and distinguished in having microscopic *setae* or bristles protruding from the terminal joints of its antennae. Of the above, *C. convexus* is shown in plate 112. Measurements and colors of the six species above have been abridged from B. W. Kunkel's *The Arthrostraca of Connecticut,* that excellent Bulletin No. 26, published by the Connecticut Geological and Natural History Survey, Hartford, 1918.

Frequent members of the tree-trunk traffic, doubtless in search of concentrations of the food animals upon which they prey, were those curious oval-bodied, high-riding creatures we call harvestmen, or daddy longlegs (plate 113), highly beneficial destroyers and avid devourers of aphids or plant lice. Few human beings stop to inquire what they may be up to, crawling about the woodland, or over our roses and other garden flowers; few wonder how they might be interrelated, when immature, with other creatures of the humus and forest litter.

Chiefly nocturnal in their wanderings, potential gleaners of any aphid-infested flower, shrub, or tree, they are among the most useful arthropods, or jointed-footed animals, in this case belonging to the class Arachnida, that large group which includes the spiders, mites, scorpions, and several other forms.

Uniquely but seemingly all wrongly designed for life in this rough and ruthless world of small animals, what with the necessity of carrying their fat little bodies beneath outrageously long legs so fragile that they break off with the greatest of ease, they are, nevertheless, equipped with bright little *turreted* eyes, and a repellent secretion to eject if tampered with.

Notwithstanding these mildly protective features, one wonders how natural selection could have allowed such absurdly long, horsehair-diametered, vulnerable appendages to evolve. Unlike those very long-legged crabs fitted with pincer claws, harvestmen have only a tiny single claw on each "foot," suitable for grasping, walking on smooth surfaces, and for walking upside down. Perhaps their stilt legs aid them in escaping enemies, although the bad-smelling and doubtless bad-tasting substance which these creatures exude when in danger would seem to be a more effective provision for defense. It has always remained a mystery to the author how these arachnids ever succeed in life without losing more of their legs than they do.

Actually, they are harmless and fragile creatures and, happily, tireless destroyers of the live-bearing and tremendously prolific sap-imbibing ubiquitous plant "lice" which have long been one of the mini-wood's inevitable insect enemies.

But it must be admitted also that, like most other animals, even aphids have their uses in nature. Through their beaks, firmly inserted into the stems and leaves of herbaceous plants of all kinds, from forest floor to sunlit canopy, sap passes into their bodies and is mostly excreted again in the form of sugar. Some scientists believe that the sap is "pumped" into the aphids automatically by pressure within the plants. In any case, the sugar syrup is eliminated in the aphids' excretia as "waste" matter in such quantities that leaves below the colonies often become coated with the substance, which then crystallizes. As stated before, the crystalline surface sometimes turns black with a mold which it supports, but the sugar's main usefulness is as food for other woodland animal organisms, both resident and transient.

A whole micro flora also flourishes on these coated leaves, and as most of us have long known, "honeydew" has become a standard food of many kinds of woodland and other insects, including flower flies, bees, and hornets and other wasps. A few butterflies and many moths also partake of the aphids' sugary offerings when they have been freshly deposited upon the leaves and before they have crystallized.

Ants the world over patronize and guard large colonies of aphids, receiving in return their "honeydew" in its clear, intensely sweet fluid state as it is freshly pumped from the aphids' protruding anal tubes. Here we have one of the classic cases of symbiosis—two vastly different species of animals mutually benefiting each other (plate 109).

Before taking up the harvestmens' activities again, a few additional notes on the interesting life-history of the Aphidae may not be out of place here, for these insects are often very numerous members of the mini-wood canopy and the stems of many plants.

As far as man is concerned, these insects do him no good, their enormous colonies draining the sap of the food plants to such an extent that the latter are often badly weakened and sometimes die, and the sticky secretions disfigure foliage and sometimes coat the windshields of our automobiles and our outdoor furniture. (See plate 109.)

In spring and summer, aphid eggs which have wintered over hatch at varying dates on the food plants chosen by the adults. *All* of these young are females, and their development is so rapid that, within perhaps ten days or less, these "stem-mothers" are giving birth to young themselves, which are believed to be hatched from eggs within their bodies beforehand. Again, these young are all of the female sex. Generation after generation of females, mostly wingless, are thus produced in a single season. But before cold weather sets in, both males and females appear at the same time, and their matings then assure fertile eggs which will all produce females in the spring again.

Fortunately, aphids have other enemies to keep them in check as well as the harvestmen which feed on them. Lady "bugs," actually little round red *beetles* with two black spots on their *elytra* (wing covers), also prey upon them in both the larval and adult stages (plate 115).

Tiny chalcid flies, not true flies, but members of the order Hymenoptera, to which ants, bees, wasps, ichneumons, and others belong, are important enemies of aphids. These parasites are so tiny that their eggs are deposited inside the aphid's body. The larval chalcids consume the

victim's vital anatomy, but without destroying the outer skin. The process anchors the bug to the leaf or stem, and frequently their brown shells may be found with a neat round hole in one side where the winged adult chalcids emerged.

As with many other pest insects, aphids have off years locally, the foliage sometimes being entirely free of them in one locality where they may be numerous elsewhere.

They are very important food items of migrating Warblers of many species, especially when these birds are moving south in the fall, when aphids, if present, are most numerous. In the mini-wood, the author observed that many migrant Warblers which normally fed in its canopy and lower stories were greatly influenced by the presence or near absence of these insects. When the aphids were abundant, the birds remained longer in each tree, passing through the mini-wood in a more leisurely manner, while in lean-aphid years, Warblers were scarce, or even almost absent during the migration periods.

It also seemed to be true that in years of aphid abundance at the lower levels, opilionid (harvestmen) traffic up the tree trunks became accelerated, perhaps thus indicating peak aphid concentrations in the canopy also.

The author believes (but has no proof to offer) that the harvestman's sight may be fairly good, despite the animal's preference for nocturnal wanderings.

Unlike the insects, whose eyes are compound—that is to say, many faceted—opilionids have a very bright-looking single eye on each side, situated somewhat back of center on top of the head. These are mounted in a raised *tubercle* or sort of stationary turret, and following along their upper contours, rows of minute, blunt-tipped tubercles may afford some protection from above.

The most astonishing thing about these opilionid eyes is that, even in dry specimens that have been stored for years in the author's study collections, they have remained as bright looking and unshrunken as in life.

Most of these are black, but occasionally a pair will look brownish, tinged with gold, while in at least one, apparently a very rare or freak case, a specimen's eyes were bright red.

Viewed through a microscope, they reminded the author of the eyes of animals much higher in the tree of life, and they seemed to be staring at the culprit observer with disconcerting accusing steadiness. The eyes also suggested the baleful stare of the giant squid, and gave the author a guilty feeling for having killed a harvestman,

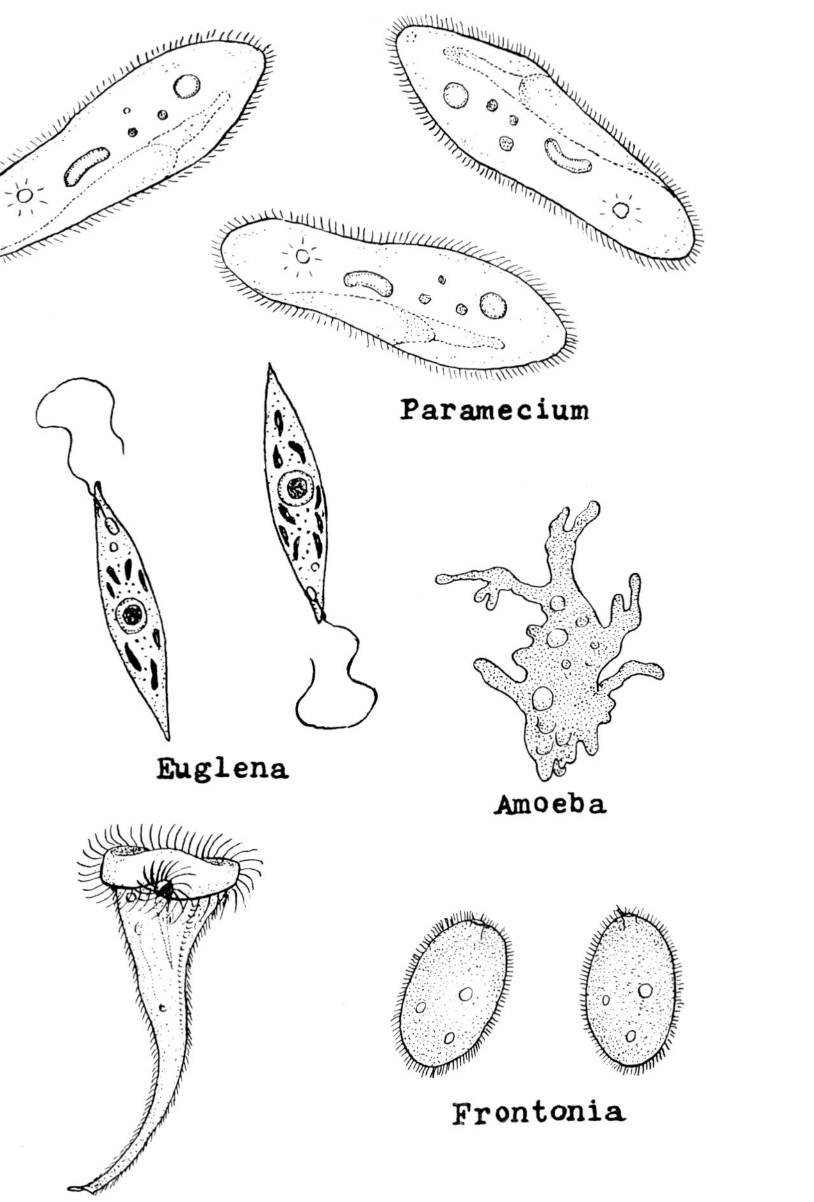

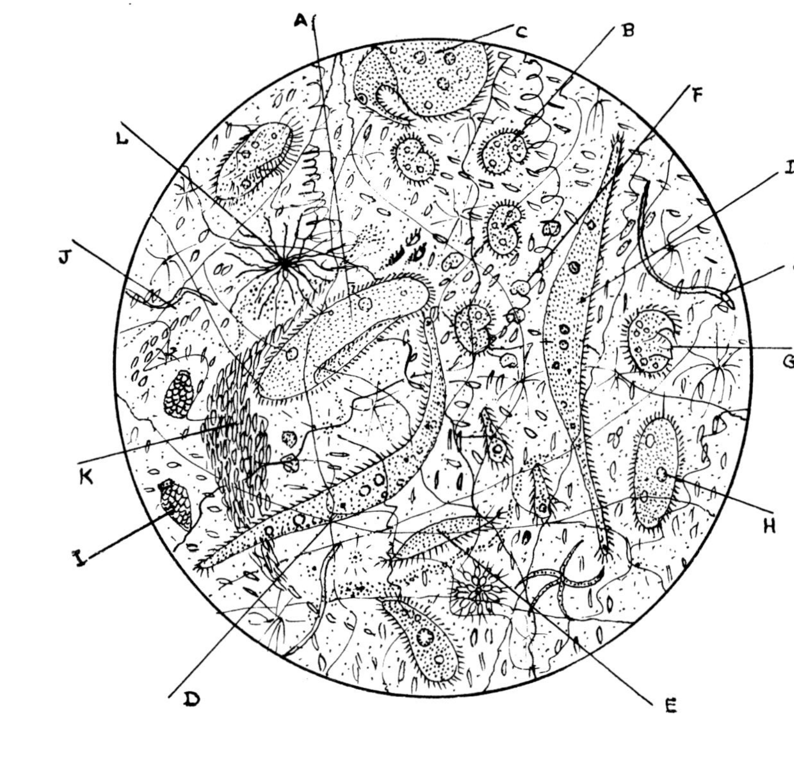

Above: Protozoa and other organisms found in the mini-wood area. (A) Slipper animalcule, *Paramecium.* (B) *Colpoda.* (C) Beaked animalcule, *Trachelius.* (D) *Lionotus.* (E) *Scyphidia.* (F) Smaller species of *Colpoda.* (G) Smaller species of *Trachelius.* (H) "Wheat kernel" animalcule, *Saprophilus.* (1) Encysted form of the rotifer *Rotifer citrinus.* (J) Enchytraeid worm and (smaller drawing) a nematode worm. (K) Massed unidentified infusoria. (L) A filamentous alga (microscopic plant). All the organisms in the circle are drawn enormously enlarged, but not all to the same scale. *Left: Amoeba, Paramecium,* and fresh-water animalcules of other genera, shown even more greatly enlarged. (Pages 72, 73, 74.)

PLATE 107

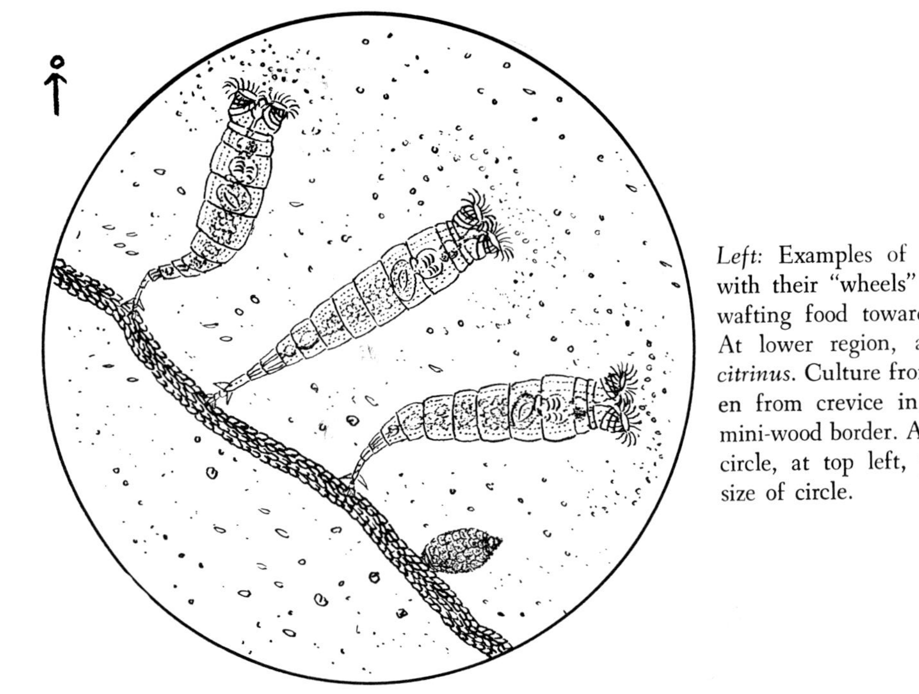

Left: Examples of *Rotifer citrinus* with their "wheels" in full motion, wafting food toward their mouths. At lower region, an encysted *R. citrinus*. Culture from rainwater taken from crevice in a rock on the mini-wood border. Arrow outside the circle, at top left, points to actual size of circle.

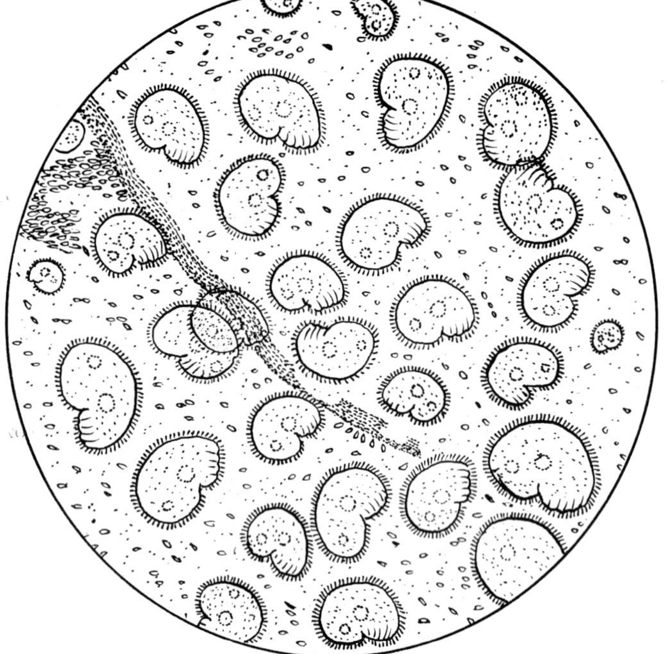

Above: Ciliates of the genus *Colpoda*, magnified 200 times, shown in a culture obtained from hay dried for five months. Solid ranks of unidentified infusoria (the "gyrating dots") have partly divided the culture, while hundreds of slightly larger organisms dart about —all in less than a drop of water. (Page 74.)

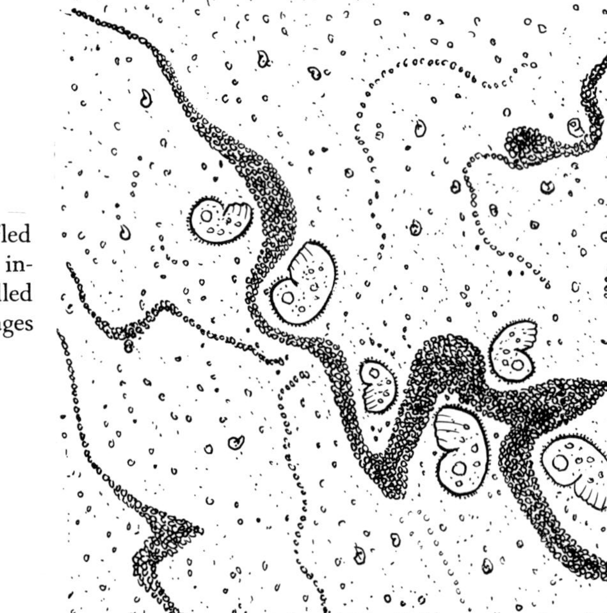

Right: Sextet of ciliates of the genus *Colpoda*, baffled in their movements by dense chains of seething infusoria, unidentified, and for convenience here called "gyrating dots." Magnification about 200 times. (Pages 73, 74.)

Above: Aphids at various ages. An adult at bottom left is giving birth to one with its legs nct yet unfolded. Enlarged $\times$ 4. (Page 77.)

Above: Typical aphid colonies on juicy stems of annual plants. An individual of the winged generation is shown at center of right-hand stem. All much enlarged. Date: August 1. (Page 77.) *Left:* Photograph, somewhat enlarged, shows how overcrowded aphid colonies become, causing leaves of host plant to shrivel up. Date: June 26. (Page 77.)

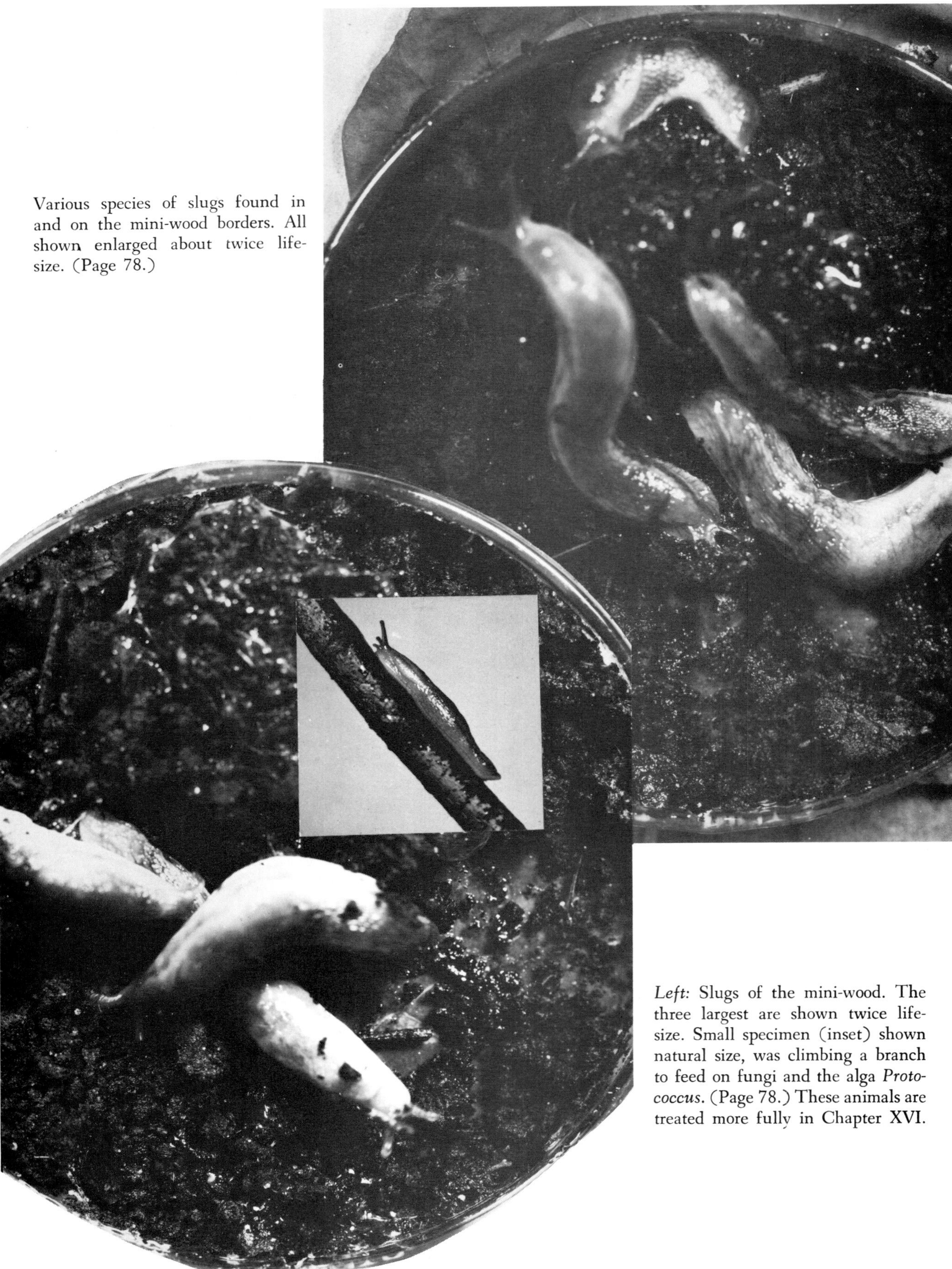

Various species of slugs found in and on the mini-wood borders. All shown enlarged about twice life-size. (Page 78.)

Left: Slugs of the mini-wood. The three largest are shown twice life-size. Small specimen (inset) shown natural size, was climbing a branch to feed on fungi and the alga *Protococcus.* (Page 78.) These animals are treated more fully in Chapter XVI.

A late fall gathering of the common woodland isopod *Oniscus asellus,* found under a slab of loose bark. There may be one or two of the species *Cylisticus convexus* among them. At bottom, two *Polydesmus* millipedes (described in a later chapter) are shown. All enlarged about three times. (Pages 79, 80.)

Plate 111

Above: An isopod of the genus *Armadillidium* rolling itself into a ball, and two of the isopods *Oniscus asellus*, viewed laterally. All about three times life-size. (Page 81.) *Below:* A portion of the same group seen in plate 111, less highly magnified and as they would look to a casual observer. (Page 79.)

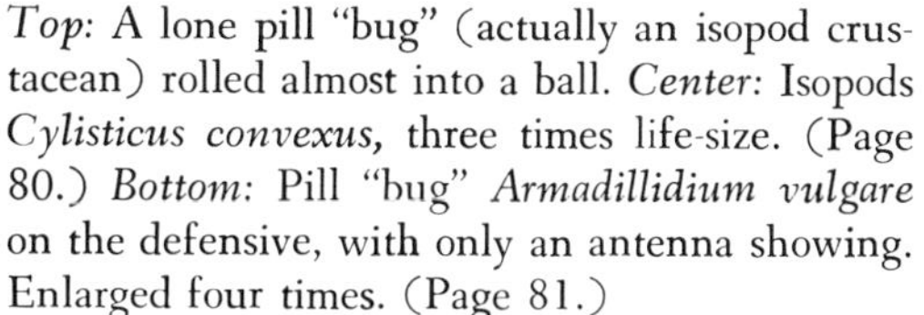

Top: A lone pill "bug" (actually an isopod crustacean) rolled almost into a ball. *Center:* Isopods *Cylisticus convexus,* three times life-size. (Page 80.) *Bottom:* Pill "bug" *Armadillidium vulgare* on the defensive, with only an antenna showing. Enlarged four times. (Page 81.)

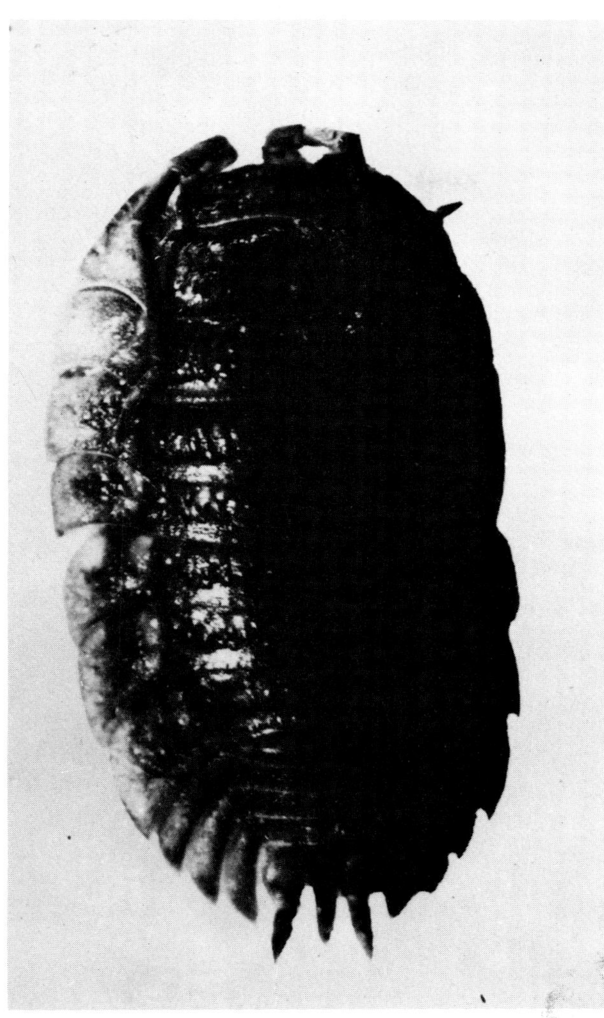

Left: Common isopod *Oniscus asellus,* usually called a sow "bug," is not an insect but a terrestrial relative of beach hoppers, shrimps, and other crustaceans. Shown eight times life-size. (Page 80.) *Above, right:* Ventral view of the same species. *Below:* Daddy long-legs courting. Male is at left. These are arachnids belonging to the order Phalangida, highly beneficial, spiderlike animals. (Page 83.)

Lace bugs (family Tingidae), which measure about an eighth of an inch in length, are astonishing insects when highly magnified, as shown. (Page 85.)

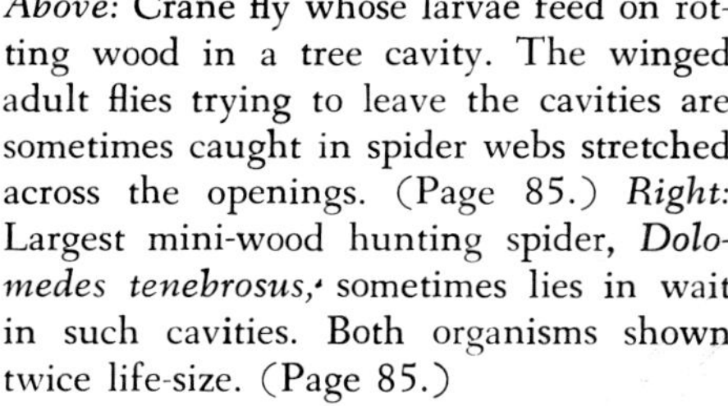

Above: Crane fly whose larvae feed on rotting wood in a tree cavity. The winged adult flies trying to leave the cavities are sometimes caught in spider webs stretched across the openings. (Page 85.) *Right:* Largest mini-wood hunting spider, *Dolomedes tenebrosus,* sometimes lies in wait in such cavities. Both organisms shown twice life-size. (Page 85.)

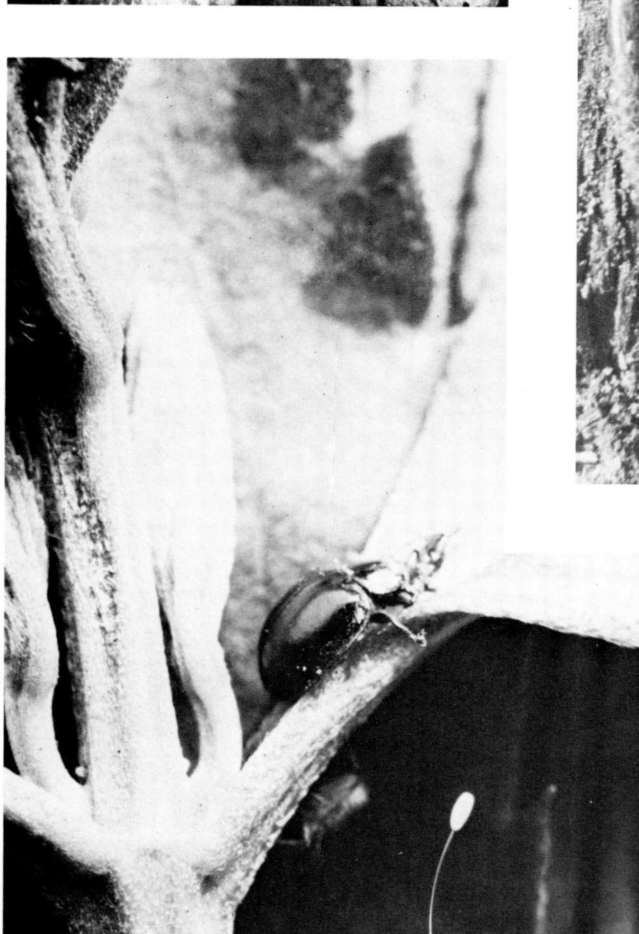

Top left: On dead trees, traffic beneath the loosened bark is often heavy. Engraver beetles (family Scolytidae) made these passages, feeding between the bark and wood. They also damage living trees. *Above:* Flat-bodied larvae of many other beetle species feed on the rotting wood beneath the bark. A tiny snail, here on rotten wood, also feeds on the algae on living tree trunks. *Left:* A lady beetle (family Coccinellidae) feeds on an aphid; and below, the stalked egg of the winged aphis lion, *Crysopa,* a great destroyer of plant life, is shown. All except top left × 2½. (Page 84.)

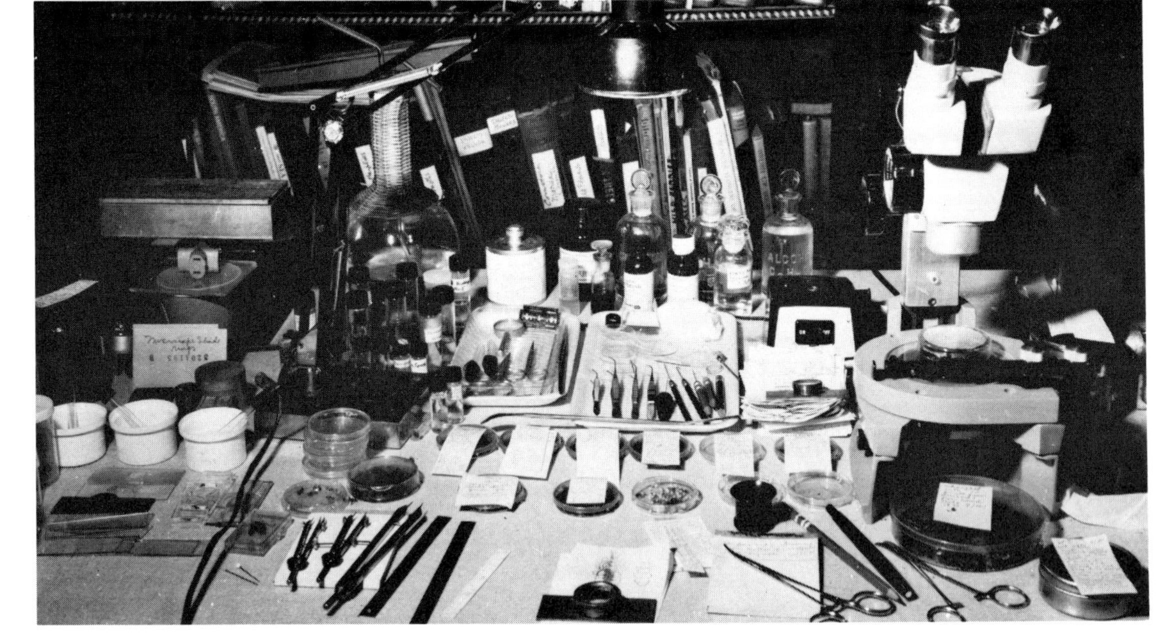

Above: A biology table set up for maintaining, microscopically studying, and measuring live micro animals of the soil. *Below*: Close-up of the plastic mini-petri dishes used in this investigation. They have well-fitting but not airtight covers, and measure (inside) 1⅞ inches in diameter by ¼ inch in depth. (Page 89.)

Above: Berlese funnel apparatus, as set up in the author's laboratory for the collection of live, minute animals of the soil. It was found convenient to suspend the lamp and shade from a light camera tripod so that the distance between the humus and heat source could be regulated simply by turning the small tripod elevator. (Pages 88, 89.)

Collembolans. *Right:* Finding it all but impossible to photograph these fast-moving, seldom still, often leaping insects from life, the author made these two pictures from mounted specimens. Here shown as they were found on fungi, they belong to the suborder Arthropleona. Enlarged about five times. (Page 90.)

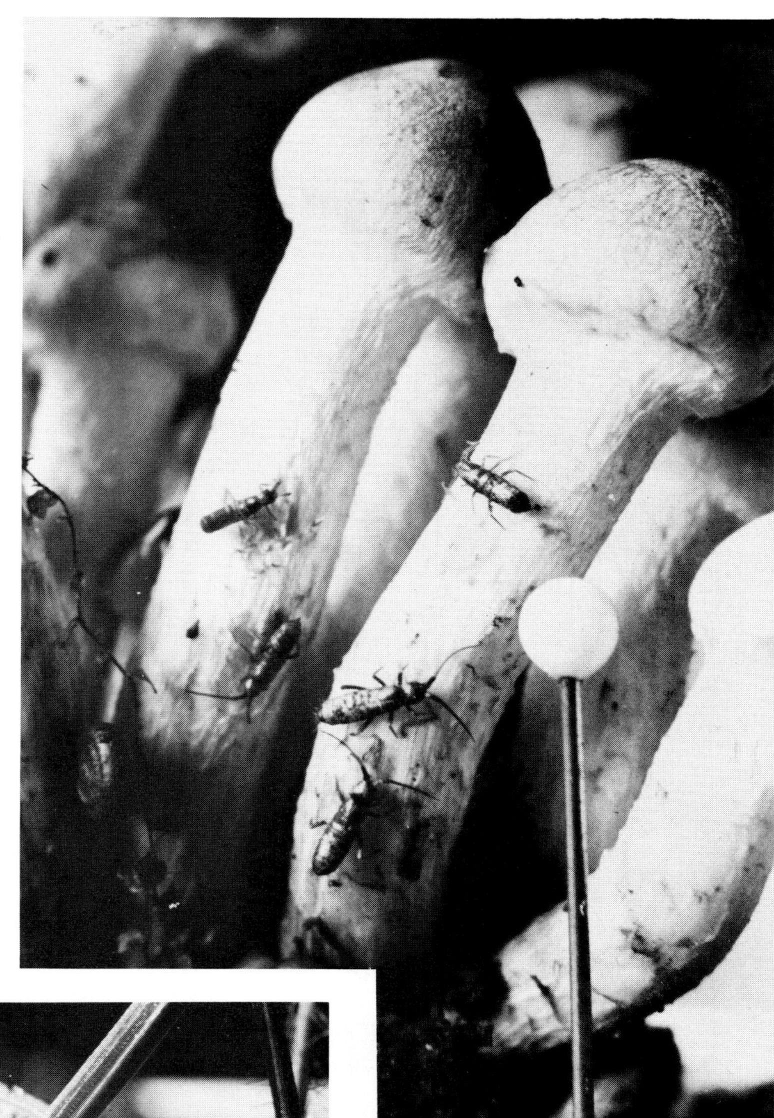

Collembolans or springtails, with coats shining like metal, are here shown with other small organisms of their world, compared in size to a five-cent piece. The white-headed pin in the upper photograph was also added for comparative size. (Page 90.)

PLATE 117

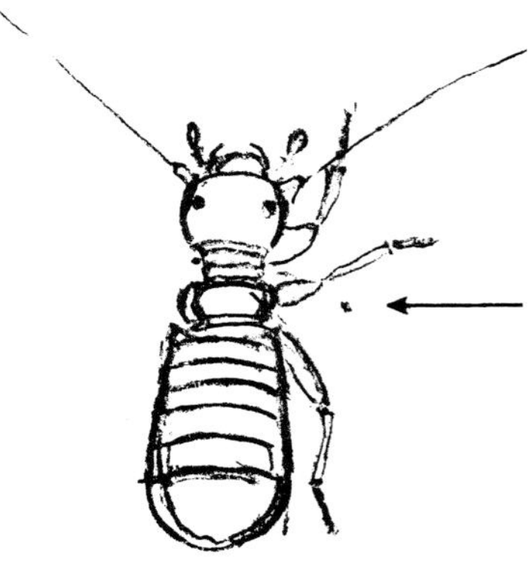

Trogiids, minute insects, lived in bark furrows. Actual size at arrow. (Page 94.)

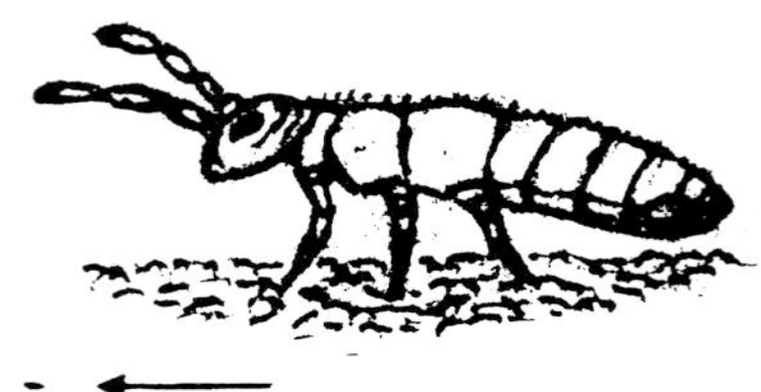

Collembolan *Folsomia quadrioculata*. A colony of these was maintained for five years on moldy paper. Actual size at arrow. (Page 91.)

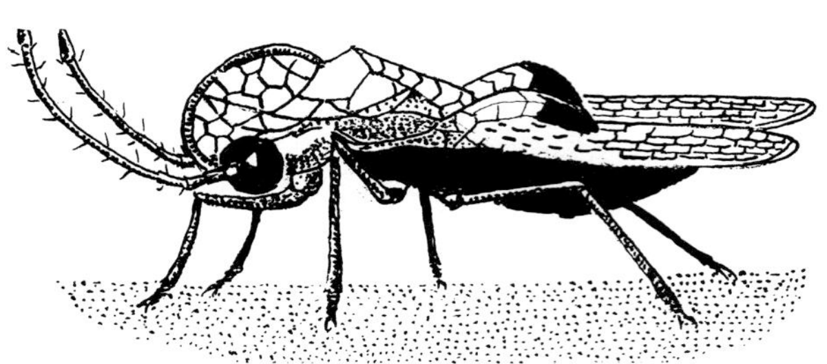

"Solarium" lace bugs (family Tingidae). *Above:* Lateral view. *Below:* Dorsal view. Enlarged fifteen times. (Page 85.)

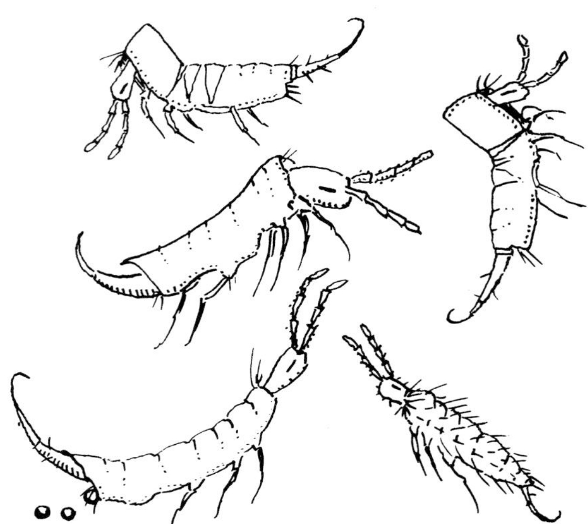

Collembolans of various forms, one laying eggs. Note extended "catapults" (furculae). Greatly enlarged. (Pages 90, 91.)

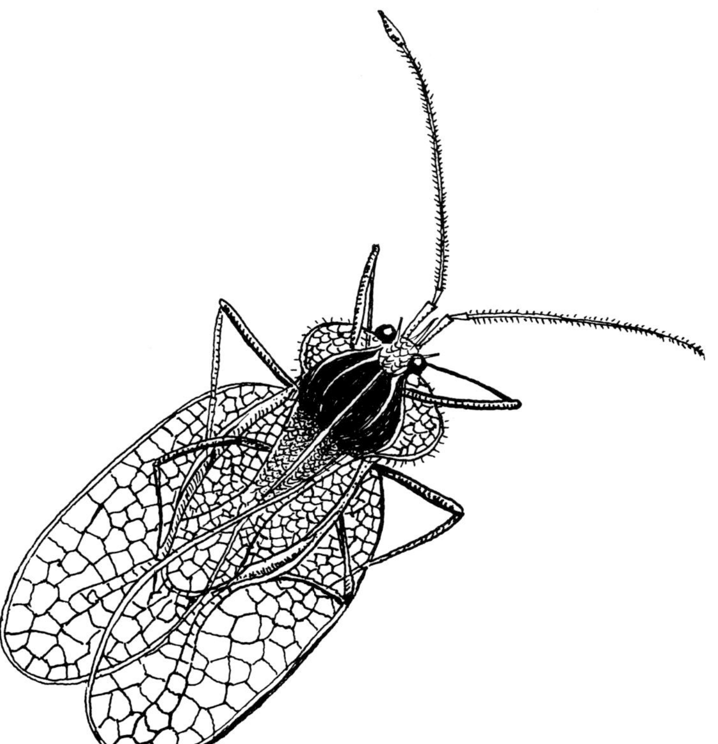

Collembolans may be black, purple, blue, yellow, gray, white, green, silver, iridescent, speckled, or spotted. (Page 90.)

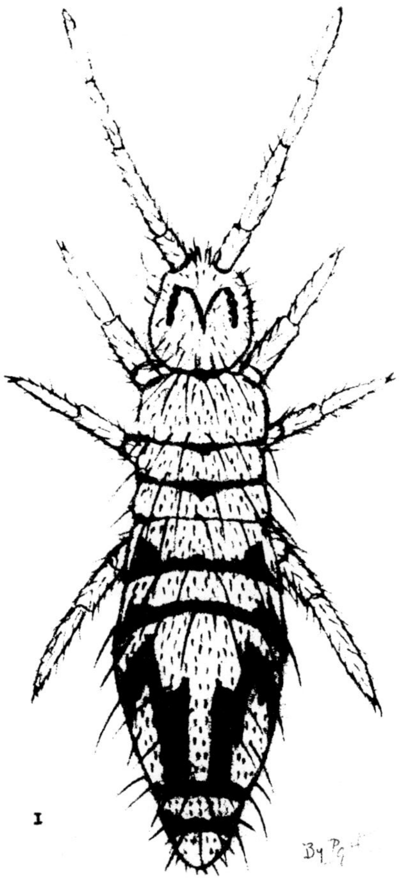

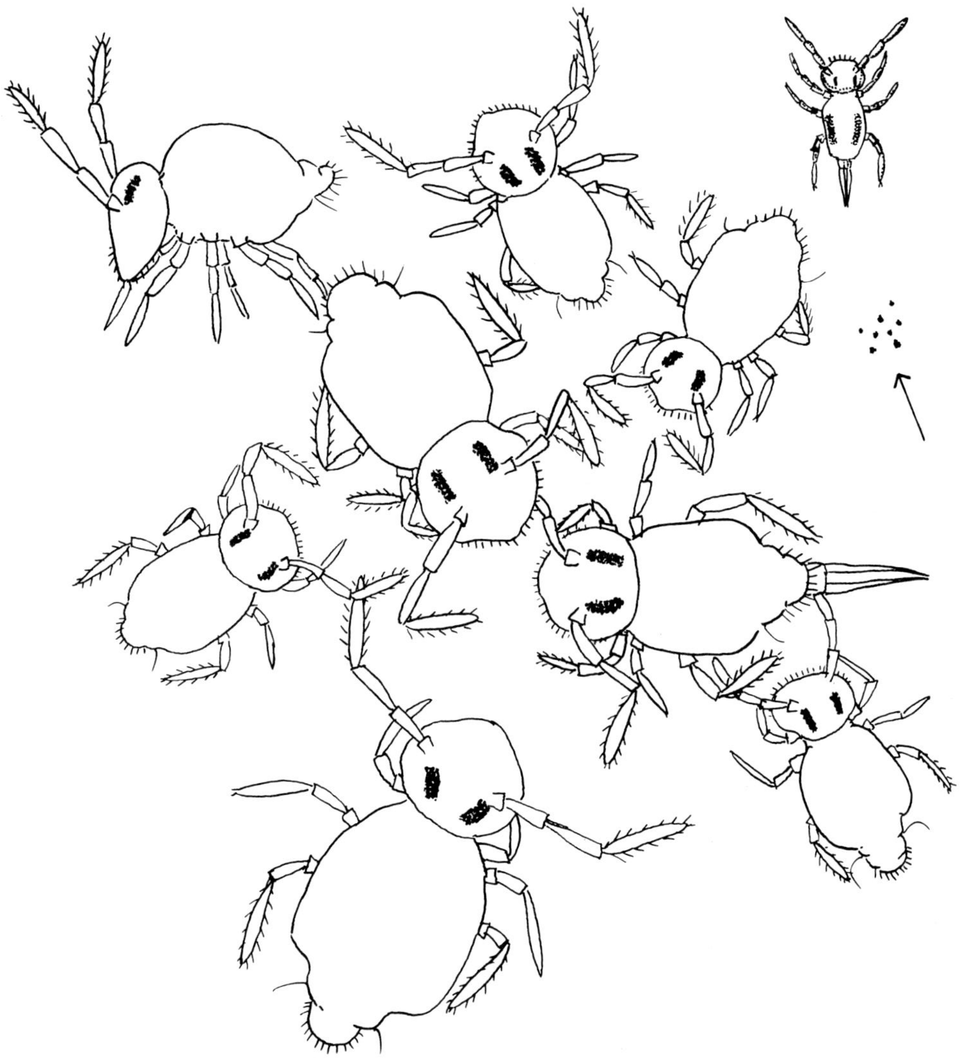

Left: Collembolan *Entomobrya multifasciata,* representative of the suborder Arthropleona, a colony of which was long studied by the author. Natural size represented at I. (Pages 90, 92.) *Right:* Group of collembolans of the suborder Symphypleona, shown enormously enlarged, two with catapults (furculae) extended. Dots at arrow (upper right) show natural size. (Page 90.)

PLATE 119

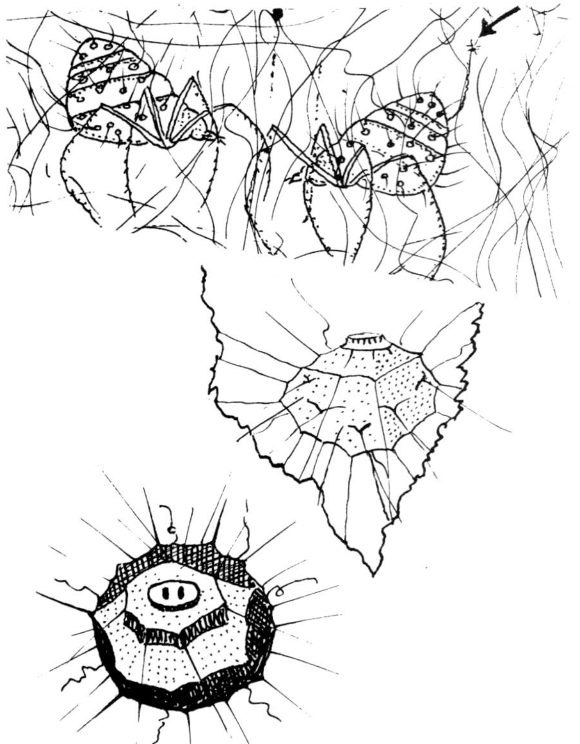

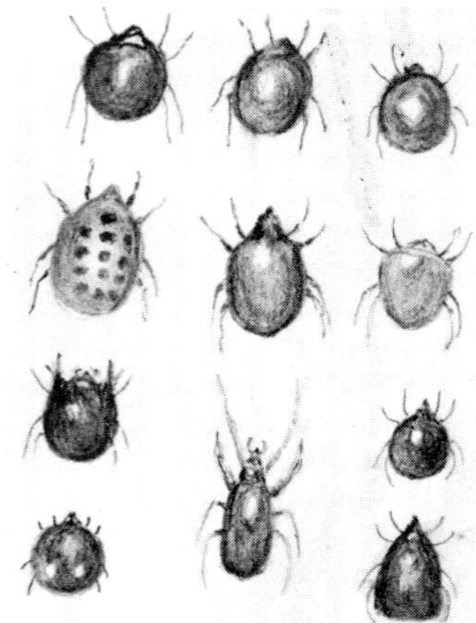

Microscopic mites from humus. Colors were (left to right): *Row 1:* Topaz; dark ruby red (high gloss); brownish yellow (high gloss). *Row 2:* Black-spotted red; plain red; bright red. *Row 3:* Black; ruby and brownish yellow. *Centered:* Butterscotch-colored, with very long forelegs. *Row 4:* White-spotted reddish brown; pale brown to milky-colored. (Pages 94, 95.)

Organisms from oak bark. *Above:* Microscopic top-shaped mites and a still smaller one riding on a single hair (at arrow). *Below:* Eggs of collembolans distorted by outgrowths of threads. All enormously enlarged. (Pages 93, 95.)

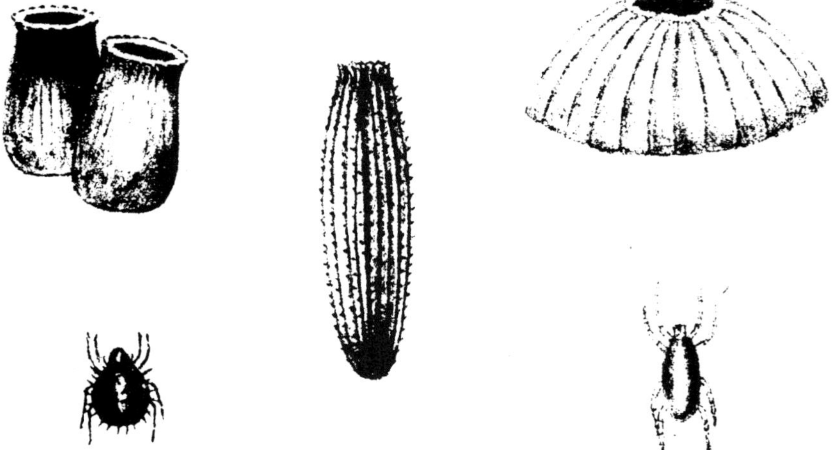

Objects found in bark crevices. Bottle-shaped, long-grooved, and "lamp shade" forms of insect eggs, and two gemlike mites. The one at right shone like a moonstone. All highly magnified.

PLATE 120

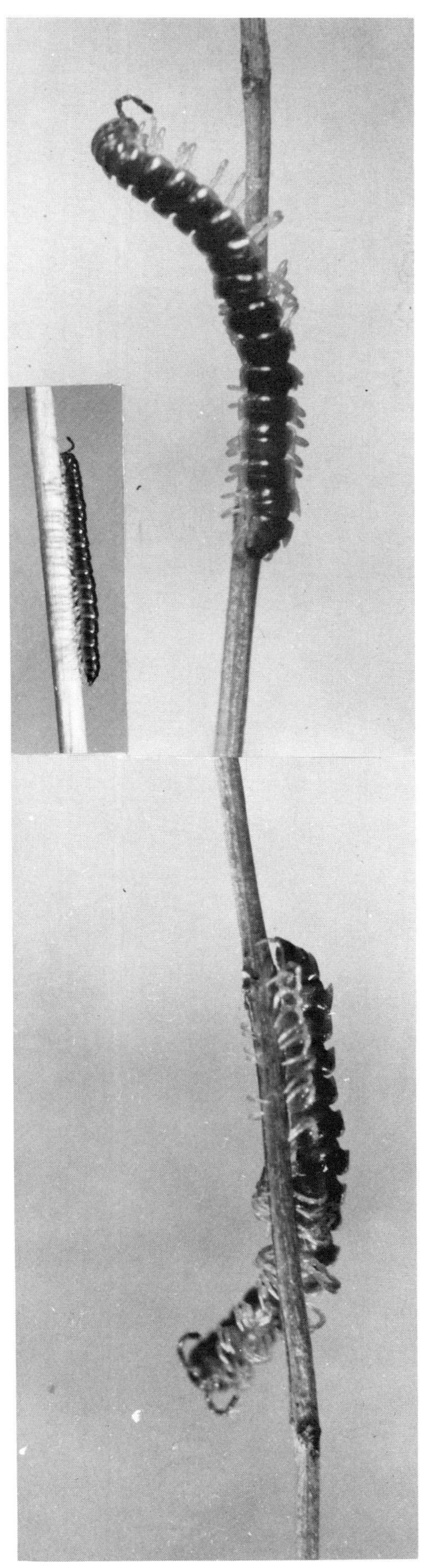

Above: Forms of some of the mini-wood millipedes (class Diplopoda), shown natural size. (Pages 97, 98, 99.) *Right: Polydesmus,* a "flat-backed" millipede which gave the illusion of being "rowed" along by fifteen pairs and three single legs on each of its sides. Inset shows the animal's actual size. (Pages 98, 99.) *Below: Julus* millipedes are the five largest, dark specimens in the photo. Shown somewhat enlarged. (Pages 98, 99.)

Centipede of the mini-wood (class Chilopoda), *Lithobius,* abundant New England myriapod, is shown here much enlarged. *Left:* On a substrate of humus. $\times$ 4. *Above:* Feeding on an oniscoid. $\times$ 2. *Below:* Eating some minute soil animal. Note elongated posterior legs, directed backward. $\times$ 2. (Pages 97, 101.)

prime destroyer of the ubiquitous sap-sucking aphids.

In the case of one very common, reddish-brown-bodied species inhabiting the mini-wood and its borders, its eggs are deposited in humus and in the soil of more open ground in late summer and early fall. Evidently they lie thus buried over winter, then in mid-May the young emerge, but whether or not they come to the surface at once, the author cannot say. They are very hard to see when newborn, as they are completely transparent at this stage and very small, so whatever may lie beneath their bodies thus becomes efficient camouflage. Under the microscope, the bodies of very young harvestmen looked much like minute bubbles. They measured a half millimeter (less than a thirty-second of an inch) in diameter, with the longest pair of legs (like transparent fine hairs) five times as long as the body.

About the only way of finding these minute opilionids, other than by painstakingly teasing out handfuls of humus with a needle, was with the funnel-separating apparatus described in chapter XI and illustrated in plate 116.

Many common spiders of numerous species must of course be included in the tree-trunk traffic, as elsewhere in the mini-wood. While many arrive aloft by the tree-trunk routes, there to stalk insect prey or spin their web traps among the foliage, it may come as a surprise to many that great numbers of newborn spiders arrive in the woodland canopies by the aerial route, carried there clinging to long streaming threads of silk which they spin out for the purpose while clinging to some lofty perch, and which finally yank them loose for these windborne journeys. The procedure has long been known by naturalists and is referred to as "ballooning."

Some woodland spiders seem to know instinctively that spinning their sticky silken threads across a cavity in a limb or tree trunk will assure them of a food supply, the ambient atmosphere of rot within the hole beckoning alluringly to various flies and other insects that habitually feed on such decomposing wood, or lay their eggs in it for their larvae to utilize.

Behind such a spider's web spun across a cavity's opening, there will sometimes be a great deal of insect life wallowing in the rotting contents, kept from drying out by captured rainwater. Flies of several species and sizes, resulting from larvae that grew and pupated within the cavity, are captured by the web as they try to get out, even large crane flies thus becoming a spider's prey at times (plate 114).

Very large woodland spiders named *Dolomedes tenebrosus* were regularly found hiding under pieces of roofing paper placed over a pile of timbers on the mini-wood border, while others were occasionally discovered in tree cavities. Moving with amazing speed on eight long legs when uncovered, they were always timid creatures, despite their ominous appearance (plate 114). These spiders spun no webs, lying in wait for their prey, or hunting at night.

Within the mini-wood, smaller spiders in large numbers were, as elsewhere, among the more important items in the diets of both transient migrant birds and those which nested and successfully raised their young within the wood or on its borders. Detailed descriptions and illustrations of the latters' nests and eggs in their natural settings will be found in chapters XIX and XX.

Among the many small arthropods found on miniwood tree trunks, and feeding on their foliage without causing great damage, were the remarkably formed lace bugs of the family Tingidae.

During the colder months, these insects hibernated in the bark crevices, and in spring and summer dwelt upon the leaf surfaces, from which they extracted sap and on which they eventually deposited batches of their sticky eggs. The curious young were often found on basswood leaves, herded together like so many gregarious sheep pastured on broad green meadows. An adult of this species is shown in plate 114.

This species and the one in the drawing (plate 118) measured three millimeters (about an eighth of an inch) in length. Although insignificant-looking to the naked eye, when viewed through the microscope, the intricacies of the latter's exterior construction were indeed astonishing.

Here was a bug above whose flattened head and thorax was something which for better description we might compare with a solarium, an elaborate dome-and-shield combination composed of chitin* and studded with many separate translucent "panes" of the same material.

It was with misgivings that the drawings of these insects were attempted, but by alternately peering through the stereo microscope and bit by bit putting what was observed on paper, and then correcting the enlargements line by line, with tedious caliper measurements, the final

---

*Chitin is a horny material, in part forming the exoskeletons or outer integuments of insects, crustaceans, and many other arthropods.

sketches gave a fair idea of how these curious bugs appeared when magnified. A third species of lace bug was also found from time to time, but with less bizarre structures than those found on basswood.

Certain boring beetles occasionally caused injury to the trunks of white oak trees within the mini-wood, seepage of the sap then setting up a whole train of interesting events around the holes made by the emerging mature insects.

As the sap oozed out from the cambium (the layer of living wood just behind the bark), a species of two-winged fly deposited eggs in the fluid, and its resulting larvae (maggots) apparently then added a digestive fluid of their own to the curious brew. Fermentation soon set in, tiny yellowish bubbles forming in the thousands and continuing to multiply until a froth was formed about the wounds on the tree trunks. As new bubbles formed, hundreds of others burst, thereby broadcasting gassy messages, tiny to be sure, but adequate to lure a host of other insects to partake of this brew automatically concocted from sap, digestive fluids, excrement, and occasional dead bodies of the larvae, bacteria, and who might say what else?

First to arrive after the flies were white-faced hornets, *Vespa maculata.* They came from a big nest of gray paper, "home made" by successive broods of worker hornets and built around a ping-pong-ball-sized nest constructed by the fertile queen for her first, "personally" raised brood in early spring.

All hornets love fermenting sap, and now much larger ones with ocher and brown bodies, measuring over an inch in length, arrived upon the scene. This was the species *Vespa crabro,* introduced from Europe and now naturalized, whose large, elongate, uneven nests are composed of brittle paper laid on in wavy bands in shades of gray and brown and cream and yellow-ocher.

These two species of hornets and several other kinds of smaller insects feeding at the wound paid little or no attention to one another, drinking at the free saloon in apparent harmony. But at length more belligerent actors (surprisingly, in the form of butterflies!) arrived. One was the red admiral, *Vanessa atalanta,* whose broods are on the wing between May and late September. The next to show up was the more modestly colored pearly eye, *Lethe portlandia,* found in the mini-wood only from late June until the end of July, and then very rarely, a butterfly always locally distributed.

Red admirals are robust creatures, spreading nearly two and a half inches, real beauties when fresh from their chrysalides, with forewings of smoky velvet banded with rich orange-red, and with many white spottings, and underwings with orange bands along their margins.

The pearly eyes are more fragile, more delicately colored, their forewings pale yellowish brown, have darker, yellow-bordered eyespots, and even lighter-colored underwings marked by wavy bands of bluish-white and brown, and the eyespots touched with white in their centers (see plate 133).

Despite their delicate wings and bodies, the pearly eyes were domineering by nature, vying with one another and the other butterflies over the odoriferous foamy sap. They were persistent and quick in their motions, but so were the red admirals. The two species just about balanced each other in aggressiveness, neither being for long intimidated by the other. They threatened, advanced upon one another, flipped their wings, and occasionally took off in short wild chases before being lured back again to the delicious stuff oozing on the tree trunks.

No harm was done; possibly all present enjoyed the excitement, so here we will leave the woodland's tree-trunk traffic.

Other butterflies and the beautiful moths of the mini-wood will be treated in a later chapter.

CHAPTER X

ARMY OF THE SOIL I

In contemplating the mini-wood, this mere half-acre of mature and in some ways aging natural forest land, four basic thoughts came regularly to mind with every visit therein:

First—always—reverence for the botanical scene, for *green* plants as representatives of Nature's supreme achievement upon which all other living things depend.

Second, the reminder that among these trees and shrubs and lesser plants, and within the humus and woodland soil, there was a vast web or network of other, minor plants, and animals, mostly interdependent, upon whose activities even the major vegetation itself was to a greater or lesser extent also dependent.

Third, gazing upward through the mini-wood foliage with wonder to the canopy and the sky, always brought the powerful reminder that this ecosystem and all others elsewhere had been brought into existence by the "horsepower" of a star, ridiculous as it may seem to apply such humanly imagined puny units of energy to that of the sun!

Most astonishing of all was the thought that our planet had found its orbit just near enough—yet just far enough from that enormous ball of fire to make green plants and all other life possible.

In the mini-wood, during the winter months, the tree trunks and their networks of limbs and branches shut out some of this solar energy. In spring, myriad freshly unfolding leaves shut out much more, and by early summer this foliage, fully expanded on both the mature trees and on those of the understory made up of their offspring, varied the degrees of shade cast, from moderate to almost total.

Microclimates, with their varying quotas of moisture, were thus set up at various levels from floor to canopy, accordingly influencing the kinds of organisms that dwelt therein. Conditions in the humus and soil were also thus modified by these and more complex factors to such an extent that, in the mini-wood, for instance, and as we have seen in the botanical chapters, the habitats offered were congenial only to a comparatively few angiosperms, ferns, mosses, and lichens.

It must be repeated here, for those who may not have read all of the earlier chapters, how important to forest land of whatever size is the annual abscission of the leaves and their accumulation on the ground. Much as the timing of this fall varies with different species, and greatly as the leaves vary in quality, they are of prime importance in the production of humus and as food for large numbers of small animal organisms. Less spectacular is the gradual fall of other spent vegetation—buds and flowers, seeds and their capsules, fruits, nuts and acorns, twigs, bark fragments, dead limbs, branches, and, sometimes, even full-grown trees. This woodland debris, together with animal matter in the form of naturally dying organisms; the excrement of mammals, birds, reptiles, amphibians, insects, and other invertebrates; and even some of the honeydew showered down by aphid colonies in the treetops is slowly or rapidly broken down and partly returned to the woodland as part of the humus. Uncountable bacteria, fungi, and large numbers of small animals of the soil, many of which we will examine presently, are some of the factors responsible for such processing.

Humus, it might be mentioned for the benefit of those who do not already know, is the organic portion of the soil, brown to black in color, and consisting of plant and animal matter in various stages of disintegration, all mixed with animal feces and mineral matter in the form of particles of area rocks already disintegrated or in the process, as described in chapter VII, on lichens.

As one digs down into a deciduous woodland such as that being investigated, the few inches of humus give way gradually or, in some places, more abruptly, to the predominantly mineral soil below, in which the larger vegetation has most of its roots. Below such mineral soils (paler-colored strata which are always much thicker than that of the humus), there is the basic rock of the region.

Some of the soil animals which contribute to the humus in one way or another, but chiefly by diminution, ingestion, and part elimination of dead or living plant and animal material, include arachnids such as the mites, spiders, and pseudoscorpions; myriapods such as the millipedes and centipedes; terrestrial isopods (the oniscoids already described in chapter IX); insects and their larvae, insect colonies, and the minute swarming woodland collembolans. Snails and slugs and vertebrates contribute also. All of them will be described and in almost all cases illustrated with new photographs or drawings made especially for this book, the subjects of course being the organisms actually found in the mini-wood.

Among the microbial populations, as among the higher animals, interrelationships may be highly involved. There may be cases of *symbiosis* such as we have encountered in other organisms in this book, in which two dissimilar organisms may live together in a mutually beneficial partnership. There may also be cases of *antibiosis* or antagonistic associations between organisms, and in which cases such association may also be called *parasitism*. Still again, there may be *commensalism*, the association between two kinds of organisms wherein one receives food or protection from the other, but without either damaging the partner by contact or benefiting it in any way in return. Among the microbial forms, the bacteria alone are capable of converting the nitrogenous "waste" matter of dead animal bodies into nitrates usable by plants. Such complex processing as this and the manufacture of humus from living and dead organic matter by living organisms will continue to operate

successfully only so long as human beings allow these forest processes to function without interference.

To tidy up a woodlot in extreme cases, such as after a severe ice storm, when even the largest limbs of some trees may crash to the ground, is one thing; but clearing the woodland floor of all debris, as in a park, may eliminate many of these animals of the soil army along with their habitats.

So much goes on among the small fry which dwell in the humus and in the disintegrating leaf-cover just above it—this stuff so often beneath *our* feet—that in order to cover the subject even as superficially as in this book, it has been necessary to devote three whole chapters to this interesting group which collctively we call "the army of the soil."

Come now on a journey of exploration with the author, down into this curious, extremely busy, and heavily populated world of microscopic and much larger (visible) organisms, a vastly diversified group, and concerning which we would dearly like to know the sequence by which came each consumer or predator consumer in the first place. Were they here before the Ice Age, and survived that? Or did all of them come slowly from more southerly regions after the glacier had melted? Here is a mystery which in all probability we will never be able to solve.

In every mass of disintegrating vegetation, in mosses and on damp earth, in the grass of our lawns and the hay of our fields, and in the furrows of bark, but particularly in every handful of rich moist humus, dwell tiny primitive and extremely active insects belonging to the order Collembola, creatures known in popular language as "springtails," and the name found very appropriate for a reason to be presently described.

Springtails are *apterous* (wingless at all ages), and they are insects of many diverse body forms and as many different colors (plates 117, 118, and 119).

Let us now observe how all of the smaller animals of the soil are collected for study.

The only satisfactory way by which those as small as the collembolans (and many other organisms of the humus) may be obtained alive is by means of a simple apparatus known as a Berlese funnel, a device so uncomplicated that the illustration in plate 116 is almost self-explanatory.

The procedure simply entails a gradual and gentle

warming of the humus and litter sample with which the funnel is charged. A 100-watt bulb in a reflecting shade suspended above the sample is the only heat source. Within the funnel, midway between the top and neck, cross wires support two wire screens, the lower one of eighth or quarter-inch mesh, and the upper one of very fine wire cloth or copper screening, and both are cut to fit closely against the sides of the funnel. A small bottle with a suitably bored cork fits over the end of the funnel neck, and into this the animals are driven as the increasing heat forces them down out of the slowly drying medium (many not surviving for long afterward if not removed from the bottle).

The collected animals may then be washed into watch-glass dishes, or into live cells for microscopic examination, and then transferred with a medicine dropper or on a dissecting needle into mini-petri dishes previously lined with layers of finely sifted humus in a dampened but not saturated condition.

Prior to the fifteen-year study of the mini-wood, and until the Berlese funnel opened up this world in detail, the auther had never fully realized what fabulous numbers of animals inhabited the humus and woodland litter.

Emphasis should again be placed on the statement that most of the animals of the humus and soil cannot tolerate dryness. Water from the media being dried out collects as a shallow layer in the bottle attachment along with the separated organisms, and while they cannot stand desiccation, neither can they remain alive for long if submerged, and they must therefore be promptly removed to the petri dishes if wanted for live study.

A single funnel full of humus taken from beneath the dead-leaf blanket in moist woodland will often contain hundreds of soil organisms. The richest hauls in the mini-wood area were taken from the first two inches of humus beneath such cover, from accumulated material under rotting woodpiles or isolated logs, and also from very moist humus under patches of skunk cabbages.

There is probably no way of accurately estimating the actual populations of these animals per square yard. Much less would it be possible to state an average number of animals occurring per acre, which would, of course, vary greatly, according to the character of the land, its vegetation, its geographic location, and the amount of precipitation it receives. The author believes it no exaggeration, however, to state that there are probably billions of soil organisms (exclusive of bacteria and other plants), even in

this particular mini-wood, or in yours, while an attempt to count the so-called one-celled organisms or Protozoa would of course be pure folly, as it would also be to attempt counting the micro fungi.

Protozoa, very animated objects, were usually present in considerable numbers in the water extracted from the humus by the funnel. We will not try to name many of them specifically, but some of their odd forms are illustrated and identified in plate 107.

As already mentioned, among the soil minutiae visible to the naked human eye and collected in large numbers by the funnel apparatus, were the tiny insects called collembolans.

Collembolans! Well, the exclamation point was not added for any foolish reason. The name has a sound suggestive of some newfound race of beings, does it not? And these tiny things, constituting one of the great animal populations of the earth, *do* remain undiscovered by most of us despite their age. This venerable race no doubt inhabited this planet for so long that, by comparison, man's wavering tenancy has been but an instant. Collembolans, it would seem, have come right down to the present time little changed from their fossil ancestors. (Plates 117, 118, and 119.)

Strange that such primitive delicate animals, creatures so soft and fragile from our point of view that the least careless handling often injured them—and sometimes killed them, just from shock—are yet so successful. In their favor in the first place, however, are their enormous numbers. They also seemed untroubled by cold, living successfully through freezing weather in the furrows of bark, or deep in the humus, a predator-rich medium at all times, but in which they have long held their own.

There were scores of species inhabiting the mini-wood and other nearby habitats. They so intrigued the author for many reasons and their lives proved so interesting that much space must now be devoted to them, but after first stating that some are harmful. At times certain species have occurred on crops in vast numbers, doing serious damage; others are bothersome on mushrooms in commercial beds, or they may suddenly swarm where maple sap is being processed into syrup and sugar. Personally, the author has had them overabundant on young lettuce plants, but it would seem that the majority of the species are harmless, while many are doubtless important, or even necessary, in the food chains of the woods.

In this book, we are interested in collembolans only as

surprising and active little entities inhabiting a part of the world so unlike the one we know.

Thriving almost wherever there is litter with damp ground underneath, and very often actually under our feet in the woods or the garden, they are nevertheless independent little animals full of energy, and go, and determination.

Technically they are divided into two suborders, the Arthropleona and the Symphypleona. In the former group the head is followed by segments called the *prothorax, mesothorax,* and *metathorax,* and several abdominal segments easily seen, while in the latter suborder the body segments are mostly fused, thus giving these species a globular apparance (plate 119).

As with other adult insects, their bodies are enclosed in exoskeletons of chitin, soft in collembolans and affording little protection. Some have densely hairy bodies, visible only with a microscope, and it was surprising, when examining certain dead specimens, to find, from their bodies scattered on the glass slide, large numbers of *iridescent* scales, scales quite different in form from the more familiar ones on the wings and bodies of moths and butterflies.

Collembolans are peaceful organisms by nature. Devoid of that wondrous thing—intelligence—on which we so often pride ourselves, when it comes to lethal weapons, collembolans possess no bombs or poison gas, or even repellent sprays with which to best or dispatch their enemies, but nature has endowed them with an extraordinarily efficient escape mechanism, called the furcula (or "spring"), from which the insects take their common name, although there are some species not so equipped.

Actually a forked lever folded back under the abdomen and held there by a catch, the furcula acts as a catapult. Instantly released at the slightest danger, the furcula is activated and, snapping down hard against any surface, hurls its owner through a parabola fifty or a hundred or maybe even more times its own length! Arriving again on terra firma after such a violent aerial journey, the tiny animal nevertheless lands upon its feet so lightly that no damage whatever results.

It is interesting to remember that nature devised this astonishing, *built-in* protective apparatus for otherwise defenseless animals perhaps millions of years before man found it necessary to think up a comparable device, for instance, the ejection seat for airplanes.

Although placed by entomologists on almost the lowest rung in insect classification, through long hours of intensive study the author has found collembolans, at least in their actions, often to be more like certain higher animals—and even some mammals—in the general care and grooming of their limbs and bodies.

Collembolan development is *epimorphic,* which is to say that the organisms emerge from the eggs closely resembling the parent, without having experienced a long-drawn-out metamorphosis. These wingless primitive insects thus enjoy a quickly reached adult life, with no need of preliminary radical intermediate phases of existence accompanied by anatomical breakdowns and reconstructions in the development of wings and stings and sundry other not really necessary equipment. While butterflies and bees, as examples, are tediously undergoing such processes in the pupal stage, the microscopic collembolans are living equivalent time already as imagoes. Thus it would seem to the author that the "primitives" are sometimes better off than the more highly organized animals.

As for the range of colors in collembolans, we find them in black, purple, blue, yellow, green, white, gray, ocher, and other shades, as well as transparent or nearly so. There are spotted and banded species, and pepper-and-salted ones, and those with brilliant metallic body finishes, while still others will be found with coats of beautifully iridescent scales.

All of them must be viewed through the stereo microscope to be thoroughly appreciated as natural creations. The spotted and mottled and variously dabbed individuals would have tickled the senses of certain kids in their Kool-Ade-acid-test phases, had they ever bothered to look at such things for pattern suggestions.

A pure white extremely minute collembolan was *Neelus albus,* so small that a forty-power magnification was necessary to observe it clearly, an abundant busybody forever on the go. Actually measuring but a quarter of a millimeter (about a sixty-fourth of an inch) in length, or the size of a dot made with a sharp pencil, they were amusing little things to watch with the stereo microscope. Apparently always highly intent and seriously engaged, whatever it was that they were doing that kept them thus continually on the move was indeed very puzzling. Oblivious of the other creatures encountered, sometimes passing right under the legs and bodies of larger animals, apparently unnoticed, they evidently had few enemies in an environment where much tougher creatures were often attacked and eaten.

With squatty fat bodies about as wide as long and, as seen in profile, with somewhat lowered, jowly little heads, they somehow put the author in mind of aging, humpbacked bulldogs with six legs.

Occurring in large numbers, sometimes in a handful or two of humus, they moved easily around or between what through the microscope appeared as great hunks or clods of soil with ample spaces in between, but which in reality were minute particles of disintegrating forest debris. (All humus seems to appear thus under the microscope.) The clods dwarfed these collembolans, as they did a score of other minute inhabitants of this cool habitat with its special microclimate.

Some of these smallest soil organisms were so light that they easily walked upon a film of water without penetrating it. Being amazingly viable also, they were able to withstand and live through the heaviest rains which soaked their realm. The best way to observe them, in fact, singly and uncluttered, was on the water film in the collecting bottle of the funnel apparatus already described.

Widely differing in appearance from these squatty individuals were species with elongate narrow bodies, and those whose prothorax was oddly upturned so it resembled an oblong, square-ended tube from which the head protruded as if from a hood (plate 118). Also, there were some with very long antennae, and, in the same habitat, many individuals with these appendages very nearly as short as their minute heads were wide.

Literally a world within a world, we might for amusement call their realm "Collembolia," a heavily populated place, yet so difficult to observe in minute detail that, during the years of this mini-wood study, the life histories of only two species were partly worked out and recorded for this book.

Maintaining these little things in colonies and in good health was not too often successful, but one species, an extremely minute form named *Folsomia quadrioculata* (plate 118), with plain or finely speckled lead-gray to blackish bodies, solved the problem themselves. How this came about was as follows:

Some small black growths about the size of marbles were removed from a rotting log in the mini-wood. Apparently dry and hardened fungi of some sort, they were placed in a small glass jar for observation.

A few days later the objects began to rot and produce considerable moisture in the jar, also moistening the enclosed data label, a strip cut from an index card, on which mold soon grew. Appearing as minute black and greenish specks, these quickly increased in number as the paper absorbed more water. In about ten days, many reddish specks of another species of mold appeared, and soon afterward, barely visible moving organisms (about the length of the smallest line that could be drawn with a sharp pencil) were observed in these pastures of microscopic plants. Viewed through the stereo microscope under high magnification, the animated black "lines" were identified as the collembolan *Folsomia quadrioculata,* many of which were apparently feeding on the mold (plate 118).

Also soon appearing here and there among the collembolans on the moldy paper were many microscopic spider-like mites of a translucent yellowish color, and as all of these organisms seemed to be living in harmony, additional strips of index card moistened with a few drops of water were added to the culture, and on these, more mold appeared and more mites and collembolans later gathered.

Thus it happened that a method of feeding, breeding, and greatly increasing a collembolan colony came about by accident, and the reader may imagine the minuteness of the newborn ones which kept appearing, doubtless having hatched from white eggs impossible to see against the still partly white paper strips.

Despite the tiny size and lowly entomological status of these collembolans, their attention to grooming and body care was a revelation. Imagine "fastidiousness" in these specks of life! Sometimes one would be seen standing on five legs while laundering the sixth with its all but invisible mouth parts. Other individuals groomed all six legs one by one, and then their antennae. In this and other ways they reminded the author of higher, more advanced animals. They seemed so "on the ball," so to speak, in their particular environment; they were fraternal creatures also, never fighting, greeting one another with quivering antennae whenever they met accidentally, but always in peace, and then each again going about its own almost continuous activities.

In order to ascertain the life spans of these minutiae, individuals and groups of three or four were transferred from the larger colony to strips of mold-grown card in plastic petri dishes measuring forty-five millimeters (about one and three quarter inches) in diameter, and five millimeters in depth. For observation, these entire small receptacles could be placed, with the covers removed, directly on the microscope stage and without disturbing the viable tiny inmates.

While many higher insects in the adult form may live only a day or two, or a week or two, certain individuals among these isolated little colonies of collembolans lived for 61, 78, 95, and, in one rare case, 127 days.

Nothing in the way of food (other than the molds which grew on the moistened strips of index card) was ever supplied to any of these colonies.

Astonishing was the fact that so many new individuals so long continued to appear in the original colony in the glass jar, brood after brood for *five whole years!* The very young were just about visible with the aid of a powerful magnifying glass, but eggs were never discovered, for reasons already mentioned.

This minute species was also found living in mini-wood humus and other debris, as well as on bark, where it could move with plenty of room to spare between the smallest furrows, or in humus, among the myriad pellets excreted beneath the dead-leaf cover by earthworms. We now know that these tiniest collembolans required only minute quantities of mold for sustenance.

Studying the much larger collembolan already mentioned, considerably more was found out about its life-history. Specimens were first found in winter, clustered together in the bark crevices under loose fragments on a white oak tree in the mini-wood.

In a preliminary brief study concerning this half-acre woodlot, in a book published some time back, the author also recorded observations on this bark-inhabiting species believed to be *Entomobrya multifasciata* (plate 119).

Measuring about two and a half millimeters in length at maturity, individuals found later varied considerably in color, but less so in the dorsal patterns. In the colony kept successfully in captivity, however, most specimens were grayish green or olive green, and with the darker markings shown in a typical example in the greatly enlarged drawing, which is quite accurate, except perhaps for some of the hairs in the insect's "clothing." On the whole, the drawing gives a good idea of how these collembolans appeared through the microscope. At best, this was a trying subject to enlarge so greatly, because of its 2.5 mm. actual length.

Our chief interest in this book concerns the living subjects and their habits, and, fortunately, the author's captive colony in this case was long lived. Its history follows.

Tender and seemingly frail, so even a short drying of their minutely hairy bodies caused them to pull in their six legs, protrude their forked "catapults" (furculae), roll over, and die, it was surprising that this species remained alive and healthy under the bark during the cold dry winter weather, when the thermometer was hovering between 10° and 30°F.

Evidently their tiny bodies were protected by some natural antifreeze in the colorless blood which bathed their whole interiors. When found, they were huddled together in little parties in tiny cracks. In a sluggish state, they were easy to capture.

By lifting strips of loosened bark and flipping the insects uninjured into a bottle by means of a fine-tipped artist's brush, a "herd" of fifty were soon taken and, in the laboratory, transferred to a small tin box containing bits of algae-coated bark and fitted with a clear glassine cover. Only greenish gray bark from the same tree upon which the collembolans had been hibernating was added, so the colony was thus returned almost at once to a natural habitat, yet in a receptacle small enough to be placed entire on the microscope stage without disturbing its living inmates.

With the bits of bark well moistened, the warmth of the laboratory soon stimulated the collembolans into active life once more, and the algae into growth. As time went on, with the colony in thriving condition, it appeared possible that bark furrows were the normal, possibly year-round habitat of this species.

Let us now look at this habitat through the microscope, into an area where the "landscape" appeared as a series of extensive parallel ridges between which were deep crevices like somber dark valleys, and much of the "land" was dotted or clothed with strange-looking "shrubs" (the algae on the bark). Here and there were larger, unfamiliar rounded plants and ultra-minute vermilion and iridescent specks one might wonder if anyone could identify.

Running about these "hills" and "valleys" (the furrows in the bark fragment) were the hairy, oddly formed collembolans, as strange to most people as would be long-extinct animals come back to life. Some were olive green, their almost transparent gullets crammed with masticated shrubbery from their realm, while others that had recently defecated were lighter green in color and distinctly patterned with dark brown. All had eight bright and shining eyes, in single rows on either side of their heads.

Suddenly startled, one instantly released its forked catapult against the ground with such force that the owner first hit the glassine cover of the observation box and, glancing off, was hurled across the several ridges, where it landed unhurt upon its feet.

Along these rough and uneven "hills" and "valleys," these creatures of the herd were almost continually on the move. Like well-fed little cattle, they only stopped now and then to browse upon choice bits of vegetation (the algal cells which dotted the bark).

Other collembolans performed head-bobbing routines (as some mating *birds* do) and antennae fencing for a second or two, all of them activities which may or may not have had sexual interpretations.

While thus observing their odd little world, as if we ourselves were shrunk to their size, a lasting impression would be one of serenity.

Three weeks after establishing this colony, a few eggs were discovered. These were deposited either in irregular short rows grouped in twos and threes or singly in crevices in the upper surfaces of the bark. They were spherical, and as smooth and lustrous as ivory billiard balls. And their size? How may one convey this to the reader when no familiar thing with which to compare them comes readily to mind? They never would have been discovered at all without a microscope, yet imprisoned within each of these ultra-minute spheres were those wonderful chromosomes and directing genes, together with exactly adequate amounts of nutrients, begotten from this woodland, with which to replicate young collembolans of the species, each with the whole inheritance of the astonishing little race in every cell of its body.

It would have required countless thousands of these eggs—more than in a whole shad roe—to cover the palm of a human hand, and were there that many available and so placed, closing the hand firmly could wipe out more potential *Entomobrya multifasciata* than bullets dispatched Americans in both World Wars! Such are our respective niches in this part of the universe.

Eggs as smooth of shell as these would have been in danger of rolling about or of being blown away, but for a strange thing which now happened. Within a short time, little points like nipples appeared at various places upon each egg. From each of these projections a delicate hairlike process then grew outward. Reaching about here and there, they soon touched at other points on the bark, thus holding fast like anchor lines, and while somewhat slack at first, they gradually tightened, thus pulling the eggs' shells into distorted contours. Some then appeared faceted; others were pulled out at one side, or at two or three places; while still others became turreted, and actually suspended in mid-air in the network of threads (plate 120).

Could this have been a remarkable example of symbiosis (a mutually beneficial relationship) between a mold and an insect's developing offspring? It did seem indeed possible, although rather improbable, that the threads were of a mold, for within the eggs the embryos were steadily growing, and mold on eggs usually indicates that such inmates are dead. Perhaps live eggs can support such a growth. The author does not know, but whatever these

threads might have been, they definitely moored each egg securely in place until the young emerged six days later.

A sure sign that emergence was at hand was the appearance of dark brownish eyespots visible through the egg shells. Instead of crawling forth from a more or less round hole eaten in the shell, as many newborn caterpillars do, the eggs split apart, and the young collembolans, *backed out,* a little at a time, waving their antennae from side to side, then resting a little after each outward movement.

Weak at first, with their legs still partly folded inward, they soon straightened themselves out, and like newborn colts, in a very short time were actually walking about and feeding themselves, but, unmammallike at such a tender age, were cleaning their legs and antennae and otherwise acting, miniatures though they were, exactly like their parents.

Newly emerged, a collembolan of this species was about as long as the first two joints of its mother's antennae, and although almost exactly like the adult otherwise, it retained a tender, pudgy look for a time which at once spelled "baby," and as with all other very young creatures, exerted a special appeal.

Completely independent from birth, these newcomers grew quickly, and with their peers, carried on within the little glassine-covered box for months, feeding, courting, breeding, dying, and revealing, by means of the stereo microscope, the secrets of their curious lives.

Collembolans may be near the bottom of the insect ladder technically, but they emerged from their eggs endowed with all of their faculties, ready to take full charge of themselves and challenge the world, accomplishments which apes and man, the highest animals on earth, cannot begin to duplicate.

Feeding habits of the collembolans varied considerably, although usually the method followed by those in the viewing box was to browse along the furrows, where they could be seen pulling off and munching minute mouthfuls of algae and rotted wood. Whether or not they actually chose these wood cells, or simply took them unavoidably along with those of the algae, the author cannot say. Cellulose components would seem beyond the ability of such tiny anatomies, unless, like termites, collembolans have those specialized organisms called *Trichonympha* which are flagellate protozoans dwelling within their digestive tracts and which might make available what nutrients there may be for collembolans in cellulose.

This same question would seem to arise with other wood eaters as, for instance, with pill "bugs" (*Oniscus*) and other terrestrial isopods, those abundant wood consumers described in chapter IV, whose millions of tiny fecal "bricks," together with feces of the mini-wood collembolans and waste matter from termite colonies and other debris-consumers are important in the processes of humus accumulation.

Collembolans are not all strictly *phytophagous* (plant feeders), as was quickly proved in observing this colony in which, surprisingly, cannibalism occasionally occurred. With great interest the author watched one such meal in detail, of course through the microscope, observed it so clearly and so highly magnified that it was almost startling.

No killing took place, however, the living collembolan simply coming upon and partaking of another that had died. First it set about devouring its relative's antennae. Next, it ate all six of its legs, reminding the observer of himself relishing the crisp appendages of a soft-shelled crab. Pulling each limb toward its mouth parts with one of its own, it would stop eating occasionally to lower its head to pick up dropped crumbs, as a dog might lick up the leavings of a heavy meal.

The most startling observation was witnessing fragments of the dead insect's body traveling through the consumer's neck passage and on down the food canal on its journey to digestion in the living animal. Watching it was reminiscent of a TV commercial depicting the passage of partly digested garbage through the chemically cleared pipes of a kitchen sink. At the same time it illustrated again, in a graphic manner, how little actually goes to waste in nature.

Through the transparent integument along the dorsal surfaces of these living collembolans, the observer also discerned definite internal waves as the slow pulsations of their heartlike organs or "pumps" bathed their more or less open interiors with nearly transparent blood. Highly animated arthropods, these collembolans, yet animals that require no elaborate networks of arteries, veins, and the myriad capillaries which are mandatory in us warm-blooded organisms at the top of the animal ladder.

At the time these colonies of bark-dwelling collembolans were under constant observation, microscopic threads about a half a millimeter or less in length, each with an opaque droplet at its tip, were discovered with the microscope. Standing up from the bark, they were believed to be mold fruits at the time and not considered to be connected with the insects' life-histories, but later it was found, upon studying the work of other biologists, that these were spermatophores deposited by the male collembolans. The threads were composed of a secretion that hardened on contact with the air, and the tiny droplets contained the spermatozoa suspended in a fluid!

Many of these are said to be deposited by a single male, and the sperms, when encountered by the searching female, are drawn into her wide-open genital orifice as she simply brushes against them, and thus the eggs are fertilized. This mode of reproduction is employed by several other kinds of animals, some of them even in the vertebrate class, but with great variation in the form of the spermatophore.

One wonders what advantages may lie in spermatophore production, for instance, in two such widely separated animals as the collembolans and certain terrestrial salamanders. It seems strange that nature, after evolving the more certain sexual process of copulation, even in many insects, would return to the spermatophore in any backboned animal.

Among the bark fragments in the box with this second collembolan colony, a habitat to which nothing was ever added by the author, except a few drops of water now and then, there were several other kinds of odd little organisms.

Very small flat-bodied insects known as booklice, belonging to the family Trogiidae, inhabited the bark furrows (plate 118), but what their function was, would be anyone's guess.

Several species of mostly microscopic Acarina, commonly called mites, appeared regularly, curious eight-legged organisms belonging to the class Arachnida, which also includes the more familiar, often enormously larger, spiders, daddy longlegs, scorpions, and pseudoscorpions, and other somewhat similar Arthropoda.

The mites constitute an enormous and greatly diversified order, as anyone who studies dead or living plant and animal material with the microscope will soon find out. In almost any kind of material taken from the mini-wood floor, mites were found in many shapes and sizes, but most abundantly in the humus itself (plate 120).

Some mites cause curious swellings on leaves and plant stems, and the young Acarina then feed and grow within these objects, called *galls*. Many kinds of galls are also caused by insects and will be treated in a later chapter. As parasitic species, mites live on birds and mammals, on insects, spiders, and even on ticks and other mites. Other species cause mange; but, as we shall see, many of them do good also, and we will meet with mites many times in these chapters.

Picture, if you can, microscopic round or oval-cut gems,

continually on the move, but on feet often so small as not to entirely spoil the illusion upon viewing some of the more colorful species found in the humus, or on plants and debris. Some of these may be rubellite pink or dark ruby red; others, rich brown or orange, coal black or dark tourmaline green. Still others are red, like carnelian agate, or red with black spots; vermilion; or yellow, like amber and honey. Many of these gemlike types (there are also numerous acarines which look like spiders) are highly glossed, and in shape almost globular, oval, or flattened like cabochons, and most of them have a spot of gleaming highlight on their bodies, thus adding greatly to the gem illusion (plate 120).

The first acarine to appear in the collembolan colony being maintained on bark was an oval bottle-green species with a luminous refraction stripe down the center of the back as brilliant as that in a fine star sapphire. These energetic mites were so small and had such short legs that they sometimes passed under the two-millimeter-long browsing collembolans without even being noticed! The second acarine to appear had a shiny brown rounded body and a barely visible triangular head; the third one was also tiny but possessed of a duller brown undistinguished body.

And now, from somewhere in these fragments of oak bark, two specimens of a species of mite never before seen by the author soon put in an appearance. Very ungemlike and curious in form, very sluggish in their movements, and pinpoint in size, these strange little organisms lived and moved among the threads of a white mold which had appeared in some of the bark crevices. Their bodies resembled five piled-up inverted bowls, with the rim of each decorated with a row of tubercles and hairs, and the "bowls" supported on very long legs as thin as the threads of mold through which the mites moved. We will call them the "topshaped mites" for lack of a better name (plate 120).

Slower in gait than any snail, they moved by dragging one hairlike leg after the other, slowed up the more at times by mold threads entangled in their own body hairs so they could make little progress for minutes at a time.

And now hear this!

A fifth, unidentified species of mite, so very minute that we will refer to it simply as "number five," came upon the scene with the top-shaped species just described. A mere orange speck of a thing with a globose body, it was merely by chance (when discovered with the stereo microscope) that it was recognized as a living animal and not some atom of inanimate material.

It came *riding* into the picture clinging to the tip of one of the top-shaped mite's longest abdominal hairs! (Plate 120.)

Here again was something new, another unsuspected or even imagined bit of life which inhabited the mini-wood. What a slim chance one would have of ever finding such an organism again in a purposeful search! Symbiont, parasite, commensalist? Who can say?

By now the reader may feel that the author has been rather far diverted from what the chapter heading implies. But this is not really the case, for the collembolans, constituting a huge group, and the acarines also, are both animals of the soil in truly fabulous numbers, and the two species of the former which were successfully bred—the first upon moldy paper, the second on bark and algae—have been allotted much space for the reason that their life histories, as far as reported, are probably quite like those of the other gregarious, and much more numerous, strictly ground-dwelling species.

The stereo microscope revealed many kinds of mites (Acarina) in the series of small petri dishes containing layers of humus taken from just below the dead-leaf cover in the mini-wood.

Here were usually numerous long-legged, fast-moving, and very spiderlike species—pink ones, white ones, pink and yellowish ones, and very minute individuals so transparent that they looked as if they were made of glass. All seemed very busy, continually on the move, some with such speed that they had to be temporarily anesthetized with ether in order to be studied at all. Most of them appeared to be scavengers, and none of those in the dishes where other, predatory organisms were also living ever seemed to be attacked. It would seem that one of their major functions was to serve as natural consumers of dead animal matter, and we will come upon them at their work many times in this book.

Very different again were tiny slow-moving mites, some of which were doubtless parasitic on insects, spiders, and higher forms.

In larger petri dishes containing greater depths of humus taken at random from the mini-wood, there were often rove beetles of the family Staphylinidae, avidly carnivorous species which fed to a great extent on other insects but which left the mites unmolested. This may have been because of a mucus or slime which many of these acarines seem to emit under stress.

Rove beetles found in the humus were mostly black fast-moving burrowers, a species of *Tachinus* five millimeters (about three-sixteenths of an inch) in length, with elongated, cone-shaped abdomens. Larger species found in the mini-wood measured up to twenty millimeters (nearly an inch) in length, but only about four millimeters in width. Their larvae were found in the dishes, but the adults were sometimes seen flying into the woods to burrow under carrion or animal feces, there to lie in wait for their prey: flies and other insects. Black or grayish black or brown, these beetles often walked with their abdomens held aloft, as if bluffing a sting. *Tachinus* of the cone-shaped abdomen sometimes captured and ate collembolans, despite their usually life-saving catapults.

Another tiny beetle, two millimeters in length, was also found preying on collembolans. Plain black except for a round white spot on each wing cover, it was discovered in association with small black snails, oniscoids, collembolans, and millipedes and centipedes, all of which had gathered in blackish debris or a sort of soil that had accumulated in narrow spaces between decaying timbers in a pile of discarded lumber within the edge of the mini-wood. In addition to the organisms mentioned, there were also a few small earthworms. All were typical animals of the soil which had crawled up from the humus below to populate this niche and its soil of sorts in the making.

It was an interesting enough occurrence in itself, but later, when the material had been collected and put in petri dishes for thorough examination, a collembolan was found which was also black and had two white spots, one on each side of its globose abdomen. The species was *Sminthurinus minutus* f. *bipunctatus,* a long but nicely descriptive technical name for so small a creature. It had only been found previously by the author in the humus of the mini-wood.

In this little community of living things from the lumber pile, it was of great interest to the observer (although maybe of no real biological significance) to find associated these two widely separated species of tiny insects, one a predatory beetle, and both with globose black abdomens bearing similar round white spots in identical number and position.

Had both been diurnal insects rather than species living in total darkness, one might have speculated that the color and pattern in the collembolan indicated a case of protective mimicry useful in the same habitat with an agile, sharp-mandibled, carnivorous little beetle. Whatever the truth may be, it is an interesting case to think about in retrospect.

A remarkable fact was that this prolific debris in the lumber pile had accumulated in a single spring and summer, *after* the pile was made. White hyphae, the mycelium of fungi resembling delicate threads, had penetrated some of the rotting timbers, but no lichen or moss spores had germinated to help the accumulation along. Under the microscope, the debris was found to consist of chewed-up wood cells and pellets of this material, fragments of insect and oniscoid bodies, and the defecated pellets of earthworms, insects, and other organisms. There were, of course, also bacteria and windblown dust in this mixture, which was not infrequently wetted by seeping rain containing additional dirt brought down from the atmosphere.

Here in this material and its inhabitants, as found, was a clear illustration of how the viable little animals of the woodland take over not only litter on the forest floor, but also *any* accumulation of debris, and by thus establishing little, growing communities, commence the work of turning waste matter into useful media almost at once.

Small living creatures are thus doing their useful jobs everywhere in the humus and soil and litter. Many times in the past and during these studies, the author stopped to scoop up soil and humus in other localities at random, subjecting it to the funnel apparatus upon reaching home, and found it always yielded micro animal forms of one sort or another and of enormous interest.

Realizing the importance of the army of the soil, it now calls for two more whole chapters on the subject. To these myriad minor animal forms, human beings and the earth itself should always feel grateful.

CHAPTER XI

ARMY OF THE SOIL II

Dwelling in the rich humus and litter of the mini-wood, and doubtless more or less numerously in undisturbed woodlands all over New England, are many interesting species of myriapods, some of them familiar, but others, although occurring here in fabulous numbers, rarely seen, and even their presence usually unknown to others than naturalists.

Technically known as the *Myriapoda,* the four recognized classes include such animals as the centipedes and millipedes, popularly known as the "hundred-leggers" and the "thousand-leggers," and many less familiar inhabitants of the humus and woodland debris. The subclasses and orders of myriapods, and also the insects, which belong to a separate class but may have descended from the former, are among the most successful groups of arthropods. Collectively, the myriapods and the insects are now referred to as the *terrestrial mandibulates.*

How confusing the popular names for organisms may be at times is here illustrated, for while millipedes have many pairs of legs—sometimes over a hundred—they never have a thousand, while centipedes may have as few as thirty, or as many as a hundred and eighty.

A millipede's body consists of many double or fused segments called *diplosomites,* each of which is enclosed in a continuous skeletal ring, or, as in the case of the *Polydesmus* millipedes found commonly in the mini-wood, in two such rings. The diplosomites may overlap to some extent, thus affording protection to these myriapods.

Centipedes are much flatter in form than millipedes, and in the centipedes the upper surface is more heavily armored, with dorsal plates called *terga,* than the underside, but centipedes' bodies lack the encircling skeletal rings of the millipedes.

The most obvious external difference which the casual observer might notice upon examining centipedes and millipedes would be the arrangement of the legs—one pair to each somite in the centipedes, and two pairs to each diplosomite in the millipedes. The common centipedes usually seen here in the northeast have rather long antennae, and their last pair of legs, which are the longest and which are directed backward, might be mistaken for the former appendages. Centipedes belong to the class Chilopoda; the millipedes, to the class Diplopoda (plates 121, 122, and 123).

The latter are slow moving, despite their extra pairs of legs. Mostly herbivorous or subsisting on debris, they have little need for speed: their bodies seem distasteful to most predators, and some of them coil up and lie still when disturbed. None of them have powerful biting mouth parts like the predatory, carnivorous centipedes which turn at once if captured and try to bite with their more or less "poisonous" fangs.

Few realize what enormous numbers of millipedes may inhabit a two- or three-inch layer of humus in an undisturbed woodland habitat, a realm of total darkness which, because it is protected from extremes of heat and cold by the annual leaf-fall, offers a microclimate very different from that known to animals tolerant of sunlight, and to which such great numbers of others have become adapted.

Millipede abundance is probably generally unrealized

for the reasons that most of them are very small burrowing animals of the humus, under or within decaying and crumbling wood, or hiding in moss and damp woodland debris. In the author's mini-wood there were often a dozen or more in a handful of the humus.

The most abundant forms varied in size from tiny things measuring two millimeters (a little over a sixteenth of an inch) in length, through many larger sizes up to twenty-six millimeters or a little more than an inch over all. The smallest of these myriapods were no greater in diameter than a horsehair; the largest, wider than the smallest ones were long (plate 121).

In the half-acre of moist woodland under examination, there must have been tens of thousands of these millipedes of various sizes and patterns, all endlessly shredding and ingesting dead vegetation and other woodland detritus of which, in the passage through their long tubular bodies, only minute percentages were processed for nutrition before the remainder was returned to the humus, diminuted and probably slightly enriched as well.

Once again we see the importance of these small animal organisms of the army of the soil, when considered collectively. The greatest oak and the smallest violet in the woodland owe much to these reducers of dead matter, as well as to hordes of springtails, oniscoids, earthworms, and mites and other minute arthropoda, some of which even utilize and further pulverize the excrement of the millipedes. In biological terms, these animals are often thought of as secondary decomposers or consumers.

Without the aid of a microscope, some of the smallest millipedes might easily have been mistaken for bits of fungus mycelium or fragments of white rootlets, whereas many other species were plainly visible to the naked eye. Many of those found in the mini-wood were interestingly patterned with colored dots or more oval markings in single rows along each side, and contrasting sharply with the body color. Some of these millipedes were very probably color varieties of the same species, and ranged from translucent white to pure white, or yellowish white to pale amber, or through shades of what the author calls "butterscotch," an easily visualized simile. Other, still larger, species were various shades of brown, and often glossy. Most all were lighter colored ventrally.

To give some idea of the many variations: an almost white millipede had five brown spots on each side; a pale yellowish white species had fifteen spots on each side; and still others, measuring only eight millimeters (about five-sixteenths of an inch) in length and of extreme slenderness, were adorned with twenty-one chestnut-colored spots on each side, beginning on the first diplosomite and ending on the fourth from the last. A graphic view of some of the spottings and colors of specimens from the mini-wood, together with their various dimensions, may be visualized in tabular form on pages 100 and 101. It might be pointed out again here that, in examining the leg arrangement in the millipedes, it will be seen that the first three diplosomites carry but a single leg on each side followed by two on each side of the remaining diplosomites.

The largest, most distinctive, and most heavily "armored" species found in the mini-wood were those of the genus *Julus,* measuring up to thirty-eight millimeters (about an inch and a half) in length, and with a total of 170 legs. These millipedes seemed to have no enemies, and doubtless possessed protective body fluids or secretions, but certain of the immature "armored" *Polydesmus* millipedes were occasionally seized and partly consumed by small black ground and rove beetles which also inhabited the humus.

For easy obtaining and observation of the latter millipedes—these were much smaller specimens measuring up to seventeen millimeters in length—it was found that cobs from which the boiled corn had not all been cut off made excellent lures, if placed on the woodland floor and covered with dead leaves.

Young centipedes kept in captivity were abnormally hungry and occasionally killed tiny millipedes, immobilizing them with their fangs and then imbibing their juices, but not actually eating their victims, tender as the exoskeletons and organs of these smallest myriapods probably were.

Millipedes possessing protective fluids, as mentioned above, are equipped in such species with repugnatorial glands from which the repellent fluid or spray is exuded through openings on certain of the diplosomites. The great abundance of even the smallest forms in the mini-wood humus would seem to indicate natural protection of some sort and, as a general rule, against most of the small predators of the habitat.

Some of the millipedes, especially those of the genus *Julus,* had the odd habit of coiling up and remaining motionless when disturbed, perhaps as a protective measure against toads or woodland frogs and salamanders, which seem to depend upon motion in an organism in order to recognize it as food.

Watching highly magnified millipedes in motion is a fascinating experience. The *Polydesmus* illustrated in plate 121, for instance, a rich glossy dark brown species with paler coloring between the diplosomites, when moving on its sixty-six legs (fifteen pairs and three single legs on each side) looked as though it were being "rowed" along as small groups of pairs of legs moved one after another, giving an illusion of *one small wave of legs* moving from one end of the organism to the other. Watching this and remembering that such power "waves" were in operation on both sides of the animal at once, the observer gained realization of what wonderful little "machines" for moving and forcing their way through the soil these millipedes really were.

The *Polydesmus* species were interesting to the author also because of their rather short bodies consisting of from eighteen to nineteen trunk diplosomites, each covered by two skeletal rings and flat "plates" or *carinae*, which produced an appearance suggestive of some prehistoric monster in miniature; as far as the age of millipedes as a group is concerned, we are indeed dealing with animals of very ancient lineage.

The millipedes of the order Juliformia were quite different from *Polydesmus* from about every point of view exteriorly, being distinctly tubular, and with the diplosomites hard and often glossy. Measuring around thirty-eight millimeters (about an inch and a half) in length and three millimeters in width at maturity, those from the mini-wood moved on 170 legs (41 pairs and 3 single legs on each side) a count difficult to make and one which suggested again that millipedes might better have been called centipedes instead of giving that name to the chilopods or true centipedes, which, as we know, have fewer legs, comparatively.

These millipedes of the order Juliformia (genus *Julus*) were equipped with *ocelli* (eye clusters) on each side of the head that were composed of many small simple eyes. How well such myriapods see is a question. Many permanently humus-inhabiting species are eyeless, whereas *Julus* with its many eyes may sometimes be seen traveling by daylight in the open, which may or may not have significance. In any case, they probably see only light and dark forms, but not as clearly as higher animals.

Appearing thus in daylight, sometimes in considerable numbers and in several localities at the same time, they have naturally been thought to be more numerous than the myriad humus dwellers. Actually they are far outnumbered by these smaller myriapods.

On rare occasions, numbers of *Julus* millipedes have been seen ascending the outside of foundations and the sides of buildings, harmless events which, however, usually bring worried calls from householders asking naturalists and museum curators for an explanation, but for which there seems to be no answer.

Many of the millipedes from the humus which measured but two or three millimeters (less than an eighth of an inch) in length were white, or translucent in the brilliant light thrown by the microscope mirror. As most of these seemed to have only three legs on each side, they were assumed to be young specimens not long out of the eggs. At this stage they might easily be mistaken by the hasty observer for the larvae of minute insects, as these also have six legs at times.

Strangely enough, during all of these investigations the author was unable to obtain the eggs of these myriapods, but it is of course well known that some species of the genus *Julus,* and possibly *Polydesmus,* burrow in the humus and there construct more or less spherical cavities which they line (by adhesive plastering) with earth crumbs. The eggs are deposited in these "nests," which are then abandoned, and the young are allowed to take care of themselves in a habitat rich in hungry predators.

As has been frequently stated—and will be many more times in this book—under natural conditions little goes to waste in nature. In one of the author's study colonies of soil animals, the dead body of a *Polydesmus* millipede had not lain thus for long before being taken over by pink-bodied acarines or mites resembling long-legged fast-moving spiders, but of course of microscopic size. Soon they had gathered in numbers and were feeding *inside* the tubular shell. When they had consumed all juice and tissue to their liking, and had scurried off to pastures new, a mold soon took over, enclosing the abandoned remains in a network of white threads from which a little later the slenderest imaginable stalks arose, each bearing stellate tufts of spore-bearing organs at the tips, and indicating that this tiny plant had fed successfully on the remains of the millipede after the mites had finished their meals.

At length, the tough integument of the dead myriapod's diplosomites gave way. The rings fell apart, bacteria

TABLE 1

DIMENSIONS, LATERAL SPOTS, AND BASIC DIPLOSOMITE COLORS OF 37 HUMUS-DWELLING MILLIPEDES* OF THE MINI-WOOD

Millimeters†	Spots and Colors‡	Basic Body Color
1.50 X .50	o	white or yellowish white
2.00 X .50	o	white or yellowish white
3.00 X .25	5 reddish brown	yellowish white
3.00 X .75	19 black	light butterscotch
4.00 X .50	14 brown	light butterscotch
4.00 X 1.00	o	white
4.00 X 1.00	9 reddish brown	pale butterscotch
5.00 X .50	13 dark brown	pale yellowish
5.00 X .50	18 reddish brown	light butterscotch
5.00 X .50	19 reddish brown	very pale yellowish
5.00 X .60	15 dark brown	pale yellowish white
5.00 X 1.00	o	very pale yellowish
5.00 X 1.50	14 black	pale butterscotch
5.00 X 1.50	15 brown	pale yellowish white
7.00 X .25	25 dark brown	light butterscotch
7.00 X .75	23 reddish brown	light butterscotch
7.00 X 1.00	o	very pale yellowish
7.00 X 1.00	13 reddish brown	honey yellow
7.00 X 1.00	22 reddish brown	light butterscotch
8.00 X .50	20 chestnut brown	yellowish white
8.00 X .75	25 dark brown	pale yellowish
8.00 X 1.00	19 reddish brown	very pale yellowish
8.00 X 1.50	26 reddish brown	butterscotch
9.00 X .50	21 reddish brown	butterscotch
9.00 X .75	30 reddish brown	light yellowish
9.00 X 1.00	26 light reddish brown	very pale yellowish
10.00 X 1.00	24 reddish brown	butterscotch
10.00 X 1.00	29 black	brownish yellow
10.00 X 1.50	27 reddish brown	brownish yellow
10.00 X 1.50	31 reddish brown	brownish yellow
11.00 X 1.75	26 reddish brown	butterscotch
11.00 X 1.75	28 reddish brown	butterscotch
12.00 X 1.00	28 reddish brown	butterscotch
12.00 X 1.00	34 dark brown	butterscotch
12.00 X 1.75	31 reddish brown	brownish yellow
13.00 X 1.00	32 dark brown	butterscotch
14.00 X 1.25	32 reddish brown	light butterscotch

---

*Includes several of each variety examined and measured. See plate 121; see also Table 2, p. 101, for dimensions and number of legs of other, mostly large, millipedes of the mini-wood.

†Small fractions eliminated.

‡Number and color of lateral spots on each side.

TABLE 2

DIMENSIONS, LEG ARRANGEMENTS, AND COLORS
OF UNSPOTTED MILLIPEDES FOUND IN THE MINI-WOOD

Millimeters	No. of Legs to a Side	Basic Color
12.00 X 1.00	30 pairs + 3 single legs	butterscotch, darker below
15.00 X 1.00	45 pairs + 3 single legs	dark brown
15.00 X 1.50	56 pairs + 3 single legs	dark brown
33.00 X 2.00	38 pairs + 3 single legs	very dark brown
38.00 X 3.00	41 pairs + 3 single legs	dark brown

no doubt moved in, and soon there was nothing left to see even with the microscope. Everything had been reused, recycled.

It has often been stated that if it were not for the birds checking insect life on the foliage, there would eventually *be* no foliage. We should remember also that, without the armies of soil animals, there would *be* no suitable humus or soil in which many woodland plants could grow. And it should be remembered also that millipedes and most—if not all—of the other small animal soil-makers were here doing their job on earth millions of years before the birds (as we know them today) put in an appearance. Millipedes, for instance, have been found fossilized in amber (ancient tree gum) of Oligocene Age, while others, related to existing genera, have been recorded from Carboniferous Age deposits.

Let us now look again at the centipedes of the mini-wood and its environs, and compare them with their slow-moving relatives, the Diplopoda, or millipedes.

Centipedes are the myriapods constituting the active class Chilopoda, avidly carnivorous little creatures, and perhaps the most to be avoided by some of the other soil animals, especially the smaller earthworms.

The commonest member of the class in and about the mini-wood and elsewhere in New England is a robust, fast-moving species of *Lithobius*. Agile, alert, determined when hungry, disappearing from view in a flash when uncovered, *Lithobius* is a dominant soil, refuse, and woodpile inhabitant (plate 122).

Around thirty millimeters (about an inch and an eighth) in length when mature, its form consists of nineteen flattened somites three to four millimeters in width and varying in linear dimensions (plate 122).

Set low on thirty powerful legs (fifteen to a side) the animal is of a rich chestnut color with a high gloss. The last two legs on either side are directed backward and seem to be useless for locomotion, but unhindered by their position, *Lithobius* travels with amazing speed on the other twenty-six. Equipped with very sensitive antennae, eight shiny little eyes in a double row on each side of the head, poison glands, and hard sharp fangs with which to quickly immobilize its prey, *Lithobius* has become a fearsome little dragon in its realm. Indeed, a close look at its details through the stereo miscroscope made at least one observer thankful there were not beasts of this nature, of, say, the length of *Brontosaurus,* alive and free, roaming the earth today!

Observing a number of them in captivity, one noticed the long, many-jointed antennae in almost constant motion and, except when the animals were resting or "sleeping," investigating and alertly probing the ground ahead. At the slightest indication of danger, the head and forward part of the body were pulled back instantly, antennae quivering, one might imagine almost as if the little animals experienced something akin to excitement.

Again, studying living specimens of *Lithobius* through the microscope, the author quickly realized with respect what marvelously evolved and constructed little organic engines for their mode of life they really are—protected by tough somite integuments and efficient poison, and perfectly adapted for speed and, when necessary, pursuit and capture of prey.

These centipedes sometimes roamed overland at night, as well as in the darkness of the forest litter, rotting logs, and woodpiles. In the daytime, they always remained hidden under loose bark, flat rocks, or refuse. When suddenly uncovered, *Lithobius* disappeared like a reddish brown streak. Difficult to capture without the aid of a long, blunt-tipped, delicate forceps, it invariably tried, once it was thus held, to employ its poison fangs while violently twisting and struggling to be free.

While it may be found in most good-sized rotting logs or in fallen tree trunks in undisturbed woodland, purposely made woodpiles left standing in the open seem to have produced favored though somewhat artificial habitats for *Lithobius*, for the obvious reasons that such piles lure and hold large numbers of edible small prey and at the same time afford suitable shelter and multiple avenues of escape. In the outdoor, uncovered woodpiles, we have odd little organic communities in the making.

The poison of *Lithobius* might cause considerable irritation, like that of a bee sting, should the centipede succeed in inserting its fangs in a human being, but its strongest urge when uncovered is to escape, and it would bite only if deliberately seized in unprotected fingers. On the other hand, in its natural environment, its poison is lethal to small soil animals, including earthworms, which are a favorite food of this myriapod. Predators such as *Lithobius* serve a particular purpose in their realm. Just as the cougar may serve to keep deer and smaller mammals from overpopulating an area and depleting its food supply, so the centipedes may help to control the hordes of certain other animals of the humus and soil.

Centipedes in the mini-wood became active under the dead-leaf cover, in refuse, and in the upper layer of humus with the first good thaw in spring. After winter dormancy spent in burrows or under other shelter, they were sometimes active early in March. Even quite young *Lithobius* centipedes, pale in color, appeared this early, probably having wintered in an immature stage.

In the author's captive colonies, consisting of several adults living together among fragments of rotting wood, young about one quarter the length of the adults also put in their appearance early in March.

Centipedes are hatched from egg clusters. The male deposits a spermatophore on a loose web of silken threads. This packet containing the spermatozoa is taken into the female's genital orifice after various courtship motions on the part of her mate, the internally fertilized eggs being deposited later in small clusters, and the young centipedes taking care of themselves soon after emergence.

The above method of sexual reproduction (these procedures following meeting of the sexes) has been assumed by the author to be similar in other species of centipedes of the order Geophilomorpha, and about which more will presently be said concerning those found in the mini-wood.

Newborn *Lithobius* centipedes have only seven pairs of legs, the additional other pairs being added successively as the animals grow. In individuals measuring only eight millimeters (about five-sixteenths of an inch) in length in the captive colonies, it was observed that the last two legs on each side (which are directed backward) seemed to serve as secondary feelers, useful as tactile appendages when the owner was backing away from danger.

Much of the centipede's internal body space is occupied by an efficient digestive tract, a rather simple, unconvoluted tube leading from the mouth to a slender esophagus, followed by the much wider midgut, which traverses the length of the body to a much narrowed hindgut and rectum. One or more pairs of salivary glands pour digestive secretions on the food particles during their long passage. The above being but the merest outline of the centipede's digestive system, for details the reader is referred to *Invertebrate Biology* (see the list at the end of the book).

Lithobius is a ravenous predator even when far from full grown. One such young one was observed seizing one of the tiny brown oniscoids (pill "bugs") described in an earlier chapter, immobilizing it with its fangs, and then feeding upon the victim's juices. Prying up one of the oniscoid's dorsal plates, this centipede appeared to be *sucking* the juice and the soft tissues thus exposed.

As already mentioned, small earthworms seemed to be a favorite food of *Lithobius*. When especially hungry, these centipedes completely consumed the worms at times, discarding only the soil which had been in the victims' digestive systems. The feces of many of the centipedes which were kept in glass jars were seen to be very black and moist, this color probably due to the small amounts of soil ingested along with the meat of the worms, upon which living food these centipedes were maintained for many months in each jar. The containers were of course well ventilated, although of necessity kept quite damp. The living inmates were supplied with strips of decaying bark for shelter, and being strictly creatures of the dark, they would turn instantly from the light as the jars were rotated.

Only small earthworms, not much longer and only half as wide as the predators, were eaten by these captive specimens of *Lithobius*. Large worms seemed to frighten them and were left strictly alone.

White worms (Enchytraeidae), which are very abundant in humus and which are treated more fully in the next chapter, were also occasionally eaten by *Lithobius* centipedes. These worms (sometimes called "potworms" for reasons unknown to the author) occur in many sizes but are all dwarfed by the earthworms, yet, together with these comparatively giant relatives, they turn dead organic matter mixed with mineral particles into the clayey excrement so valuable to the growth of plants. In eating the enchytraeids (plate 124-6), however, the centipedes evidently leave this useful material unchanged, or maybe enriched by the passage through their digestive systems.

Many *Lithobius* centipedes inhabit woodpiles, living in the spaces beneath bark strips loosened as the logs begin to decay. Such habitats in the mini-wood were also favored by terrestrial isopods *(Oniscus)*, beetle larvae, ants, and small earthworms, thus offering a ready food supply as well as excellent shelter for the centipedes.

In view of the fact that, while these centipedes are carnivores, they do not have to *chase* their food organisms, and also the fact that they appear to have few if any predator enemies themselves, one wonders why nature endowed them with such great speed.

In one of the dozen colonies of mixed myriapod species kept in observation dishes, a *Lithobius* centipede attacked a slow *Polydesmus* millipede. This was on one occasion only, and may have been because of the absence of more suitable game or because small worms were not supplied often enough by the author. It proved, however, that these somewhat soft-underbodied diplopods may be utilized as food by centipedes when other more desirable organisms are absent. (*Polydesmus* is shown in plate 121.)

Seizing its victim in its poison fangs and puncturing the integument just in back of the head, the centipede then chewed a hole through the chitinous skin, and was watched through the microscope apparently enjoying the released colorless blood before consuming parts of the body and leaving it in several pieces.

Now, again, the pink, long-legged acarines (mites) followed the centipede, reminding the observer of hyenas rushing in for their meager gleanings after the lion has eaten to repletion.

Finally, as we have seen before, bacteria and fungi completed dissolution of the *Polydesmus* remains. Once again, nothing had been wasted or lost to the woodland humus, here represented by the organisms in the observation dish.

In colonies of eight and ten, *Lithobius* centipedes, purposely confined together upon damp fragments of bark, surprisingly all seemed to live in apparent harmony. Fed only eight times on small worms of various species collected from the humus, most of them remained vigorous and healthy up to five and six months, while a single individual had been kept for 225 days when it escaped and was not seen again, despite a diligent search of the laboratory. This was an unfortunate accident ending a possible record of longevity for the species. Many times the author had watched this centipede with admiration as it subdued an earthworm thrashing from side to side when first attacked.

To illustrate once more the extent to which *natural* conservation may go at times, there existed in the mini-wood community certain microscopic mites with highly convex, glossy yellowish bodies, whose strange habitats were the *edges of the mouth parts* of these *Lithobius* centipedes! Here they doubtless picked up a living on smears of escaping juices or uneaten particles of tissue as the chilopod cut up and chewed its prey, and in return for this sustenance, the mites served as living "napkins." Technically another case of symbiosis, it seemed particularly interesting to the author.

It seemed equally interesting and remarkable that well-armed and aggressive little creatures such as these centipedes lived so peacefully together in the confined quarters of the glass jars. Even in midwinter, when for considerable periods worms for their maintenance were not available, these chilopods never attacked one another, although ravenously hungry, as attested by the way they instantly attacked worms when these latter finally were supplied.

Other very interesting species of centipedes, or in some cases perhaps color varieties of the same species, were found in the mini-wood dwelling in the humus or under rotting logs.

Like *Lithobius*, all of these possessed fangs capable of ejecting poison with which to quickly immobilize or dispatch their prey.

A much less often found species of centipede, measuring around twenty millimeters (thirteen-sixteenths of an inch) in length, bore much flattened tergal (dorsal) plates which in predominant color closely resembled butterscotch lozenges, a simile often to be used in this work. (Some may laugh at such a simile, but in a book of this kind, using familiar things for color comparison seemed to the author better than employing the more cryptic terms used in technical descriptions.)

This centipede was a member of the order Scolopendromorpha, an animal with forty-two legs (twenty-one on each side), and although a burrowing species, was equipped with four *ocelli* (tiny eyes) on each side of its head.

Only a few of these centipedes were found, and, at long intervals, in the mini-wood, and it is possible that they are rare elsewhere, as most of those in this order are tropical. Altogether a graceful little animal, it moved with the greatest of ease, almost as if "flowing" in and out among the humus particles.

Amorphous as this medium may look to the naked human eye, when viewed through the stereo microscope, it is seen to be made up of great numbers of separate clods and pellets thrown together helter-skelter and interspersed with unidentifiable fragments of animal and vegetable matter, as well as mineral grains which are particles from disintegrating rocks. Some of the latter are colored; others are clear quartz. Magnified, they become rocks themselves, while the clods and pellets and other items in the mixture will be seen (in fresh humus) to be coated with mere films of water, deep enough, nevertheless, for protozoans to swim about in unhindered.

The investigator with a microscope will also be surprised to find that plenty of space, and therefore air, exists among the clods and pellets and other particles, and in these the animals of the soil army, whether herbivores or carnivorous predators, move and "breathe," eat and breed, and do what else they must to maintain themselves.

Moisture is an absolute essential in this realm. If the humus dries out, its fauna die unless they can escape in time into temporary habitats where moisture still prevails. In the observation petri dishes, the inmates died very quickly if the medium was not kept properly dampened.

On the other hand, some animals of the humus (which often becomes temporarily rain saturated) may withstand submersion for surprisingly long periods. In using the funnel-collecting method already described, a layer of water extracted from the humus usually accumulated in the collecting bottle attached to the apparatus. Submerged in this water, certain acarines (mites) lived for hours, some of them secreting an apparently protective mucus about their bodies. Some millipedes did not drown easily either, but centipedes, which mostly live in sheltered situations, quickly died in water. Oniscoids and abundant white worms, campodeans, and several other soil animals (to be described later) lived for varying periods submerged, while many springtail species and some of the mites were so small and so light that they simply hopped about the collecting bottle without ever breaking through the surface film of water.

Rather slow-moving, long, and very slender centipedes found in the mini-wood belonged to the order Geophilomorpha. Species in this order may have from 31 to more than 180 pairs of legs (1 leg on each side of each somite). Specimens found by the author measured up to fifty-three millimeters (a little over two inches) in length, and barely two millimeters in width. Their bodies consisted of forty-nine somites, a pair of legs to each, with the posterior pair directed backward and not used in walking (plate 123).

These *blind* centipedes, when in motion, seemed to be gliding along, so smooth was their progress—a delight to observe—with only their antennae guiding them, as a seeing-eye dog guides a blind human being.

In color, this species was light yellowish brown, with a much paler narrow line along its central dorsal surface. With the microscope, this was seen to be the animal's heart, a long tubular organ just below the surface of the somites, and through which the almost colorless blood was being moved by regular and clearly observed pulsations.

Patterns on either side of this lighter dorsal area consisted of distinct dark markings, seen only with the microscope and as more or less half-moon-shaped. On these animals there were also darker areas visible in places which, when magnified, were seen to be fine individual lateral markings. The head and antennae, and the last somite with its appendages, were a somewhat darker glossy yellowish brown than the rest of the body in this species, while all of the walking legs were very light colored and unmarked. Highly magnified, the hard fangs were seen to be tipped with jet black.

Imagine now, if you can, what this creature would be like had it been fashioned by evolution in feet instead of millimeters. Imagine a horrendous dragonlike creature fifty-three feet in length and two feet in width, a nightmare thing rushing at you on ninety-eight powerful legs, poison fangs spread two feet apart, ready to grab you before you could run ten yards. It is indeed a good thing for many of the other animals of the soil that they do not possess intelligence, or eyes to see as ours do!

Most beautiful of the geophilid centipedes found in the mini-wood, but very rare in the experience of the author, was a chilopod fitting the description of the former, except for color. Of a solid coral red shade, it was a delight to look upon, but very difficult to find. Its smooth, steady glide upon not quite a hundred legs illustrated perfected organic coordination in an animal supremely confident of its antennae. Little thicker than a coarse shoe thread, it normally moved in a straight line, fully extended or with its body gracefully undulating, according to the roughness or smoothness of the substrate. Vulnerable, from our point of view, easily injured in the forceps, this centipede was, nevertheless, a carnivorous hunter, which the author never observed being itself attacked (plate 123).

Myriapods "breathe" through tiny ports, technically called *spiracles,* situated laterally on the somites or diplosomites, and which are the openings to elaborate systems of tubes or *tracheae* by which all parts of the body are reached. The coral red geophilid centipede, for instance, was equipped with a pair of spiracles on all but the first and last somites. The blood of most of the soil animals, and that of the great class of insects, including those living in the soil, is aerated by such tracheal systems.

During the concentrated hunts for these centipedes, the author was not only rewarded but astonished on rare occasions to find specimens that were piebald!

Measuring ten millimeters (about three-eighths of an inch) in length, and less than a millimeter in width, exactly two-sixths of their bodies anteriorly were of a rather pale butterscotch color; the next three-sixths, coral red; and the posterior sixth, again the yellow candy color. Very unusual myriapods!

There were still other mini-wood centipedes in this extraordinarily slim category which measured from seventeen millimeters when immature to thirty millimeters (somewhat over an inch) in length when adult. They had eighty legs (forty to a side), were the color of light opaque amber on most somites, and darker on others.

Whether or not these various specimens were separate species was not determined. In the opinion of the author, some of those which seemed identical except for color were most probably simply color varieties of the same species.

The piebald individuals, on the other hand, may have been very interesting mutations. These latter specimens lived only a month in captivity, refusing to eat the tiny white worms offered. Not expecting such a tragedy, the author found, by the time their dishes were opened again for inspection, that they had disintegrated under the onslaught of spider mites, and no others were found and preserved or photographed.

Magnificent objects for viewing through the microscope, many tougher centipedes were kept healthy in the smallest petri dishes on thin sifted layers of moist humus, being fed on the smallest white worms available. As mentioned above, only the mutant rarities died before they could be properly studied or photographed.

It was a happy fact for the observer, however, that these long, slim centipedes did not try to escape over the rims of the dishes when the covers are removed; thus they could be watched for considerable periods even under strong illumination.

Like so many other animals of the soil army, centipedes were found to be fastidious creatures. Often they were seen stopping other activities to lick or brush a leg here, a limb there, or to remove some bit of foreign matter from their bodies. At other times they would bend back an antenna to thoroughly cleanse each joint of this all important appendage. An individual with an injured leg on a central somite turned its head and half of its body repeatedly backward to the exact location of the trouble, then, with all the apparent concern and care shown by much higher forms of life, tried persistently to straighten out the fractured appendage with its fangs.

Often the author wished that everyone might see some of these small creatures of the woodland for themselves, acting as they do in such normal ways, even when in captivity. One gains new respect for the army of the soil and what it means to our forests and ourselves.

Perhaps the cleanliness exhibited by these small fry was their most appealing quality. Merely realizing that these creatures—many of them barely visible even to the best of human eyes—seemed to care about themselves to such an extent, was all that was needed to bring a permanent salute from the author, be their "toilets" consciously performed, or merely instinctive.

Obtaining photographs of normal, living myriapods, that is to say, specimens that had neither been purposely chilled, or otherwise tranquilized for the purpose, was one of the many problems to be met in the preparation of this book.

It was solved with some success with the optical apparatus on hand by selecting as subjects (especially where fast-moving chilopods were concerned) animals living in long-established mini-habitats prepared in jars and petri dishes. Such individuals were accustomed to captivity and therefore calmer than recently obtained specimens, and the covers of the receptacles could be removed with comparative safety for short periods if very cautiously lifted.

Working rapidly, the selected animals were first photographed by flashlight at a distance of 1¾ inches. The color-slides thus obtained were then projected and photographed directly enlarged (in black-and-white) on 5 x 7 super Panchro Press, Type B, Eastman cut films, from which the illustrations were printed, all work except that of processing the Kodachromes being done in the author's laboratory.

Another strange and very small centipedelike creature seen by very few people, but a fairly common inhabitant of the humus in the mini-wood and elsewhere in suitable media, was a myriapod of the chilopod class Symphyla, an animal called *Scutigerella* (plate 124).

These pure white, blind chilopods measured about four millimeters (under 1/6 of an inch) in length, and well under a millimeter in width. When mature, they had twelve pairs of legs, but only six or seven pairs in juvenal stages. Their long, delicate, twenty-seven-jointed antennae, closely resembling strung shining beads, were in constant quivering motion whenever the animals moved about. Their manner of locomotion was peculiar, being a curious sort of go-and-stop, go-and-stop gait, or series of short advances with equally brief pauses between stops, the animals all the while testing the way ahead with their sensitive antennae. Their pure white color made them easily visible to the naked eye against the black color of the moist humus, and it was observed with the microscope that very recently hatched young ones, measuring a millimeter or two in length, progressed with the same jerky motions as the adults.

The male *Scutigerella* is known to be equipped with a pair of spinnerets on the next-to-the-last somite which have to do with very delicate silklike stalks produced outside of the body, at the tip of each of which a droplet of sperm, called a *spermatophore,* is affixed. Many forms of them are produced by various members of the soil army in which there is no actual sexual intercourse between the males and females of the species. In such cases, there is no coition; instead, the male produces the spermatophore and the female finds it and takes the sperm-bearing package or capsule or droplet into her genital opening herself, or as in the case of *Scutigerella,* into her mouth, where the ova are fertilized one by one before being deposited in the humus.*

Scutigerella might be easily mistaken by the uninitiate in soil biology for another white or yellowish white inhabitant of the humus called *Campodea.* Campodeans are primitive insects, not myriapods. Although, like other insects, they have only six legs, their body form and color, and their long, many-jointed antennae, resembling minute beads, do make the two animals look much alike at first sight, but careful examination will reveal that campodeans have a pair of "tails" which are absent in *Scutigerella* (plate 124).

Technically known as *cerci,* they appear to be as sensitive as the similar-appearing antennae at the other end of their bodies. A *Scutigerella* might be mistaken for a campodean that had lost its cerci, as they sometimes are lost during the animal's passage through the funnel-collecting apparatus. (The campodeans are more fully treated in the insect chapter, XIII.)

The author believes that *Scutigerella* sometimes preys on campodeans, the supposition being based on the fact that in petri dishes in which both kinds of animals were kept together on shallow layers of humus, the campodeans were on rare occasions found partly dissected, or with their bodies collapsed, as if they had been sucked dry. *Scutigerella* was never actually observed in this predatory act on this species, and in the dishes in which the latter alone were kept, most of them lived in good health for several months. By close watching, however, it was discovered that *Scutigerella* definitely preyed upon the smallest *brown* pill "bug," *Metoponorthus pruinosus,* described in an earlier chapter, which was so common in the mini-wood humus. The predator was observed attacking the unresisting isopod, immobilizing it, apparently with its fangs, and then chewing an opening between two of the victim's platelike

---

*See Friedrich Schaller, *Soil Animals,* in the book list at the end of this volume.

segments. It then seemed to be sucking out the juices, not actually chewing the tissues. *Scutigerella* in a petri dish with campodeans is shown magnified in plate 124, in a photograph taken from life.

Because of its extreme fragility (softness), the author found it difficult to transfer living *Scutigerella* specimens uninjured from the collecting apparatus shown in plate 116 to prepared petri dishes. In employing this method of obtaining soil animals, however, a shallow layer of water extracted from the humus almost invariably collects in the retaining bottle on the end of the funnel, and from this the animals may be rinsed into a watch glass by means of a fine wash bottle or a medicine dropper. Once safe in the watch glass, *Scutigerella* could often be rescued, before drowning, on the tip of a very fine dissecting needle (bent at a right angle), and then very gently induced to crawl from the needle onto the humus layer in the petri dish. As such rescues must be accomplished with the aid of a microscope and are often time consuming, these very fragile creatures may sometimes appear to be dead when lifted onto the needle, but if they show the slightest signs of life, even the twitching of an antenna, they will usually recover completely once safely on the humus and dried off to some extent once more.

CHAPTER XII

ARMY OF THE SOIL III

We come now to those worldwide members of the soil army belonging to the phylum Annelida, the common earthworms, the far less familiar but also abundant white worms of the humus, and certain other wormlike animals found in the mini-wood but belonging to the phylum Platyhelminthes, the flatworms, a large lower group.

Most of these worms play a part in breaking down the dead vegetation or litter of the forest floor, and thus, of course, in building the humus. Unpalatable and tough as some of this raw material may be, earthworms in particular are capable of coping with it, feeding on dead leaves, including even such tough and acid leaves as those of the oaks, and reducing this harsh material in its passage through their bodies to substances partly useful to other animals of the soil and partly to the woodland's living vegetation.

The word *worm* may conjure up unpleasant thoughts in the minds of many of us—visions of squirming fish bait or night crawlers, or, again, of noxious parasitic species and other enemies of man, but as in other animal groups, among the hundreds of species of worms there are the good, the very good, and the bad ones.

Let us then at the outset dignify the highly beneficial common earthworm with its technical name and status, *Lumbricus terrestris,* oligochaete of the order Opisthopora.

Classic in the study of biology, great numbers of these worms are preserved and sold by scientific supply houses to schools and colleges, where, in class study, they undergo dissection to illustrate invertebrate anatomy and demonstrate bilateral symmetry.

The species most employed in this kind of study and the worm most familiar in our gardens and lawns and woodlands and, in fact, almost wherever there is moist earth or vegetable debris, is *Lumbricus terrestris,* as just mentioned (plate 125).

As everyone who once studied biology in school knew, but which few remember, the body cavity of the worm—the coelom—consists, stating it simply, of two tubes, the outer one forming the body wall, and the inner one the wall of the long digestive tract. This system consists of a muscular pharynx or swallowing apparatus and glands to moisten the food; an esophagus; a food-storing crop; a gizzard wherein the ingested food is milled into fine particles; and a long intestine, where, as in other animals, absorption of nutrients and other functions are carried out before the remainder is rejected in the form of feces.

Some may also remember that the earthworm's circulatory system consists of two main vessels. One of these, dorsal to the digestive tract, carries the blood from the worm's numerous segments forward, and the other vessel, ventral to the gut, carries it back. Students have often been surprised to find that annelids have no definite heart, the blood apparently being circulated by dilations of the dorsal vessel.

It is all vastly more complicated than outlined in the paragraphs above, and earthworm anatomy in detail is beyond the scope of this book.

The extreme anterior end of the worm is called the *prostomium,* and just posterior to this, in the first true segment, called the *peristomium,* the mouth is situated.

In feeding, the fleshy pharynx, situated just behind the mouth, is eversible to some extent, and may be employed to grasp such food items as leaf stems or fragments, bits of grass and other mostly dead vegetation, which, thus secured, can then be withdrawn into the animal's burrow to be consumed in safety (plate 125, center; note mass of worm-excreted soil pellets).

The food is first stored in the crop, where it is doubtless softened by secretions not already added from other glands before being moved to the muscular gizzard, an organic threshing machine which then pulverizes the material, rendering it in part digestible; the remainder, mixed with much ingested soil, is ejected as feces or "casts" (plate 125, top). These familiar objects are sometimes found on otherwise bare ground in the hundreds, especially after a rain. Leaf stems and grass blades protruding from the ground indicate where earthworms have been feeding at night.

Feeding and casting is what the big night crawlers and other worms are doing on the surface after dark. In a single year, the earthworms, taken all together, turn over and somewhat enrich vast amounts of soil, their castings often including much clay brought up from considerable depths, eventually to become mixed with more nutrient substances. In a very gradual manner, stones and other surface objects are thus partly buried while the millions of worm burrows aerate the soil and bring rainwater directly to the roots of vegetation.

Enormous quantities of dead leaves are gradually consumed by *Lumbricus terrestris*. During one September, as a test, the author raked freshly fallen leaves into a deposit seven yards long and a yard wide, and averaging eighteen inches in depth. Left thus, wired off and untouched for a year, the mass had been reduced in depth to an average of five inches by the following September 15.

While some of the loss had of course been due to natural compression, to rain and snowfall, and to other kinds of soil animals, its bulk in the main had been consumed by a concentration of earthworms. Just below the wet and lowest remaining leaves there were literally millions of pellets replacing the more familiar surface "casts" which worms leave in such numbers. Under the worms in plate 125 (center picture) the form of these pellets may be clearly seen.

The softer, evidently much preferred leaves of ash and birch, sassafras, chokecherry, basswood, and sugar maple disappeared completely, and those of the spicebush nearly so.* Only the tough, resistant, and probably less nourishing foliage shed by the beech and oak trees remained mostly intact after a full year. Leaves of the red maples were about 50 percent consumed by the worms and other soil animals. All of those of the three most resistant species *do* finally disappear, given additional time, as everyone knows who has studied the annual leaf-fall.

In late fall, the author found that the earthworms became concentrated in greatest numbers at the bottom layer of the leaves, hence the vast numbers of their pellets. Composed of cellulose, soil, and other indigestible matter, they become ready-made little packets for further exploitation which are utilized by numbers of lesser soil organisms, including bacteria and fungi, the latter two kinds of plant forms aiding in the final breakdown and amalgamation of the remainder into the woodland's nutrient soil.

Earthworms are *hermaphrodite*, both male and female reproductive organs occurring in the same individual. When two earthworms copulate, sperm from each one moves into a seminal receptacle in the other. When the fertilized eggs are ready for deposit, the *clitellum,* a bandlike object or girdle lighter in color than the rest of the worm, secretes the wall of a cocoon containing the eggs, which is slipped over the worm's anterior end and left in the humus or soil (plate 125; cocoons are shown in plate 126). Young worms, which are almost identical with the parents, take care of themselves from the day of emergence from the cocoons.

Most worms are quick to react when uncovered, and their return into the burrows is aided by micro setae or bristles, objects to be seen, only when highly magnified, as very short spines protruding but slightly from many of the segments of the animal's long body. As the worm's muscles contract, the setae evidently act as spike holds on the sides of the burrow, thus aiding withdrawal.

Inhabiting saturated woodland debris, or living among dead leaves which retained thin water films during the spring, small very *pink* earthworms occurred which were probably of a different species. Early in the year there are many such wet habitats in a deciduous woodland: near springs, or in swampy areas where skunk cabbage and jack-in-the-pulpits grow prolifically. Although the majority of the larger worms seemed to drown quite easily,

---

*Notwithstanding the aromatic quality of the spicebush's leaves, they are the preferred food of the spicebush swallowtail butterfly, *Papilio troilus,* for its caterpillars. Neither did the leaf secretion seem to bother the earthworms.

this pink oligochaete was apparently a semi-aquatic species, at least seasonally.

When extended, a typical example measured twenty-five millimeters (about one inch) in length. Periodically, as it progressed, it protruded an exploratory pharynx. Thrashing about from right to left, or scouting straight ahead, it continually protruded and withdrew this bit of its anatomy as if testing whatever might be ahead or on either side; doubtless it was thus able to find what to its simple taste was delicious amongst the soaking dead leaves and other decaying vegetation.

At the same time, and in the same habitat with the curious "proboscis" worm, microscopic crustaceans were found darting about in the water films held on the surfaces of the rotting leaves.

Bearing a single eye in the center of their heads, these animals were of the genus appropriately named *Cyclops*. Familiar to every biologist, the minute creatures progressed in jerky fashion within what we may imagine to them were ample lakes. Sometimes these micro animals appeared in pairs, one clinging behind the other, as horseshoe crabs appear along the Long Island Sound shores at spawning time.

Where these *Cyclops* came from each early spring in the mini-wood, how they fared during summer droughts, and what their function was in this habitat, the author cannot state, but it is logical to suppose that these little crustaceans may be valued food items for some other minute creature of the teeming mini-wood fauna.

Often the female *Cyclops* was found carrying an ovasac on each side, protruding from the first abdominal somite, and through its transparent walls numerous eggs could be plainly seen with the aid of a microscope and water-cell slide. Before the partial drying out of these earlier wettest woodland habitats (that is to say, before the hot summer days arrive), batches of *Cyclops* eggs, if detached, must remain viable but unhatched until favorable conditions return once more, sometimes maybe not until the following spring, when these little animals become common again.

In the soil of the author's vegetable garden, dark olive green earthworms appeared rather suddenly during the studies being made for this book. Later, they turned up here and there about the mini-wood borders, and while they may have been introduced in the soil of some pur-chased plants, such as tomatoes or other vegetables, it is also a possibility that their color might have been due to their diet, which seemed to include the bright green alga *Protococcus,* but not dead leaves, a staple of the commoner earthworms.

These worms stretched out into extraordinarily long creatures at times. Freshly dug, quite robust-looking examples, when dropped into a smooth earthenware crock, extended themselves in feeling for a way of escape into animals 84, 107, and 128 millimeters (over 3, 4, and just 5 inches) in length, and only 1½, 2½, and 3¼ millimeters (1/16 to about ⅛ of an inch) in diameter.

Upon being touched or grasped, these worms wriggled violently, sometimes at once deftly tying themselves into intricate knots, a reaction seemingly of no value, but rendering them easier for a Robin or other bird to carry away! Within a minute or less, they would unwind themselves, stretch out as soon as released, and then search energetically for a way of escape again. At such times, when fully extended, the color of their bodies lightened very much, becoming pale yellowish olive, and the clitellum or girdle in which the egg cocoons are formed standing out in opaque white and contrasting with the rest of the segments.

No lateral body setae (bristles) were discernible with the microscope, which seemed odd in a deep burrowing species, in which cases such structures protruding from the body aid other worms in rising in the excavated passages.

When released in numbers in a clean glass jar, these worms thrashed about at once, at the same time secreting a mucus and churning it into a froth of viscid bubbles in which they soon became partly hidden, a possible protective reaction released by the unnatural condition of many individuals thrashing about together.

In captivity, under normally reproduced conditions of ample acid soil having on its surface fine green moss and the alga *Protococcus,* described in an earlier chapter, these oligochaetes lived well enough for a time while under observation, moving about at night, and apparently feeding on the green stuff supplied by extending and withdrawing their wide, bluntly fan-shaped pharynges, the soft eversible part of the digestive system in these worms. Up to publication time of this book, the author had been unable to locate an authority to whom to send specimens of these worms for technical identification.

* * * * *

When keeping microscopic watch upon some small fragments of dead and rotting leaves taken from a saturated area in the mini-wood in mid-March, a tiny, well-camouflaged wormlike animal was discovered that had not been seen before by the author.

Its body, glistening like polished glass, was dark yellowish brown, like the leaf fragments on which it was found. On the central dorsal surface, a blackish line extended from just short of the anterior end to almost the posterior end. A dusky, indistinct line could also be discerned on each side of this plainer one, which under high magnification was seen actually to be composed of minute blotches, spots, and distinct V's of black pigment difficult to discern through the body highlights.

Somewhat sluglike when at rest, it appeared quite robust for its short length, but upon being disturbed, when its hiding place was exposed, it quickly extended to five or six times its resting length, thus becoming a comparatively slender creature thirty millimeters (more than an inch) long as it glided out of sight again beneath another leaf fragment.

Watching through the microscope as it moved about in its wet habitat (within a mini-petri dish) the animal's anterior end was seen expanding and contracting, continually changing in contour, but always, as it became fully expanded, forming a fan-shaped probe feeling out what might be ahead and at either side, yet periodically being *completely* withdrawn. A startling thing to observe, the action gave the impression of a deflatable object being suddenly sucked into the body, followed by an illusion of it being rapidly pumped up and out again. The fully extended fan-shaped end, when thus probing ahead and from side to side, was so suggestive of a standard household cleaning appliance that the author often thought of the animal as the "vacuum cleaner worm" (drawing, plate 126).

Preferable, for purely personal use, had the author been better informed, would have been to write a properly descriptive Latin binomial as a temporary tag until a published technical identification might be found. It can be stated, however, that this creature of the mini-wood was a free-living flatworm (in this case with a tubular body), phylum Platyhelminthes, class Turbellaria, and probably a species of the genus *Placocephalus*. The sketches in plate 126 show how it looked in various attitudes, moving and at rest, the pale dorsal lines representing highlights.

There are a great many species of Turbellaria in the world, the majority apparently aquatic, and obviously, as we have seen (and will see again in this chapter), not all of them have flat bodies. Flatworms are unsegmented organisms much used in the study of biology and noted for the ability to regenerate both ends of the body when that of certain species is cut in two.

Many are predaceous animals, often, if not always, secreting mucus or slime in subduing their victims, and they are peculiar in having their round or slitlike mouths situated on the ventral surface of the body, usually near the center. One would expect to find the mouth in the fan-shaped "inflatable" anterior end of the flatworm under observation, but this was not the case, which would no doubt be baffling to an uninformed student or other observer.

This turbellarian exhibited another strange capability at times: that of extending its bluntly rounded posterior end into an elongated tubelike probe. An eyeless creature, it would seem that both ends of its body have been evolved into tactile equipment, extendible at will, and possibly as useful to the owner as antennae and cerci are to other kinds of animals. Observing this flatworm through the microscope, it was most interesting to see it extend one end and then the other, from its five-millimeter length when at rest into an animal five or six times as long when on the move and "exploring" its surroundings.

For the 137 days (March 6 through July 21) that this animal was kept under observation, it could not be determined what it was using for food. Apparently still in good health at the latter date, it died very suddenly, first shrinking to an object five millimeters in length, as shown in plate 126c.

This specimen had apparently produced no eggs or egg capsule, and no offspring had appeared during its life in captivity. For a considerable period, no identical specimens were found, but at length a somewhat similar species was recovered from black humus which had been dug at the base of the large angular glacial boulder shown in plate 2, the recovery having been made along with many other animals of the soil army with the funnel collecting apparatus described earlier.

Its body resembled that of the former turbellarian minus the fan-shaped extendible anterior. This specimen glistened also, its surface as smooth as polished glass, but of a dark gray color. The "head" end (it might so be called in this instance, as it seemed to have a central eyespot) was bluntly rounded and employed merely as a probe as the animal moved about among the dead and decaying leaves. Its posterior extremity was similarly blunt and rounded.

When fully extended, this interesting find measured sixteen millimeters, or approximately five-eighths of an inch in length, but only one to two millimeters in breadth.

After many days of confinement in a mini-petri dish containing bits of dead and rotted leaves, it commenced to leave a milky trail behind it wherever it moved. Then, as the hours passed, this secretion greatly increased in volume, and the creature then became motionless, finally lying embedded, as if in a shroud, in this mucus. Twenty-four hours later this turbellarian had completely disintegrated and, it seemed at first, leaving only the milky mass. But now the mucus itself began to shrink, and as it disappeared altogether, it left exposed (as revealed by the microscope) a curious oval object or cocoon two millimeters in length and less than a millimeter in width and, measured from base to the top of its domelike form, a little higher than wide (plate 126d).

This was a flat-bottomed capsule with a neat narrow rim running all around its base, and the whole object was so nearly transparent that it could be looked into and its contents partly discerned when placed under the stereo microscope. The date was June 15. It contained six tightly packed in objects, resembling wheat kernels in shape, and yellowish white in color. After seven days (left undisturbed in its wet surroundings), the inmates had become black at one end and a lifelike pink in the center, but remained the original yellowish white elsewhere. From the time this capsule or cocoon had appeared, on June 15, until the eggs or, possibly, embryo flatworms had developed some color, took just a month. During this period, with the greatest of care, the object was examined from day to day under the microscope, actual contact being made only once or twice with a very fine dissecting needle to remove ultra-minute strands of a white mold which had started up from a leaf fragment near the capsule, whose inmates seemed to be developing nicely.

But then very suddenly, came the tragic end to what might have afforded new and valuable biological data. On July 30 the sides of the capsule fell away, and in place of the young turbellarians which had been expected with no little excitement on the author's part, *hundreds* of extremely minute parasitic worms called *nematodes*, organisms which will be described more fully later in the chapter, poured forth from the devastated and barely discernible remains of the capsule and what could not be said with certainty to have been the offspring of this flatworm rarity, which the author believed to be *Rhynchodemus terrestris,* an organism named by O. F. Müller, and described by L. H. Hyman in 1939.*

On a cold morning (March 26) with the frost just out of the ground in a few places in and around the mini-wood, two specimens of a much larger terrestrial turbellarian were found entangled with one another on the surface of the ground under a slab of bark. These organisms were active despite the low temperature of their habitat which then registered 38°F.

Removed to a petri dish containing soil from where they were found, they lived from March 26 to the first week in July. Then, as in the case of the turbellarian described before, they suddenly disintegrated, but in this instance leaving no cocoon, or fragments of themselves from which in some cases in flatworms new individuals are regenerated.

These were robust animals but very soft. Fully extended, they measured fifty-seven and sixty millimeters (about two and a quarter inches) in length, and in this state becoming much slimmer than when at rest, as shown in one of the photographs in plate 126. In color they were a rich shade of yellow, with a purplish longitudinal line traversing three-quarters of the central dorsal surface. They were equipped with eversible pharanges, but as shown, much less broadly fan shaped than was the probelike object put forth by the "vacuum cleaner" worm described earlier.

Here in the mini-wood, most surprisingly, were these curious organisms, a form of Turbellaria never before encountered by the author anywhere in over sixty years of field work as a naturalist! The species, which was identified as *Placocephalus Kewense,* an animal said in some cases to reach 250 mm., was figured by Ward and Whipple in their classic volume.†

The mini-wood humus, as elsewhere in deciduous forests, was the natural habitat of several kinds of worms and wormlike organisms in addition to the species already mentioned. In employing the funnel apparatus, these soil animals were obtained from most all fillings of this device,

---

*Libbie H. Hyman, "New Species of Flatworms from North, Central, and South America" (Washington, D.C.: Smithsonian Institution, U.S. National Museum, Proceedings, vol. 86, no 3055, 1939).

†Henry Baldwin Ward and George Chandler Whipple, *Fresh-Water Biology* (New York: John Wiley and Sons, 1918).

some as rare single instances, others in great numbers, the latter (white worms), however, varying numerically according to the richness of the humus and apparently to a considerable extent to the plant life in the location tested.

Novel forms with baffling features may sometimes surprise the investigator. One never knows what may wriggle through the strainers.

One such, taken from humus that was very black, rich, and wet, was a slim white species twelve millimeters in length. Its "head" end was equipped with an eversible lance-shaped food-gathering probe or pharynx which could be extended to an extraordinary degree. One-third of the body length consisted of twelve segments resembling closely strung glossy beads with a deep constriction between each two, while the rest of the worm was a smooth tube, undistinguished and bluntly rounded at its extremity.

Another species, measuring sixteen millimeters (approximately five-eighths of an inch) when extended, and then but a half millimeter in diameter, had a suckerlike anterior extremity, and some distance back from its "tail" end, a constriction encircling the body so deeply that it seemed as if the animal must suddenly break in two. This of course, it did not do, even when caught up in a struggling tangle of more normal looking species.

Occurring in the mini-wood humus in great numbers were very small white worms of the family Enchytraeidae. Important members of the soil army, they are seldom noticed members of the class Oligochaeta. Like the earthworms of the same class, they have more or less tubular bodies without true heads, and all of those obtained from the mini-wood appeared to be vegetarians, or detritus feeders. Many resembled bits of transparent plastic thread.

It was not unusual to obtain a hundred or more Enchytraeidae from a single funnel full of rich humus, and it was a very interesting fact that the greatest hauls of these worms were taken from humus where in May and June and July large skunk cabbages had completely hidden the ground over an area of several yards in each direction, but where, by the latter part of September, not a vestige of these bulky plants could be seen, even their thick central veins and stems having completely disappeared from the woods (plate 124-6). Taken with the worms in large numbers were the small brown oniscoids described in chapter IX; thus it is quite possible that these two organisms may aid to a considerable extent in leaf disposal of these plants in such a thorough fashion. The fact that the ground was often saturated where the skunk cabbages grew so pro-

fusely would not be an inhibiting factor in the case of Enchytraeidae, for a great many of the species are naturally aquatic in habit.

When, as often happened, water collected in the retaining bottle of the funnel apparatus trapped enchytraeids when present in numbers, they usually became entangled in writhing balls of struggling individuals which could then be lifted en masse with an L-shaped tissue teasing needle and thus be transferred to petri dishes for microscopic examination. All sizes were found, from truly minute specimens up to worms several centimeters in length. They were mostly viable creatures and, as already mentioned, able to withstand long submersion, an adaptation serving them well during long rainy spells, when the humus became saturated (plate 124).

The author found enchytraeids to be acceptable food items for various species of centipedes; when very small, they were preyed upon (occasionally) by tiny soil animals such as the pseudoscorpions, which are arachnids of the order Pseudoscorpionida, and by the small, somewhat centipedelike myriapods named *Scutigerella,* all odd soil inhabitants already covered or shortly to be described in these chapters.

In regard to the actual numbers of enchytraeid worms inhabiting the author's mini-wood or yours, probably no one could say with any degree of accuracy, for such populations must vary greatly, according to the makeup and the amount of moisture in the forest soil. Judging roughly, from those that have been collected from single quarts of mini-wood humus at times, and where this medium was richest and black, there were doubtless hundreds of thousands of them populating such areas within this half-acre alone. As for their numbers collectively, suffice it to say that, represented by hundreds of probable species, they have populated the seas, the fresh waters, and the land all over the world where conditions were favorable. Terrestrial forms, together with the earthworms, are important in the economy of the woodlands because of their ability to ingest the partly decomposed organic matter in the humus, to mix it with mineral matter, as we have seen, and to reject (in their feces) the indigestible materials in the form of a mixture useful to plants and to other animals.

Vast numbers of enchytraeid worms are bred and sold annually to fanciers of tropical fishes, for which they make ideal food for both young and old fish. In any mostly undisturbed deciduous woodland community, these worms are doubtless regular food items of several species of carnivorous beetles and their larvae, and less often of centipedes and pseudoscorpions, as already noted.

Before leaving the worm fauna of the mini-wood, some space must be devoted to what may constitute the most fabulously numerous group of all, the Nematoda, small to microscopic free-living and parasitic worms, so numerous and distinct that they constitute this phylum by themselves. Bluntly rounded anteriorly, and sharply to even needle-pointed posteriorly, transparent or nearly so, they move by more or less rapid contortions of the whole body, thus "whipping" themselves from one place to another (plate 124 at 4 and 5).

Some ten thousand *species* are known, while *thousands* of others doubtless remain to be named. They are the most viable of the worms, living in fresh and foul waters, in inland and coastal mud; by the hundreds of thousands in rotting fruit and fungi; by the million in humus, soil, and sand; and in highly acid solutions, including vinegar, where, known as "vinegar eels" they sometimes spoil the product with their vast numbers. How they get into such mixtures is anyone's guess. As for the soil, evidently they have *always* been its inhabitants, probably thousands of millions to the acre!

In every funnel full of moist humus from the mini-wood, and as often in decaying vegetation, and frequently as parasites of both animals and plants, the nematodes in varying densities were only easily observed with the microscope, but occasionally some were large enough to be seen with a hand glass.

In many instances they appeared as if made of glass, often taking the shape of a new moon, but much more pointed at one end. Often in continual motion, their forward progress was, nevertheless, very slow, or sometimes nonexistent, their method being to bring both ends partly together on a single plane, then jerking them apart again. This manner of locomotion is alone diagnostic of the nematode organisms in general. When thousands of them were thus in motion in a few drops of water, the sight was one of utter confusion.

That the reader may gain a better idea of their ubiquitous (worldwide) as well as "backyard" distribution, the author has taken the liberty of quoting from that great work *Fresh-water Biology* by Ward and Whipple, and already mentioned in this chapter. In this volume of 1111 pages and 1547 text figures, the nematodes are beautifully illustrated in microscopic detail.

Quoting these authorities on the subject of nematode distribution:

They occur in arid deserts, at the bottom of lakes and rivers, in the waters of hot springs and in polar seas. They were thawed out alive from Antarctic ice by scientist members of the famous Shackleton expedition. In the examination of beet seeds imported into the United States, several species were found. The tap water of even well-conducted cities often contains them. Their microscopic larvae even more readily than the adults are transported from place to place by an exceedingly great variety of agencies; they are carried by wind, by flying birds and running animals. They float in all the waters of the earth, and are shipped from point to point throughout the civilized world in vehicles of traffic. Sometimes the eggs and larvae are so resistant to dryness that if converted into dust they revive again when given moisture, even after as long a period as a quarter of a century. There are beneficial nematodes, though knowledge of this phase of the subject [1918: date inserted by the author] is in its infancy. Some feed exclusively on their injurious brothers. Others devour beneficial micro-organisms. Their adaptations in these respects appear to be similar to those of insects.

Nematodes sometimes occur in uncountable numbers within hosts very small in themselves. In such cases they may multiply so rapidly and congregate in such vast numbers, moving en masse rather than "whipping" about individually, that only a microscope can resolve the moving bands and other figures into separate entities.

To cite an example, a species was found in the mini-wood feeding within hard black nodules of a fungus growth, the plant itself feeding upon a decaying log. The growths in a glass jar were observed for several days, and suddenly hundreds of thousands of nematodes poured forth from the nodules.

Why they performed this mass exodus was not explained. It might have been due to overpopulation and lack of food, although plenty of the fungus seemed to remain uneaten and not too badly damaged. If they were fleeing fetid surroundings, it was a condition which they had no doubt brought about themselves.

Hundreds of them collected within single drops of

water which had condensed on the glass walls of the jar, and in which they formed medusalike balls. Masses of others collected elsewhere, their anterior ends sticking out in all directions and waving from side to side like so many tentacles. Within some drops of water, groups of nematodes swirled about in an aimless state of confusion, while literally tens of thousands of others formed long regiments, a few millimeters in width, all interlocked and advancing in snakelike ribbons up and down the sides of the glass container. Highly magnified, these bands reminded the author of migrating elvers or baby eels when in the glasslike stage.

After a day and a half they began to break up into shorter lines, oddly shaped figures, and dense blobs. This was a sign that their end was near, and in another few hours, after more of such aimless maneuvers, all of them died, but leaving networks of their dessicated remains as evidence of this great death march of animals, which perhaps, like the famous lemmings, victims of overbreeding, were, en masse and in panic, rushing to meet their fate.

It may be imagined from these observations how many other fungi (under natural conditions) might have been infested in the mini-wood from such a horde as this, or as a result of even a few out of these thousands reaching suitable hosts in which to deposit their eggs. We see also, once more, how so-called waste matter becomes food for other living things, in this case the fungus nodules, now spent, but with their spore-releasing, species-insuring mission successfully accomplished before infestation. Furthermore, had the exodus occurred in natural surroundings rather than in a laboratory glass jar, the million or more nematodes that died would also have been utilized as food by still other micro organisms.

It is of course difficult to say with accuracy whether nematodes like those just discussed are parasites, scavengers, or both. But a great many other species, which gain entrance to plants and animals (many by means of eggs taken in food), may become true parasites, feeding on their living hosts.

Nematodes cause serious diseases in man and in other mammals and lower animals, and may seriously damage crops and other foodstuffs. But knowing that ten thousand species have been recognized and that at least as many thousand more may eventually be known, it is fairly safe to say that the great majority of them must indeed be comparatively harmless, and hundreds of others quite possibly beneficial in one cryptic way or another.

* * * * *

Living down in the black humus of the mini-wood, mostly well below the layer of dead leaves and litter, were much more sparsely distributed and highly interesting tiny crablike creatures called *pseudoscorpions,* arachnids of the order Pseudoscorpionida (plate 127).

Unknown to the vast majority of humans beings, perhaps chiefly because they are minute creatures whose intriguing forms and ways of life may only be observed in detail with the stereo microscope, they are confident little predators in their habitat, armed with poison glands and strong *pedipalps* (pincer claws for capturing prey), looking extraordinarily like those of many crabs.

Pseudoscorpions walk on four pairs of legs, holding the pedipalps, which are the second pair of appendages, wide apart, like a person with outstretched arms. These appendages are actually four-jointed, terminating in much thickened, sometimes almost bulbous "hands," each with a movable finger, forming hard pincers called *chelae.*

Those found in the author's mini-wood measured from two to four millimeters (less than and up to a little over an eighth of an inch) in length, with pedipalps spreading wider than the body length (plate 127).

The smallest of these, obtained from the mini-wood humus, were reddish yellow anteriorly, gradually shading into a browner hue posteriorly, while the underside in this (unidentified) species was white tinged with yellow, a conspicuous feature when the animal walked with the *opisthosoma* (abdominal) portion elevated.

The four pairs of walking legs were pale yellowish white, as were the rather bulbous pedipalps with their chelae or crablike pincer claws tipped with rich reddish brown.

The microscope revealed numerous sensory hairs called *trichobothria* and *setae* here and there on these claws. The chewing mouth parts, called *chelicerae,* were equipped with minute teeth and micro combs with which these tiny creatures were actually observed cleaning and grooming their soiled fingers after eating.

While some pseudoscorpions are equipped with one or two pairs of eyes, those from the mini-wood appeared to be blind.

Interesting, after one's initial astonishment and pleasure at seeing pseudoscorpions alive for the first time, were the unhurried movements of these little animals and their slow, one might almost say casual, advances, but with pedipalps outstretched and at the ready to capture unwary prey. These stalking movements contrasted sharply, however, with their instant back-stepping when danger threatened ahead, as when the author confronted them with the tip of a fine needle and their instant reactions were like those of the typical small jumping spiders.

Pseudoscorpions with bodies measuring about two and a half millimeters in length and also found in the mini-wood humus and leaf litter, were of a pale yellowish color, including the legs and pedipalps, except for the pincer claws, which were tinted reddish brown, becoming richer and redder at their tips. Almost the same size as the first ones described, these may have been simply color variations of the same species.

Many will be surprised to learn that these tiny arachnids may construct silken shelters or hideouts by means of silk glands, a pore and spinneret situated on the chelicerae. Some species hide in these shelters with their pedipalps protruding, ready to grab their prey in their pincer claws. The females, who carry their eggs and developing embryos outside of the body, also hide in these silken shelters.

Food items include collembolans and doubtless other minute insects, and very small enchytraeid worms.

Watching the feeding process through the microscope, one is reminded again and again of the procedure of the much more familiar crabs, arthropods of another great class, the Crustacea, and animals hundreds or thousands of times larger than the pseudoscorpions.

The pincer claws first passed the food, pulled into rough pieces or whole, to the chelicerae, whose business then seemed to be to mince it into minute fragments and probably mix it with digestive fluids. Some species, maybe all of them, are known to have a digestant in the mouth which liquefies some of the food before it is swallowed. In some cases observed by the author, the animals seemed to be chewing without success and finally discarding these unmanageable fragments. Those from the mini-wood did not feed very often, and after having a good meal, seemed to take no further interest in other living things for several days. The observer will soon realize also that, unlike crustaceans such as shrimps, crabs, and lobsters, whose feeding methods they might seem to mimic, pseudoscorpions have no antennae but must depend upon sensory hairs with which to feel their way about, test other objects, and sense enemies.

Collembolans, those tiny insects described in chapter X, seemed to be a most suitable food for these arachnids. In one of the mini-petri dishes containing sifted humus, ten very minute white collembolans, and a single pseudoscorpion, the collembolans had all disappeared in ten days, and although the act was not actually observed, it was assumed that they were eaten by the predator.

While it is of course well known that pseudoscorpions possess poison glands, and a tooth for administering it situated on one of the pincer claws, the author could not ascertain whether those from the mini-wood humus were immobilizing their prey before reducing it to fragments or were killing it outright with their pincer claws.

Minute as these creatures are, and as primitive in the scale of living things as they may be, they nevertheless perform ritual dances and other courtship motions, and their methods of transfering sperm to the female's genital orifice are remarkable. One sometimes wonders how Nature came to evolve such strange sexual routines for some of her tiniest organisms, which, it would seem, might do better in their habitat with simple copulation.

Following courtship antics, the male (as we have seen with some other small animals described in these chapters) deposits one or more spermatophores. Each of these consists of a stalk, sometimes very simple, or in other cases more complicated. All support at the top a "package" containing the spermatozoa. This object may be globular, cup shaped, or of more intricate form.

In one species, after the sexes have met and the male has been stimulated to deposit a spermatophore, he backs away from it, and the female advances and then stands over the "package," rupturing it and thus releasing the sperm, which are then (with the aid of the male) forced into her seminal receptacle!

That various and more involved sperm "packages" occur, and that different insemination procedures are followed by other species, has led scientists to believe that evolution from simpler processes not involving mating in any form and toward more conventional procedures may be here represented.

The whole subject is of very great interest but too complicated in detail for a book such as this one, except for brief mention, as in this chapter on the army of the soil. Those who wish to know the whole story about pseudoscorpions are enthusiastically referred to a recent outstanding little book on the subject, a book that is very well illustrated as well as lucidly written. Covered are the life histories of these odd little creatures of the woodland soil and (less often in the author's experience) tree trunks. It includes the extremely interesting details concerning the spermatophores and courtship, and the female pseudoscorpion's great concern for, and care of her eggs and embryos, a wonderful story in itself.*

Unfortunately, too few people experience the pleasure and excitement attendant upon the discovery for themselves of an organism previously unknown to them except perhaps through pictures or reading. Fascinating it is, when peering through a stereo microscope, to come suddenly upon such a creature, barely visible, or wholly invisible to the naked human eye, which yet may seem to have the ways of an animal from a much higher phylum.

Thus it was when examining humus taken at random in all parts of the mini-wood that the author kept making these little "discoveries" which soon brought almost embarrassing realization of how little he had known about the army of the soil previously.

To cite an example: long after the portions of this book pertaining to the mini-wood millipedes and centipedes (classes Diplopoda and Chilopoda) had been completed, the author, while searching the humus for additional pseudoscorpions, found minute white organisms never encountered in any previous microscopic examination.

Measuring approximately half a millimeter (about a thirty-second of an inch) in length, these soil animals might best be described as suggesting rather shortened, *humpbacked* centipedes, to which they were doubtless distantly related, but the five tergal or dorsal "humps," each one saddling and partly covering two of all but the last of the eleven body somites, seemed to link these myriapods as closely to the millipedes, and, notwithstanding the nine single legs on each side, in an arrangement more like that of the centipedes.

The animals proved to be myriapods belonging to a class distinct from the Diplopoda and the Chilopoda—the Pauropoda, creatures minus hearts and blood vessels, and lacking respiratory systems, oxygen being exchanged through the skin surface and the blood being aerated in spaces between the tissues called *lacunae*. Unfortunately, attempts to obtain adequate photographs of these pauropods were unsuccessful.

In biological language, the word *anamorphic,* applied to an animal, means that it develops from a simpler to a more elaborate stage as it matures. The term applies to the pauropod just described, a species of *Pauropus* whose

young emerge from the eggs with only six legs and three tergal plates, but which acquire two additional "humps" and twelve more legs as they mature.

These animals of the humus appeared to be fitted for a carnivorous diet, but whether this was the case or not, or what indeed they did feed upon, was not ascertained, nor were other details concerning the life history observed by the author.

Much more interesting and easier to obtain were other pure white minute inhabitants of the humus belonging to order Diplura, primitive insects of the genus *Campodea* (plate 124).

Soft-bodied and very delicate, sometimes dwelling under rocks as well as in the humus, these are the kind of organisms from which, it is believed, the more advanced orders of insects may have descended. Rightfully, their place in this book should be in the next chapter, along with the mini-wood abundance of other insects, but they have been included here because they are so typically animals of the soil army.

Easily maimed, even when with all possible care being flipped into a vial with the aid of the most delicate sable brush, it was soon found that the funnel apparatus so often mentioned was most satisfactory for obtaining living campodeids for study, several of these minutiae sometimes coming safely through the sieves from a single humus filling of the funnel. Washed into a watch glass under the microscope, they could then be "rescued" and transferred on the bent tip of a dissecting needle to the humus in a mini-petri dish. Nearly drowned at times, these tiny things would usually recover completely, groom themselves, and then, in the great majority of cases, remain active and healthy for *months,* apparently requiring only a degree of moisture in their habitat, and feeding on detritus or other minute items in the humus supplied.

Never trying to escape when the dishes were uncovered, they could be observed at any time directly under the stereo microscope.

Full-grown specimens measured up to eight millimeters (slightly over five-sixteenths of an inch) in length. Very slender, very soft, pure white to yellowish white, low slung on six rather short legs, they were ever on the move. Eyeless animals of the soil army, they bore long, delicate antennae whose nineteen joints, viewed highly magnified, resembled clear glass beads (plate 124).

At the posterior ends of their bodies they were equipped

---

*Peter Weygoldt, *The Biology of Pseudoscorpions* (Cambridge, Mass.: Harvard University Press, 1969).

with another pair of appendages or "tails" called *cerci*, somewhat shorter than the antennae and the joints bearing bristles.

The cerci were carried spread wide apart in the form of a continually trembling V. These, it proved, were more easily broken off in passing through the funnel apparatus than the longer antennae, although the latter suffered also at times. In either case, campodeids survived sometimes even thus maimed, for considerable periods. Uninjured individuals could drown in the collecting bottle, however, as easily as the maimed, and needed to be rescued promptly. Very young individuals (campodeids pass through no elaborate stages or metamorphosis) were as viable in humus but perished as readily in water as their elders.

In observing the cerci, it was obvious that these "secondary feelers" were of great importance, continually in motion, often being waved from side to side, touching objects here and there, and at times spread far to the sides, or being bent forward up over the body. As tactile appendages they seemed to be as useful as the antennae, particularly when their owner was moving backward, which was almost as often as forward, doubtless a valuable habit in a predator-rich environment.

Such soft-bodied defenseless creatures as this probably rely on speed as well as on their tactile appendages, for when living in normal particle-formed humus, they were often observed slithering out of sight in an instant. Whatever their combined methods may be, they must have proved efficient through the ages, as evidenced by their numbers in the soil army of today.

These may be blind and wingless primitives, but they are organisms beautiful in their simplicity and in their ghostly outward design as they move with such zest in their black and moist world.

Watching them through the microscope and observing their fastidiousness in the care of their bodies and appendages, seeing them groom their delicate legs and cerci and apparently *licking* each glossy bead of each antenna or leaning back to reach parts farther away, one realized how important is cleanliness among even the lowliest animals with which we share the earth.

Their care of the bristles protruding from the cerci indicated that these microscopic "hairs" encircling or otherwise located about joints of these appendages were highly sensitive tactile units in themselves.

What campodeids were eating was not definitely determined, but as their mouth parts appeared to be simply con-structed and not very strong, it was assumed that they were vegetarians, or detritus feeders. Whatever their food, they found it themselves and evidently in ample supply in the finely sifted humus in the petri dishes.

Although a great many were collected uninjured and maintained for considerable periods in small groups, no eggs were ever observed, but very young campodeids appeared from time to time in the dishes. This was very satisfying, as these newborn individuals were seen to be exact replicas of their parents in miniature, and their presence proved that conditions supplied within the little dishes were satisfactory for successful breeding. Mature campodeids often showed excitement upon meeting, touching and quivering their antennae. Probably some of these contacts were in the nature of courtship procedures, but although we know that male campodeids produce stalked spermatophores bearing globular sperm packets, if such were deposited in the petri dishes, none were ever located during the author's long hours of observation with the microscope.

Somehow campodeid behavior seemed to suggest to the author a confident, not to say carefree, sort of life as the animals went methodically about their business, ever active, long antennae continually waving, quivering as they contacted things as if in pleasant expectation, and their sensitive cerci doing likewise at the rear. Delicately probing this and that, or one another, they went their restless way in total darkness, a condition, however, meaningless to them as it must be to all little animals without eyes, and many of which are so completely successful and such numerous members of this great soil army.

While such members as the campodeids must have many enemies in their densely populated habitat, being soft and succulent, and no doubt good eating, it was a curious fact that in the many mixed soil communities kept under constant observation by the author, only two other organisms were ever actually seen feeding on these primitive insects. One of these was a species of pseudoscorpion; the other, another species of insect.

The latter, a curiously formed minute ant, orange-yellow in color, and in contour resembling that of the tiny, slender, and much elongated rove beetles living in the same habitat, seemed to be a solitary-living creature, and no colony, as of other ants, was ever found in the humus. Like the miniature rove beetles, it was a streamlined little predator, nicely evolved for life among the humus particles.

Only two other species of ants were occasionally obtained from the humus by the funnel apparatus. Both of these were also orange-yellow insects and considerably larger than the attacker of campodeids. One had very long legs and extra lengthy antennae. All three ants, together with the carnivorous rove beetles just mentioned, doubtless filled important roles in natural population control in their environment.

For students and others who may become interested in studying these soil animals, the simple technique followed by the author in rescuing living campodeids and pseudo-scorpions, geophilomorph centipedes, millipedes, and other small fry which came through the funnel apparatus, will be repeated in a little more detail here.

Required will be a stereo microscope, a fine-tipped small wash bottle or medicine dropper, a watch glass or two, extremely fine-tipped tissue-dissecting needles, a delicate touch, and plenty of patience.

The shallow layer of water which accumulates in the collecting bottle of the apparatus is rinsed with the collected animals into a watch glass on the microscope stage, and this can best be done with a medicine dropper.

The animal to be rescued is now very delicately approached with a bent or L-tipped teasing needle.* By gently penetrating the water film and moving the needle tip beneath the living, often floating subject, then raising this very slowly, the minute creature will often avail itself of this chance to leave the water, crawling "aboard," one might imagine almost "thankfully," although many attempts at rescue may sometimes be necessary before achieving success.

Wet and bedraggled or half drowned, the animal must often be shaken or otherwise induced to leave the rescuing needle for the properly prepared mini-petri dish with its finely sifted layer of moist humus. The humus layer should not be saturated in these covered receptacles at any time. The animals thrive when moist, but if the humus in their dishes is too wet, it will grow lethal mold.

As already mentioned, campodeids and some other soil animals often recover very well after submersion once they are on "land" again. Certain Collembola, however, die very quickly, while other minute species are so light that they walk on the water film without breaking through. Millipedes stand submersion better than centipedes, and the

minute species of ants succumb much sooner than the vigorous predator rove beetles.

In preparing for a colony of *Campodea* or *Scutigerella* or other soil minutiae, a circle of filter paper is cut to exactly fit into the bottom half of the petri dish. This paper is thoroughly moistened, after which *dried* humus just deep enough to completely cover or black out the paper is applied by means of a 60-mesh (60 to the inch) laboratory sieve. Such a fine mesh is necessary, as large humus particles in the dishes would allow the minute animals to burrow in and remain out of sight. When not under observation, the dishes are kept dark by cardboard or black paper covers.

The procedure just outlined is necessary for the reason that damp or wet humus will not pass through a 60-mesh sieve, and even when the medium is used dry it must be *rubbed* through the screening. The tiny particles of dry material become properly moistened but not saturated upon hitting the moist filter paper, and by this method a very thin, evenly distributed layer in each dish is possible. A few drops of water applied with a medicine dropper from time to time will maintain proper conditions in the colonies. An occasional additional thin layer of humus, sifted on top of the original layer, is all that will be needed to maintain a colony of detritus feeders.

In terminating these chapters on the Army of the Soil, it should be mentioned again that counts or estimates of the numbers of the various organisms found in the author's half-acre of woodland, or even in a square yard of it, probably could not be made with any degree of accuracy. This would doubtless also hold true of most any deciduous woodland area, the hundreds of samplings made showing that media taken from apparently identical locations often contained many identical species, but in very dissimilar percentages.

Among these, such comparatively large member organisms as the earthworms, the *Lithobius* and *Geophilus* centipedes, the larger millipedes, and possibly the isopods (all of which various animals are easily visible to the naked human eye) might be numerically estimated at times, but when we come to the microscopic mites, the tiniest species of millipedes, the nematodes and enchytraeid worms and many other, still smaller, organisms, we might as well try to count the overhead stars in some small area of the Milky Way.

It is well to remember that, with few exceptions, these

---

*An extra-fine needle may be made by removing the head of a O- or OO- gauge insect pin, inserting this end in the eraser of a pencil, and then bending the pointed end to the desired angle.

animals are working for our own good in the production of humus and soil, as one superbly organized society, and for their raw materials are to a great extent dependent upon the vegetable litter supplied by the trees and shrubs and smaller herbaceous plants of the woodland community, much of which is shed in the form of a continuous gentle shower, and at other times almost as a deluge (which, of course, is the comparatively sudden annual leaf-fall).

And be it also always remembered that these woodland plants and animals are doing what nature intended of them, *only*. Never has the Army of the Soil polluted its environment, never has it tampered with or dumped experimental poisons into its habitats, nor has it or the plants of such woodland communities ever contributed to the degradation of the atmosphere, which we as human beings are so conceited as to speak of as "our own."

CHAPTER XIII

INSECTS I
MOTHS AND BUTTERFLIES

It would be quite possible to write five hundred pages about the insects which were found in and on the borders of the half-acre woodland investigated.

A number of these, including the tiny to microscopic species of collembolans, the campodeids, and some other minute or invisible-to-the-naked-human-eye forms have been covered in earlier chapters, where, in arranging this work, they seemed fittingly to belong in the overall story of the mini-wood community.

The reader may get some idea of the many orders and the great variety of species which have occurred in the area in question by a close study of plates 128, 129, and 130, illustrating part of the author's *seasonally* arranged collection. Represented here are those which have bred within the mini-wood, or have appeared there occasionally or with annual regularity, some of the latter showing up on the same, or close to the same, date from year to year. There are a few which have shown up but once, for reasons not easily explained. While many other species are illustrated in the fourteen additional insect plates, even collectively they offer but a suggestion of the possibilities for further work, as species new to the collection continued to appear after the book had been completed, and will no doubt continue to do so in the future, for as long as such a vigil may be kept, thus indicating what a study in depth of the entomology of even this small area might entail.

Aside from the numbers of odd, small, and inconspicuous insects, there were many larger, often beautiful, sometimes spectacular things which appeared in these woods mostly by night, species seldom seen or even imagined by many people who also own New England woodlands, and some of whom, we might say, dwell in or very close to the same habitat with these creatures.

In every woodland, much of the insect life hides away during daylight hours or, relying instinctively upon natural camouflage, may choose in many cases to remain motionless but in full view on the bark of trees and on dead and living foliage (plate 131).

Outnumbering all of the *plainly visible* insects, both in orders and families, and in number of species captured within or on the borders of the mini-wood, were moths, technically, Heterocera, or one of the two subdivisions of the order Lepidoptera—the scale-winged insects—to which the butterflies or Rhopalocera belong in the other subdivision of the order.

Whether such flying insects are moths or butterflies, is, as most of us know, quickly determined by examining the antennae, which in the moths are merely hairlike appendages or appear in the form of feathery plumes, while the antennae of the butterflies are hairlike but end in club-shaped enlargements or knobs.

Many moths fly by day as well as by night, whereas the butterflies seem to be strictly diurnal and sun loving.

Not so long ago, but mostly before the years of study covered in this book, it was quite easy in fall or in winter to find cocoons of our largest northern moths. There was *Hyalophora cecropia,* a great reddish brown and grayish brown species with large half-moon spots upon wings which spread nearly six inches. These was *Antheraea polyphemus,* mostly yellow-ocher, with four big eyespots and a five-inch spread.* There was the lovely pale green collector's delight, *Actias luna,* also with five-inch wings, and splendid, out-curving tails up to two inches in length; and there were the much more numerous, smaller, dimorphic moths, *Callosamia promethea,* the females reddish brown and the smaller males much darker, some at a distance looking almost black (plate 131) and whose silken wintering shelters containing the pupae were once a common sight dangling from the twigs of spicebush and wild cherry trees.

From many localities, these splendid moths have gone with the wind, or perhaps we should say with the DDT. Likewise, and becoming even rarer, there was the great pale yellow, pale lavender-and-brown-spotted *Eacles imperialis,* in which the female's wings spread five and a quarter inches; and still rarer, even in years gone by, we occasionally, on red-letter days, found the brown and gray and richly yellow-spotted royal walnut moth, *Citheronia regalis,* another species with a five-and-a-quarter-inch wingspread (plate 131).

Once these fine moths were attracted by the lights in our houses and delighted us when we found them in the morning resting and clinging motionless on our screens or walls. All of these species, now preserved in the author's collection, were found within the mini-wood area, but none of them had been seen there for a decade before the completion of this book.

Compensating for their loss, however, are today's most spectacular survivors, the *Catocala* and *Euparthenos,* or underwing moths, a group outstandingly beautiful and at the same time protectively patterned to a most remarkable degree. During the fifteen-year period of the mini-wood study, exactly fourteen species of *Catocala* moths and one underwing of the other genus were recorded and collected there.

The forewings of a *Catocala* moth, camouflaged to imitate bark, nevertheless vary greatly in their exact markings, not only from species to species, but also quite often *within* the species as well. Various grays and white, various browns and black, are the only colors in these larger, forward pairs, but so wonderfully have they been evolved through natural selection, with wavy lines, spots, and more intricate light and dark figures, that, when they are characteristically folded, rooflike, over the insect's body and under-

wings, the camouflage is so complete when the moth is resting on bark or other darkly weathered wood that it is truly remarkable (plate 131).

Catocalas are, nevertheless, very wary moths. The least disturbance near the resting place—a slight jar or vibration, in particular—instantly sends the insect off through the woods in erratic and extremely rapid flight. It is then that the brilliant red, yellow, or white black-banded underwings at once become conspicuous. The flight is hard for the human eye to follow. In a zig-zag course of instantaneous flashes, the moth selects a distant tree trunk or other dark woody object, and swooping slightly upward, comes to rest again; then, upon folding its camouflaged forewings, it blends instantly with the bark. And so, as we look with astonishment, even if only from a few feet away, the moth seems to have vanished from the woodland.

The theory has long been held that the highly colored and often striking underwings baffle bird or other animal predators when all color vanishes as the moth terminates its flight. The idea at first seems logical, admitting that the same insect may easily fool a human being. But the theory becomes porous when we inquire about such potential predators. What birds or other animals would they be? Owls, it is true, eat many insects at night, but then these moths are flying, and would owls be able to see colors in the dark? And as for bats, it is doubtful if they see colors at all. Perhaps the flying squirrel might add such a moth to its diet, should it run across one on a tree trunk (a mere speculation on the part of the author, however without a single verifying observation). We must remember also that there are many *Catocala* moths whose underwings are black.

But these moths doubtless fitted into a food chain somewhere in this well-established mini-wood community and in a way not detrimental to the *Catocala* population as a whole. Probably their large, oddly shaped, hairless caterpillars were eaten by birds and other animals high in the oaks and other woodland trees, where these larvae themselves feed on such foliage.

The caterpillars were very hard to find, being mostly colored in grays and browns. Many that are known also have humps, or a "saddle," somewhere on their long, doubtless succulent bodies. The fourteen species of Catocala moths and one underwing of another genus recorded within or on the borders of the mini-wood during the years of this study (twelve of which are illustrated in plate 132) were as follows:

With black underwings very narrowly bordered with

---

*These two moths were formerly *Samia cecropia* and *Telea polyphemus.*

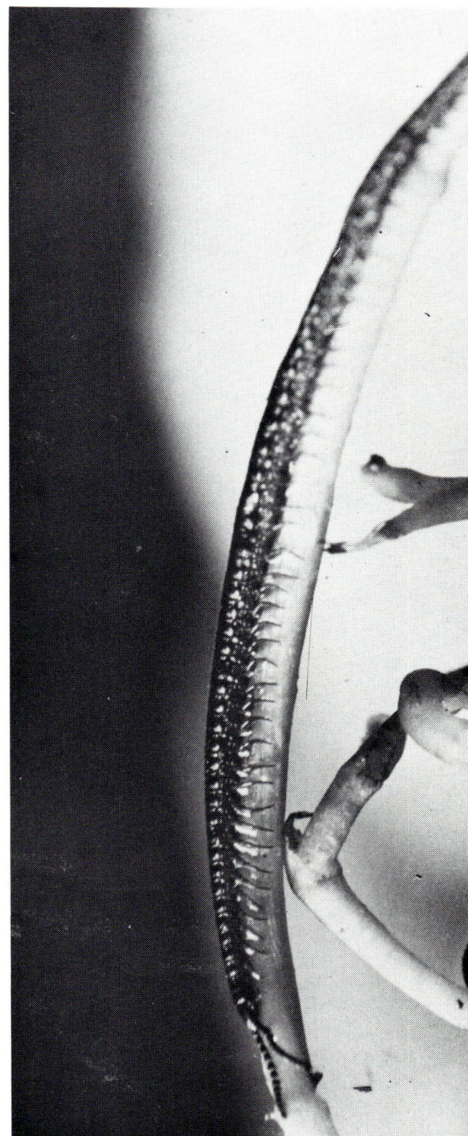

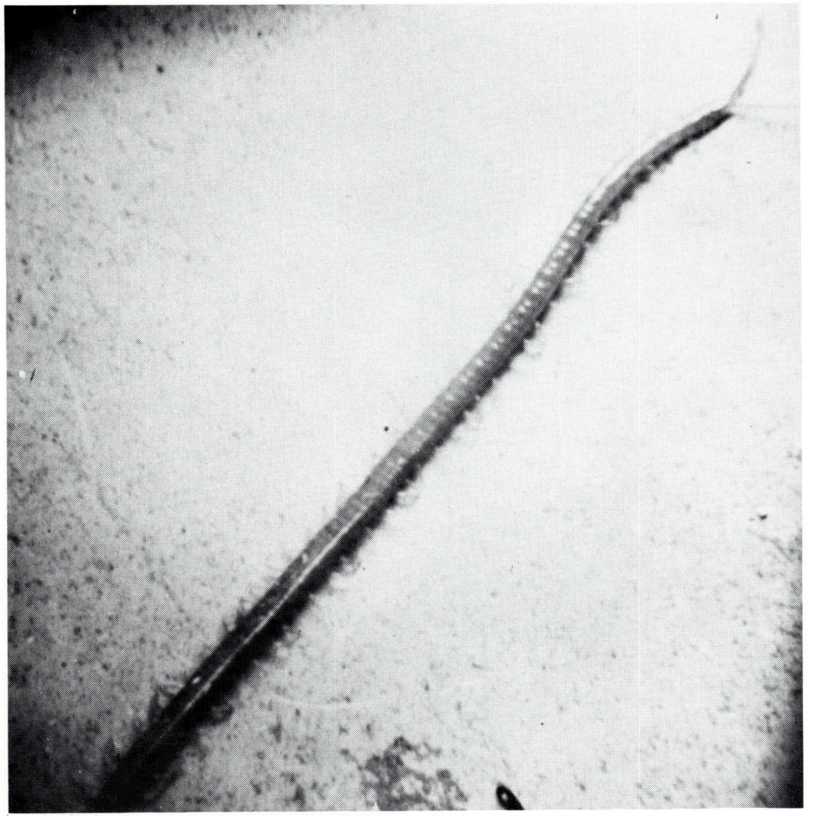

Top left: A mini-wood centipede of the order Geophilomorpha, a blind species, shown emerging from the humus with forty-three of its forty-nine pairs of legs visible. $\times$ 5. *Above:* Same species ascending a rootlet. $\times$ 4. *Left:* A delicate and rare coral-red geophilid centipede found in the mini-wood humus, shown on the move. $\times$ 3. (Pages 104, 105.)

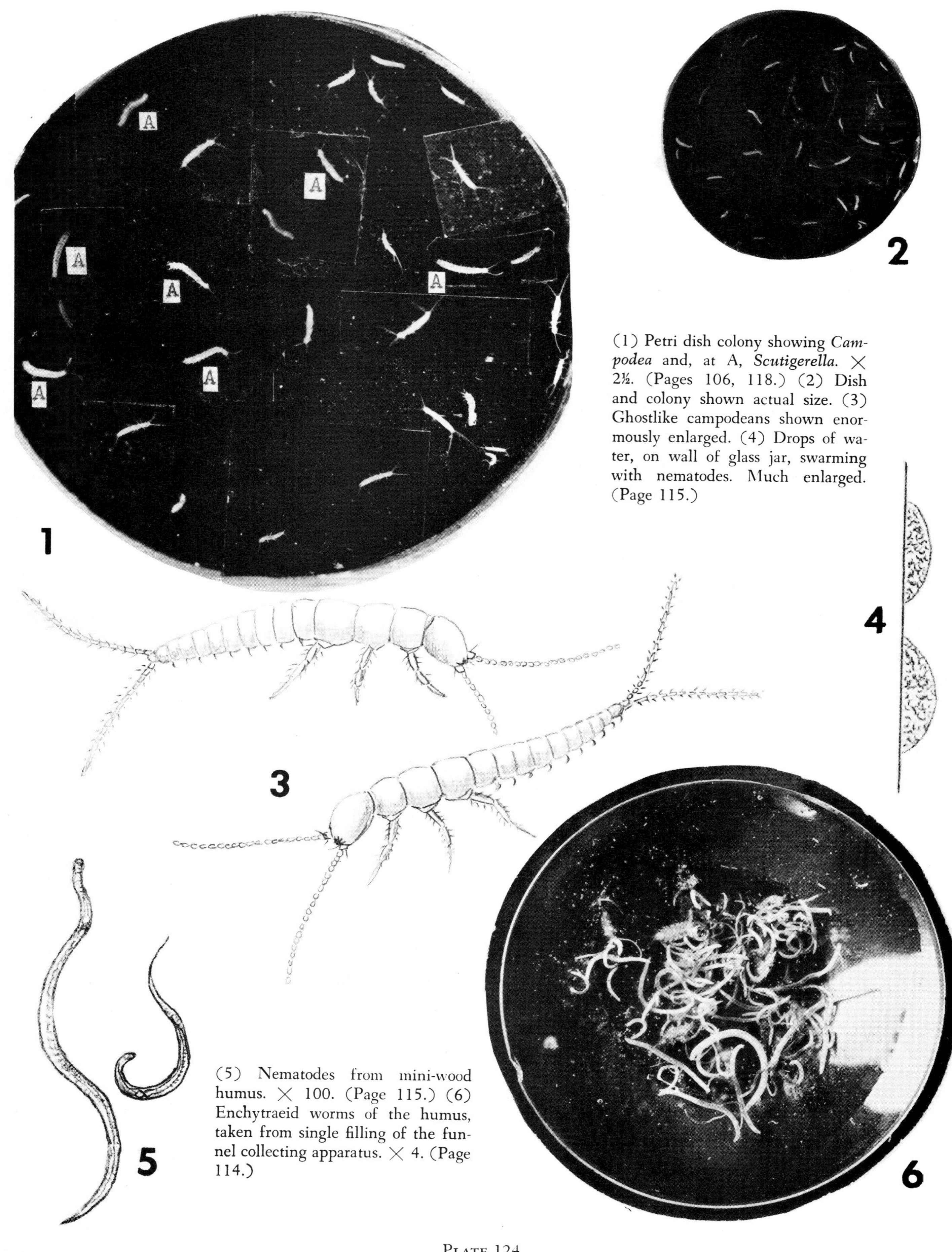

(1) Petri dish colony showing *Campodea* and, at A, *Scutigerella*. × 2½. (Pages 106, 118.) (2) Dish and colony shown actual size. (3) Ghostlike campodeans shown enormously enlarged. (4) Drops of water, on wall of glass jar, swarming with nematodes. Much enlarged. (Page 115.)

(5) Nematodes from mini-wood humus. × 100. (Page 115.) (6) Enchytraeid worms of the humus, taken from single filling of the funnel collecting apparatus. × 4. (Page 114.)

The common earthworm, *Lumbricus terrestris,* a highly important species of the soil army. *Top:* Worm has ejected a "cast" consisting of clay and other soil materials from its digestive tract. (Page 110.) *Center:* Worm at right is about to extend its fleshy pharynx and grasp a leaf fragment. *Bottom:* Earthworms, hermaphrodite animals, about to copulate. The clitellum (bandlike object), which secretes the egg cocoon wall, shows clearly on left worm. (Page 110.) All × 2. See also plate 126.

PLATE 125

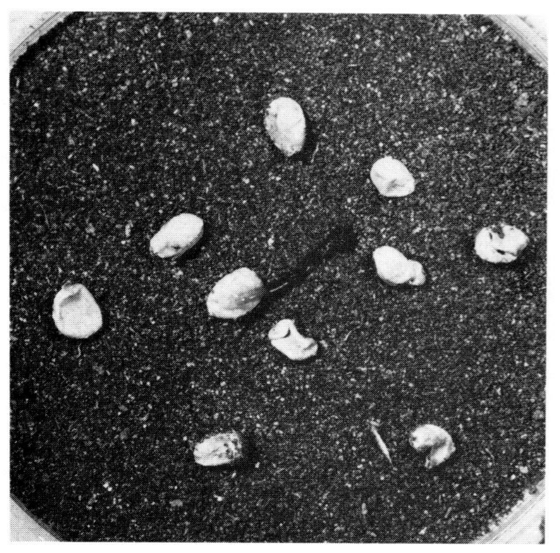

Cocoons whose walls are secreted by the earthworm's clitellum. Each holds several eggs, but usually only one worm emerges (see cocoon in center). Natural size. (Page 11.)

Free-living flatworms (Turbellaria), showing fan-shaped anterior "probes." See caption at bottom right for name and other data. (Page 113.)

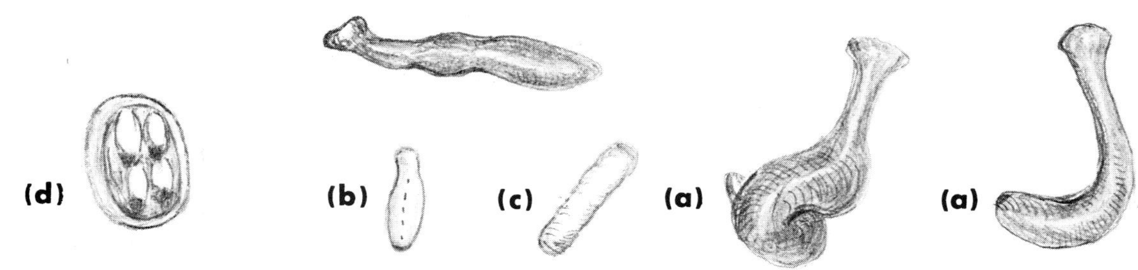

(d)   **(b)**   **(c)**   **(a)**   **(a)**

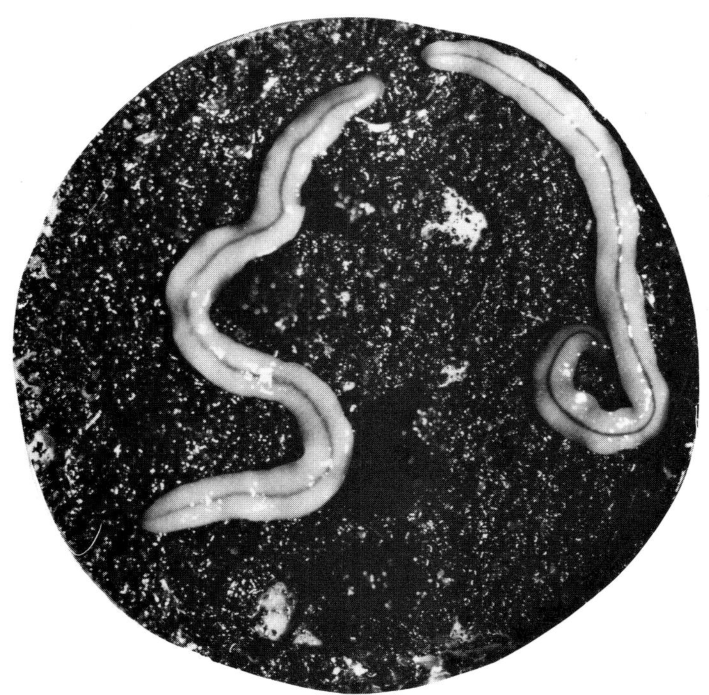

Sketched above: Turbellarian (*Placocephalus?*) (as described on page 113). (a) On the move (partly extended). (b) At rest. (c) Just before disintegration. (d) Egg capsule or cocoon left by a similar species upon its disintegration. All are greatly magnified. Photograph from life (at left) shows two turbellarians, *Placocephalus kewense* (Mosely), more than twice life-size. (Page 113.)

Silhouettes of minute pseudoscorpions. Even though their presence is unknown to the vast majority of people, they are among the most interesting and colorful tiny predators of the soil army. *Left:* The animals mounted in a microscope cell, shown twice life-size. *Above and below:* Pseudoscorpions from the mini-wood humus, $\times$ 19 and $\times$ 22. (Page 116.)

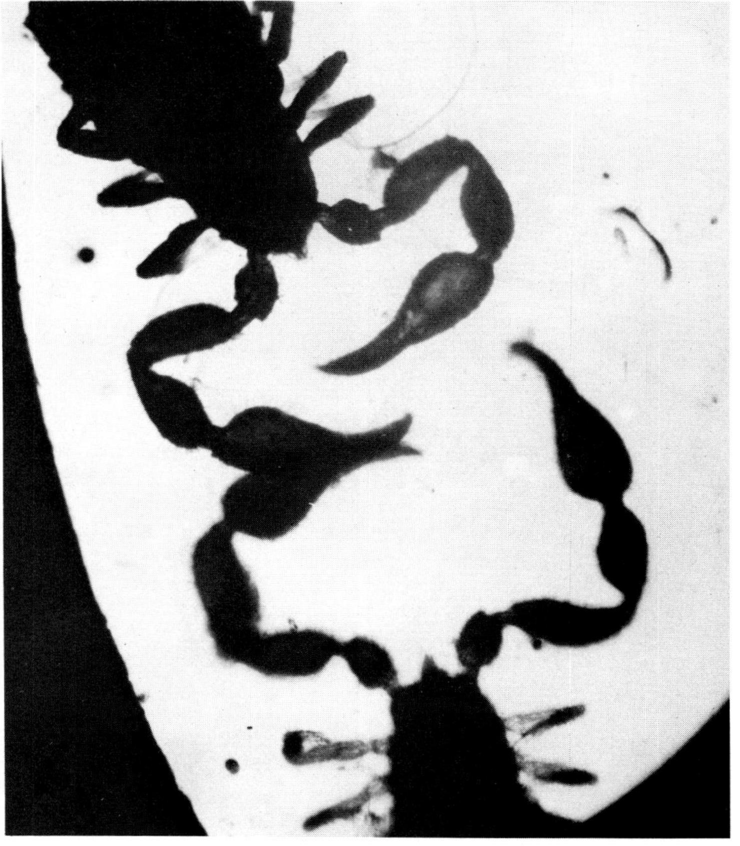

Holding their pedipalps ("arms") wide, pincer claws at the ready while hunting or contacting one another, they are very crablike in their motions. Shown $\times$ 22. (Page 116.)

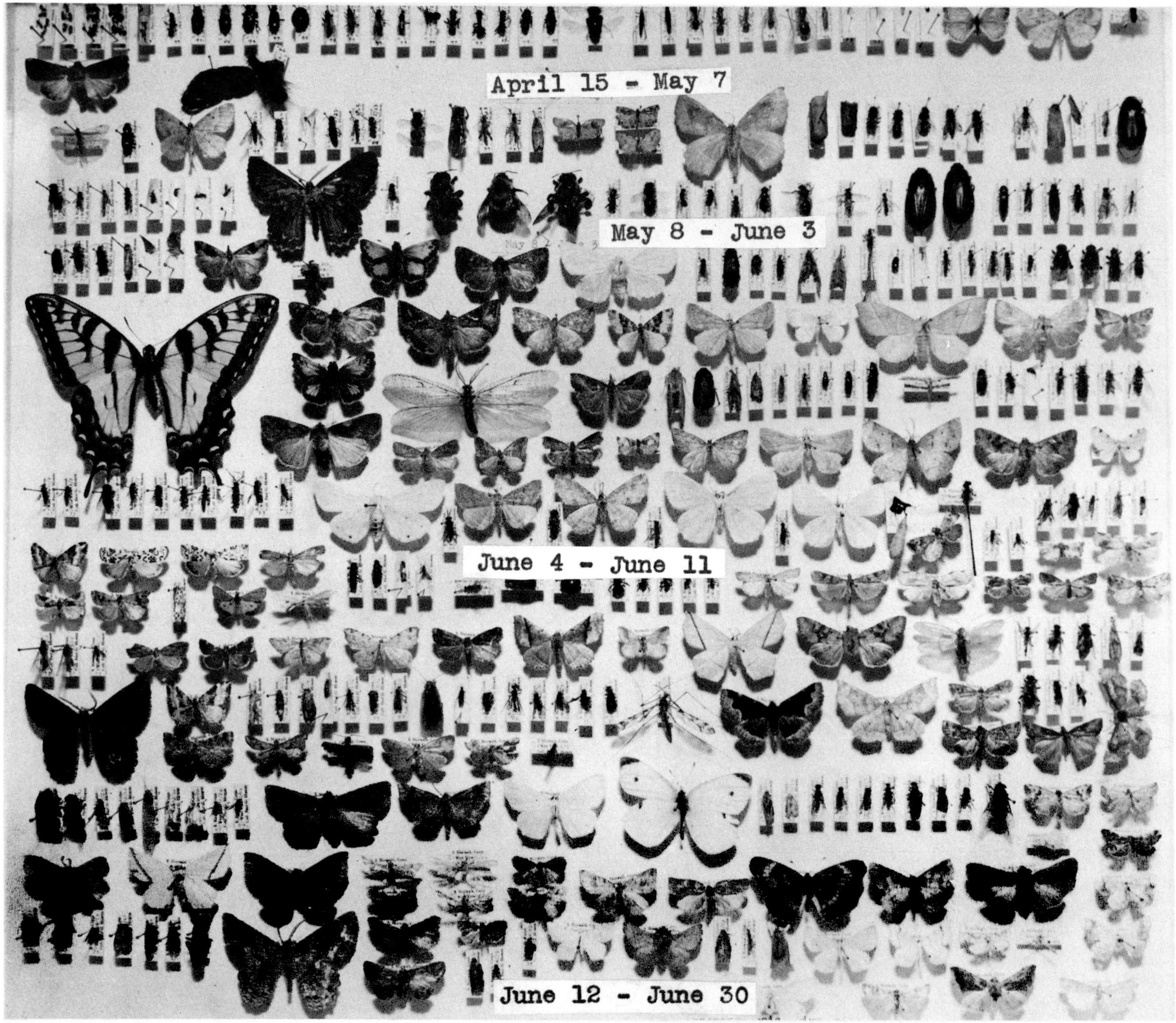

Seasonal insect collection. This, and plates 129 and 130, give some idea of the variety of insects which occurred in, or on, the mini-wood borders during only *one* season of intensive collecting. More exact periods are indicated in the photographs. (Page 123.)

PLATE 128

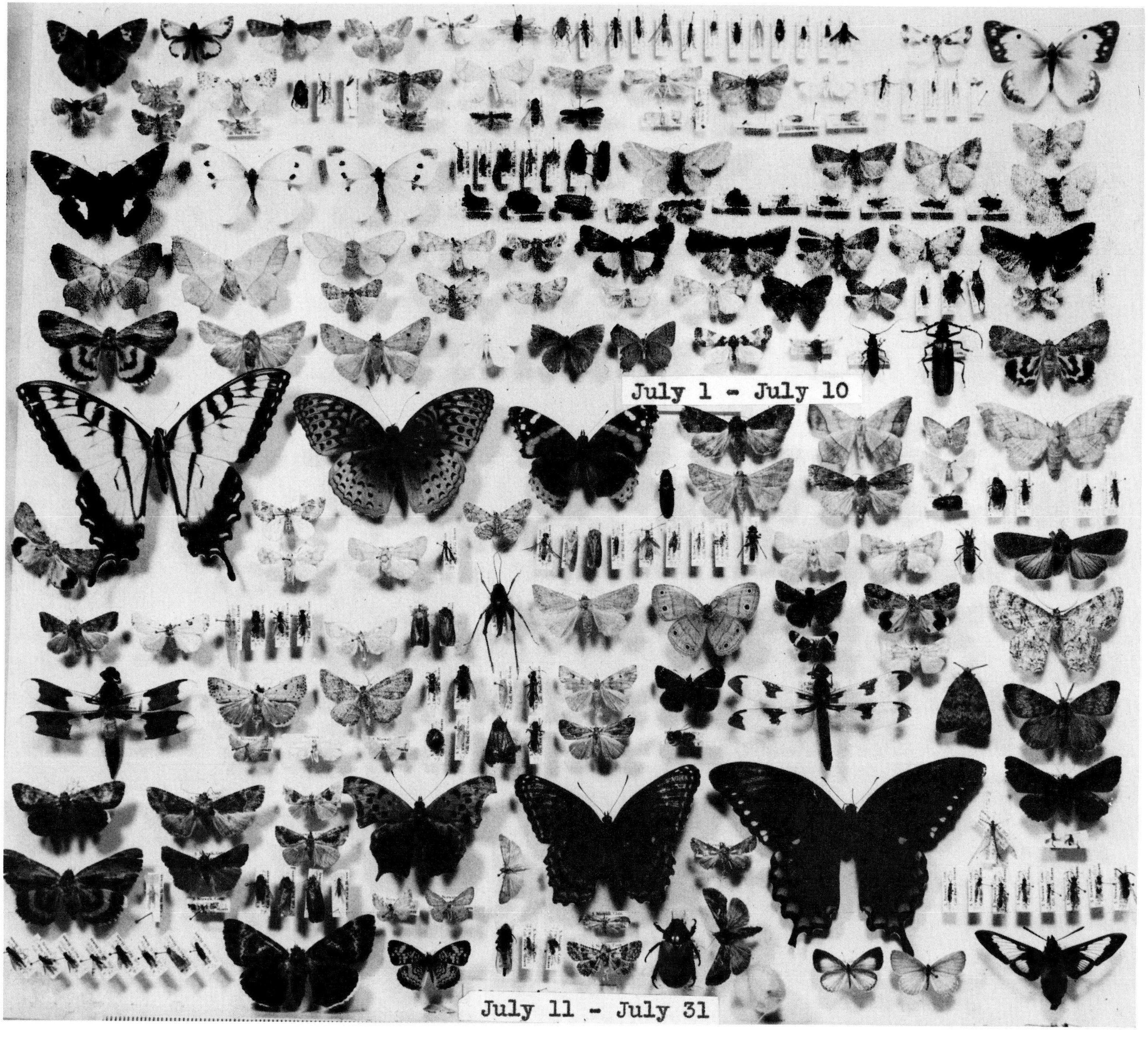

Seasonal insect collection (continued). Duplicate specimens in this and in plate 130 illustrate second broods of the same species (see also plate 128). Entomologically inclined readers will know the names of most of the specimens represented. (Page 123.)

PLATE 129

Seasonal insect collection made in a single year (continued). Periods when specimens were obtained are indicated in the photographs. The one-foot rule indicates actual size of any specimen. (Page 123.) See also plates 128 and 129.

PLATE 130

white, *Catocala retecta, C. robinsoni, C. agrippina,* and *C. andromeda.*

With yellow, black-banded underwings, *C. neogama, C. palaeogama, C. amica, C. grynea, C. antinympha, C. fratercula,* and *Euparthenos nubilis,* the only underwing of a different genus recorded in the mini-wood.

With red, black-banded underwings, *C. ilia, C. parta, C. cara,* and *C. ultronia,* the commonest species in the area.

Like so many other species, these moths lived only a few days in the adult form, and it is indeed a satisfaction to be able any day to look upon brilliant and well-preserved perfect specimens in the collection, creatures that long ago would have been reduced to dust, following their meager allotments of time as splendid living imagoes.

About 275 definite species of moths were taken by the author in or on the borders of the mini-wood during this study. Numbers of these would be dismissed as "just millers" by unobservant human beings, but a closer look might alter such a viewpoint. Most of these moths have distinction in one way or another more worthy of attention, and all of them have technical (Latin) names somewhere duly recorded in the universally understood and distinguishing terms of science.

Dozens of the tiniest species carry exquisite patterns on their wings, formed of precisely laid-on scales or micro "shingles," in many color combinations in addition to black and white and plain glossy white, and including gleaming silver and gold, copper and buff; some have a peacock iridescence that even a rainbow cannot duplicate. So exquisite is the maillike sheathing on some of these smallest moths that, when first viewed through the microscope, it causes one to gasp with astonishment.

The colored "dust" which rubs off on one's fingers when a moth or butterfly is handled consists of hundreds or thousands of these minute "shingles." Each such scale terminates at its broadest end in two or three tines which hold it in its position, overlapping another scale in the pattern. There must be millions of scales, as well as great numbers of colored hairs of various lengths, sheathing the wings and bodies of our largest moths and butterflies.

As many of the smallest moths found in or near the mini-wood were rather weak fliers, it was assumed that they were not deliberate transients, being either breeders whose larvae fed on the abundant variety of vegetation available, or, in some instances, might have been blown into the area. The close observer knows that in many cases the same spe-

cies may be collected at the same dates and places year after year, a fact which would seem to strengthen the supposition that these are local breeders, as often as not.

Despite the fact that hundreds of these insects have been technically named and described, there is still very little known about most of their life histories, many of them (in the larval stage) being minute leaf and stem miners seldom found except by chance.

Occasionally, however, we may come upon an unfamiliar caterpillar which is not easily raised to the imago (adult) form as we may more often raise some of the larger species. To be able to describe the stages (from egg to moth) of a tiny species whose life history was previously a blank is the worthy goal of many an entomologist.

Many caterpillars are controlled at least to some extent by internal parasites. Extremely small ichneumon wasps and insects known as chalcid flies deposit their eggs in the bodies of these larvae by means of specialized short, greatly elongated, or often curved *ovipositors,* many of which are as fine or finer than a horsehair. Ultra-minute ichneumons and chalcids deposit their own eggs in *single* eggs of other insects, including those of moths and butterflies, and still other species parasitize wasp larvae and pupae and also the egg clusters of spiders. So tiny were certain parasites observed by the author that over two dozen in the winged adult stage issued through a mere pinhole in a moth egg, while over four hundred came forth from a single wasp pupa (plate 132).

Some of the largest ichneumons are wasplike insects equipped with long, horsehair-diametered "tails," which include a remarkable ovipositor and a "drill" capable of penetrating inches into dead wood through which eggs are deposited on other insect larvae feeding therein. There are also parasites *of* parasites (known as *hyperparasites*) and also parasites of the hyperparasites!

Notwithstanding the fact that parasitism is one of nature's most efficient methods of species control, it was sometimes disappointing to have an ichneumon wasp rather than a fine butterfly or large moth issue from a chrysalis or cocoon.

The Ichneumonoidea and Chalcidoidea constitute the two largest superfamilies of the Hymenoptera, which also includes the ants, the bees and wasps, and the less well known sawflies. In the two superfamilies, we have hundreds of species which occur in as many forms and sizes, and which collectively are probably the world's most useful natural controls over other insects, both harmful and otherwise. DDT killed legions of these useful wasps.

Insects are quick to exploit the natural or accidental appearance or occurrence of forest food materials. A dead animal of any kind is rapidly consumed by the young of blowflies and other dipterous insects. Even dainty butterflies sip the juices of decay. Rove beetles, which are insect tigers, with elongate bodies, short square wing covers, and ravenous appetites, hide beneath carrion and animal excrement and feed upon the other attracted insects.

Most interesting of all the carrion consumers are the large black beetles with big red or yellow markings on their *elytra* or wing covers. They are able to locate dead-animal matter with surprising promptness, and upon arriving at the game, immediately dig a hole beneath it into which it is then lowered and completely buried for their own use. The author witnessed the burial of a Scarlet Tanager within the edge of the half-acre woodland. The bird was completely interred in the daylight hours of one day. Due to the work of these beetles and other yellow-patterned species, dead creatures are seldom found lying on the woodland floor. These carrion beetles, a natural sanitary corps, constitute the family Silphidae.

Here again we see in the activities of these arthropod groups which gather at carrion, the natural conservation of organic "waste" taking place; and when these dead bodies are those of the smaller birds, or the woodland mice, or the still smaller shrews, all of whose delicate bones soon soften and disintegrate, mineral matter useful to the forest is thus also returned to the humus in a recycling process which is no doubt accelerated by the carrion beetles.

Another type of exploitation observed in the mini-wood, and entailing an entirely different group of insects, concerned the seeping fluid from cracks and abrasions near the butts of certain white oak trees, a liquid which usually fermented. Such "working" sap exerted a powerful lure upon many insect species, as recounted in more detail in chapter IX.

Insects indeed seem to be attracted to sap wherever it may be found. When the migrating Yellow-bellied Sapsuckers stop to drill rows of drinking holes in willows and maples and other trees, hornets and many flies take a share. Occasional moths and butterflies also sip the oozing fluid after the Woodpeckers have gone elsewhere, and sometimes such vertebrates as Starlings and gray squirrels find these holes and drink at them.

Rather rare butterflies found drinking the oak sap, and with nothing else present to account for their appearance, were the pearly eyes, *Lethe portlandia* (plate 133), whose green caterpillars bearing two curious red horns on their heads feed on grasses not found in the mini-wood.

The mourning cloak butterfly, *Nymphalis antiopa,* so resplendent in purple, with rows of blue spots behind broad yellow wing borders, delights us because it is one of the first to appear in spring after having hibernated in a tree cavity, outbuilding, or other shelter. The author once found one with its wings (with black undersides) shut tightly above the abdomen and head, hanging wings down under the eaves of the garage.

The hop merchant, *Polygonia comma,* also appears in early spring about the same time that the mourning cloak first flies. The latter's reddish brown, black-spotted wings, when folded straight up as it alights, seems to transform the insect into a dead leaf with instant perfection, only a very close look then revealing the little silvery *comma* marks from which the buterfly takes its name. Both of the above species are shown in plate 133.

Both butterflies hibernate in the fall, those which thus winter over being the late-emerging individuals from waning summer or early fall broods of caterpillars. The author found freshly emerged hop merchants as late as the end of August, and mourning cloaks well into September, both in and about the mini-wood.

Several species lay their eggs on nettles, plants which have not been found in or on the mini-wood borders. The hop merchant is one of these, according to other writers, and the author found none of these odd caterpillars. Holland, in his classic *Butterfly Book,** states: "They vary greatly in color, some being almost snow white," which leaves one wondering why he did not tell of their other, more normal colors.

Mourning cloak caterpillars, preferring the leaves of elm, willow, or poplar, found a ready supply of the former leaves in the fine mature elm shading the author's house, while the hop merchants must have found their food plants well beyond the area.

The blackish, red-legged, and very spiney caterpillars of

____

*W. J. Holland, *The Butterfly Book.* (New York: Doubleday, Page & Company, 1902). Many reprintings since original edition, and still available through occasional book lists. Book has forty-eight plates in color, and black-and-white illustrations.

the mourning cloak butterflies fed unmolested by birds or other animals in this elm, and the curiously spiked chrysalides were occasionally found suspended from the thick end of clapboards on the house not far below the tree's overhanging branches (plate 135).

Two of our largest butterflies, the strong flying tiger swallowtail, *Papilio glaucus*, and the spicebush swallowtail, *Papilio troilus* (plate 133), were frequent visitors about the borders, and less often within the mini-wood. Here in the North, the vast majority of the "tigers" are of the striking bright yellow and black-banded form, and only rarely observed are the very dark gray individuals (always of the female sex), in which the upperside black banding on the forewings and the stripes on the underwings may be nearly obscured. Fully expanded, the wings of one such dusky form spread just four inches.

The forewings of the spicebush swallowtail are basically black, with small rows of bluish and yellowish spots close to the borders. The underwings, also black, each bear a single, diffused broad greenish or bluish band, and are bordered with half-moon-shaped greenish or bluish marks, and one orange spot located on their upper surfaces. Specimens collected in or near the mini-wood which had bluish banded wings were all females.

The first tiger swallowtails emerged from chrysalides about April 30 and were followed by other and larger individuals as the season advanced into summer, the last ones having been recorded in mid-August. These late-summer butterflies deposited eggs; from these the larvae that emerged and succeeded in pupating before cold weather set in wintered over in that stage to become the first spring butterflies of the following year.

The preferred food plant, and the one at which the big females were often seen hovering and depositing their eggs one by one on the leaves, was the wild cherry or choke-cherry, a tree common within the mini-wood and about its borders.

The full-grown caterpillars of the tiger swallowtail are among the more extraordinary looking of the lepidopteran forms. Of a dark, mottled green, robust, with the head-end rising and then sloping forward like the front of a 747 plane, and equipped with a pair of brightly ringed *false eyes*, they have a not-to-be-tampered-with appearance which no doubt is of value in frightening off would-be predators. A curious raised semi-ring or collar, posterior to the "eyes,"

adds to the odd appearance of this fat caterpillar (plate 135).

The spicebush swallowtail, *P. troilus*, as its common name implies, prefers spicebush leaves on which to deposit its eggs, but it also sometimes chooses sassafras. In and out of the mini-wood, a few appeared in mid-May, but the majority emerged later—through June and into July—and the last (females), in September.

The caterpillars are as spectacular as those of *P. glaucus*, being similarly shaped and dark green shading to lighter green below, but with large double ringed eyespots, part blue and part black. They also have black-rimmed yellow spots with blue centers on the first abdominal segment, and black dots with blue centers on the remaining segments. The chrysalis is shown in plate 136.

Both of these big butterflies—the tiger swallowtail and the spicebush swallowtail—seemed irresistibly attracted to the blooming phlox in the author's flower beds, where they were easy to observe at close range. The sweet-scented nectar-rich blossoms exerted a calming, almost tranquilizing effect upon them, causing both to forage for long periods and forget their customary erratic flights. At the phlox they tarried to probe each flowerlet, sucking the nectar through their long uncoiled tongues.

A delicate little butterfly named the hickory hairstreak, *Strymon caryaevorus*, proved to be a regular warm-weather member of the mini-wood community (plate 133G).

Measuring but an inch and three-eighths in width with wings fully expanded, it blended splendidly with the over-all color of the leaf carpet when flying close to the woodland floor, although it was oftener seen sunning itself upon horizontal leaf platforms.

Plain grayish brown on the upper-wing surfaces, it was more attractively patterned underneath, the hind wings here marked with an orange eyelike spot on either side of a patch of blue. Delicate wavy lines side by side in brown and bluish white were also conspicuous where they crossed the underwings. From about midway between the orange spots on the borders of these wings, the tiniest imaginable swallowtails could be seen in movement in the slightest air current. How these tails remained unbroken was a mystery, and sometimes they still remained, even on old and tattered individuals.

The caterpillar of this butterfly remained unknown to

the author, despite the long search for it on the various oaks and the one pignut hickory in the mini-wood, upon whose foliage it has long been reported as feeding. That the species bred within the half-acre area seemed proven by chrysalides recovered, from which the butterflies emerged successfully. These were found beneath objects lying on the woodland floor—pieces of peeled-off bark or a fragment of board; in one instance, two chrysalides were found under a piece of roofing paper that had probably been blown into the area.

In general outline the pupa resembled that of an ordinary baking bean, except for a slight constriction which marked off the abdomen from the thorax of the developing butterfly. Close inspection revealed the form of the legs of the insect, as well as its antennae, molded on the exterior of the pupal skin, as is usually the case with pupal moths and butterflies. Those of the hickory hairstreak measured only three-eighths of an inch in length, and were dark brown before the butterflies issued. From two of these, found under a piece of bark beneath the pignut hickory, on the leaves of which the caterpillars probably fed, the imagoes emerged on June 15 and June 27.

These frail looking, neat little insects are alert and even comically pugnacious toward other members of their species, often giving harmless chase to intruders into their chosen territory.

This innate pugnaciousness in butterflies has always seemed to the author to represent a form of jovial bluff, for these insects have no possible means of hurting one another. It is the fun of the chase and the excitement of the thing, to zip off thus in pursuit of one's own kind or a rival or possible mate, or, better yet, to succeed in putting a bigger butterfly than one's self into retreat with all possible haste.

Past masters at this sport are certain members of the family Hesperiidae, commonly known as "skippers."

There was one species in particular which appeared each year with great regularity on the mini-wood borders. It made a rollicking sport of the chase, darting at every butterfly and, sometimes, at other kinds of insects which came within its apparently sharp vision. This was a little skipper that always impressed the author with its dashing spirit, with its compound eyes, large for its size, and with its stoutly proportioned thorax evolved to house the powerful wing muscles for its special kind of existence. This was the zabulon skipper, *Poanes zabulon,* a rich yellow and black-winged insect with a spread of about an inch and a quarter (plate 133H).

Annually selecting its territory close to the woodland border, it also selected two or three special watching and resting places, to one of which it usually returned after each chase or other sortie. While at rest between these dashes, which were almost too fast for the human eye to follow, it had the peculiar habit of spreading and flattening its underwings while holding the forewings open at an angle above, thus displaying its black and bright yellow patterns to the best possible advantage, probably to catch the eye of any passing member of the opposite sex.

Despite these regularly displayed, rather aggressive traits, zabulon was a tame and trusting butterfly where human beings were concerned. Often, each season, as the author sat in the sunlight near one of the skipper's regular resting places, it would suddenly shift to the head or a purposely outstretched finger of the amused observer. Then, in a confident manner, it would sortie and return to the same newfound vantage point, as if the author were naught but another harmless inanimate object.

From about May 26 until September 20, individuals appeared each year, all following the behavior just outlined; and no individual lived for more than two or three weeks, as far as could be ascertained.

Another, similar skipper, the "whirlabout," *Polites vibex,* (plate 135), followed the same behavior pattern as the zabulon skipper (plate 133H), tame, but vigorous, and appearing at the same time of the year.

A third species, the larger silver-spotted skipper, *Epargyreus clarus,* powerful of flight and conspicuously marked, was also an occasional visitor along the mini-wood borders (plate 133P).

Moving through the mini-wood at times in search of nectar, like a tiny bluish shadow, was the eastern-tailed blue, *Everes comyntas,* already described in the chapter pertaining to the area's wildflowers. (See plate 133, top right, single photograph.)

Butterflies and moths have been evolved with a wonderful variety of colors and patterns which doubtless serve to protect them, although in ways not always easy to explain. To further insure their success as an order and make successful coition more certain, the last abdominal segment in the males has been furnished with paired *claspers* with which to grasp the female's abdominal tip above the entrance to the *bursa* or seminal receptacle.

The form of these claspers may be roughly visualized by holding one's hands, palms together, and with all fingers straight up and touching for their full lengths. Then, with the thumbs representing the two wings of a hinge, open and close the hands.

So firmly do the claspers hold during the mating process that occasionally one may see the female of certain species carrying a male below, head down and with his wings tightly folded. Indeed, the walls of the claspers may also be equipped with a series of ridges which make his grip more certain during mated flights.

Purely transient butterflies that occasionally drifted through the mini-wood, or tarried for a bit in response to some subtle scent or to dip experimentally into the nectaries of some unaccustomed flower, included the great spangled fritillary, *Speyeria cybele,* a strong, swift-flying species, which appeared sometimes by June 15, and was followed by more in July and August, and, rarely, by one or two in early September. Its wings, which spread three inches, are orange-brown with black spots, crescents, and other dusky markings, while the undersides of the underwings are gorgeously spangled with silver, brightly polished, startlingly beautiful! (plate 133K.)

Eggs of *cybele* were doubtless deposited on the leaves of violets, the customary food plants of this genus, but although a long search was made by the author, no eggs or larvae were ever found on any of the species of violets which grew within the mini-wood or close by its borders.

The red-spotted purple butterfly, *Limenitis arthemis astyanax,* was another of these fine transient species which were occasionally seen drifting through the woods or feeding nearby. With a three-inch spread of dark purple forewings, each showing a row of bluish-white border markings, and underwings each with two rows of blue markings separated by a band of black, they are truly beautiful insects. The numerous red spots from which the species takes its common name occur on the undersides against a sort of smokey brown background on all four wings (plate 133L).

A related, somewhat similar butterfly of rare occurrence was distinguishable at a glance by the presence of a broad white band on each of its neat dark purple wings. This was *Limenitis a. arthemis,* seen but twice in the area under study. Its common name, the white admiral.

The red-spotted purple was recorded at various times between June 9 and September 10. The rarer, white-banded species probably had a similar flight period. Fond of decaying fruit, both butterflies were lured for close observation by placing mashed pears on the ground in a sunny situation.

Delicate little gray wood satyrs, *Euptychia cymela,* butterflies with two dark eyespots on each wing; larger insects called graylings, *Cercyonis pegala alope,* bearing yellow patches with two black-ringed blue eyespots on each forewing; and the painted lady, *Vanessa cardui,* patterned with yellowish-orange, black and white above, and displaying yellowish-ringed black eyespots set in gray and white on the undersides of the hind wings, were other interesting visitors in the mini-wood or about its borders (plate 133M [underside], -N and -O).

Less often observed were the colorful common sulphur butterflies, *Colias philodice,* with bright yellow black-bordered wings; and the always rare (in woodland settings) alfalfa butterfly, *Colias eurytheme,* a more or less similar-appearing insect, with black-bordered more or less orange (rather than sulfur-colored) wings (plate 133Q and -R). Both of these butterflies are more typical of the meadows and open country.

The introduced and abundant European cabbage butterfly, *Pieris rapae* (plate 133S), a creamy white species with two black spots and a black tip on each forewing, was seen annually from the first warm days of April until late September, there evidently being more than two broods. In midsummer they were often observed depositing eggs singly on separate leaves of the author's cultivated spider plants, an interesting fact and substitution in this area where growing cabbages are a thing of the past. (The spider plant is *Cleome spinosa.*)

The green, hairless, and spineless caterpillars of this species have been destructive to the cabbage crop over much of the country elsewhere, but the pungent odor of the spider plants, which bloom right up until frost, proved diverting to this butterfly and evidently suitable for its eggs.

Two large species, the familiar monarch butterfly, *Danaus plexippus,* and its "mimic," the smaller viceroy butterfly, *Limenitis archippus* (belonging to the same genus as the red-spotted purple), were frequent visitors in the area under observation.

Both of these butterflies have the same basic orange-brown wing coloring, both have striking jet-black veins in all four wings, which are also black bordered with *similar* white spotting, but which will be seen to be quite different upon close inspection. The viceroy is also easily distinguished from the monarch by an unmistakable diagonal black line somewhat back from the black-and-white-spotted border of each underwing. The monarch has a wingspread of up to three and a half inches; the viceroy, three or a little over (plates 133T and U).

The old theory maintains that the monarch is so bad tasting that predators leave it strictly alone, and presumably the viceroy, which might be *good* eating, gains immunity, or at least protection in some measure, because it so closely resembles the former species, and thus it is termed a "mimic."

The theory does not always hold water, however.

Experimenting with monkeys at the Bruce Museum, the author found that Joe, a *Rhesus* male well along toward thirty-six years of age when this was being written, accepted monarch butterflies but ate the bodies only after tearing off the wings.

Much more significant, however, is the fact that our native Crested Flycatcher, which nested many times in birdhouses in and outside of the mini-wood, regularly fed its ravenous offspring on several kinds of butterflies, including red admirals, mourning cloaks, tiger swallowtails, and at times the monarch, as documented in the author's motion picture films.

Dragonflies of the largest size were also fed to these young birds. In *all* cases (butterflies included) the insects were fed to the birds *whole,* the nestlings somehow managing to get wings and all down their eager gullets and into their stomachs. The adult Flycatchers did not first remove the insects' wings, and no wing fragments were found within the nest boxes upon examination directly after such feedings.

It is now believed by some that the monarchs may become more or less distasteful as food for the birds, according to the species of milkweed upon which the butterfly's larvae fed, the plants having various chemical qualities.

Experiments have been reported in which birds that were induced to eat monarch butterflies vomited soon afterward. Milkweed latex, a sticky pure white fluid within the stems and leaves, is very bitter to the human taste buds. A species of this plant, the northeastern milkweed, *Asclepias syriaca,* grows abundantly not far beyond the mini-wood area, and if the monarchs fed to the young Flycatchers were of such local origin, it would indicate that here at least this particular milkweed which is so common in Connecticut contains no bad properties for birds in its sap. (The plant referred to is illustrated in the botanical portion of this book, in plate 47.)

CHAPTER XIV

INSECTS II:
CATERPILLAR EXPLOSIONS

Looking at this bit of woodland, it was a comfort to know that the bulk of its greater and lesser arboreal members—that is to say, its species of trees and shrubs—had long triumphed over winter storms, summer winds and lightning, at least two hurricanes, most plant diseases, and other hazards to which plant and animal communities, large or small, are subject.

Seventy or more feet in height, its oldest trees informed one that these things must be true, but at the same time, as the author lived along with this beloved mini-wood through the winters and looked with delight upon its spring and summer riches of foliage and bloom, a false sense of overconfidence for its future safety was for a time engendered.

Unfortunately, the forest's tranquillity may be rudely upset at times by foliage consumers which may seem more avid than ever before—caterpillar species which, unnoticed by most of us as they increase in numbers over several years, suddenly reach the zenith of a cycle, horrifying the uninitiate by their sheer numbers and an unaccustomed steady pattering sound, like gentle rain, caused by finely fragmented green foliage converted to frass (droppings) by the insects, resulting in steady lessening of summer shade as the infested trees are defoliated.

Two of the most damaging of these insects have been the imported—and now naturalized—gypsy moth, *Porthetria dispar,* and, of late, an inconspicuous native all-white "miller," commonly known as the elm looper and technically as *Ennomos subsignarius,* whose caterpillars have become indiscriminate defoliators of trees and shrubs and

also many kinds of smaller plants, if the avid insects happen to drop on them (plates 134 and 135).

It was a curious fact that, while the gypsy moth had been numerous in New England for some time, and this included parts of Connecticut, only recently did it reach somewhat alarming numbers in the mini-wood and in other parts of Norwalk, and in New Canaan and the surrounding country. It first appeared in the mini-wood (noticeably, because of the numerous males suddenly flying) in 1968, increasing, one might say, right under our noses for a number of years, and mostly during the time when adequate spraying was no longer possible because of its lethal effect on other organisms. The author was fortunate in making an interesting discovery concerning the gypsy moth caterpillars and the adults in the mini-wood.

Within this area, many birdhouses had been maintained, watched over, and cleaned after any species of bird or mammal had used and abandoned them. In almost all instances in which hole-nesting birds had taken over these boxes, they had finished nesting activities by the middle of July, a date which coincided with the most active time of the moths' rapidly maturing caterpillars, large numbers of which were then nearing full growth.

It was discovered, upon cleaning out the recently used boxes, that gypsy moth caterpillars, sometimes in large numbers, had been coming down after feeding all night in the canopies of the various oak trees and congregating within these boxes by day, regardless of the often foul condition in which they were left by the birds.

Many of the full-grown caterpillars now *remained* in

the boxes and there transformed to the pupal state. Larvae and pupae were found there in numbers, often crowded together, the latter sometimes separated from the caterpillars just by a flimsy network of silken threads.

As the summer advanced, the larger *flightless* black-speckled white females commenced emerging from their chrysalides, but never leaving the boxes thereafter. They were soon followed by emerging smaller males, who then mated with the larger insects, so to speak, on the spot!

Other males from other parts of the woods were also attracted by the supply within the birdhouses, and soon the ripe females were forming curious flat cushions of yellow woolly material or scales (it would seem from their own bodies), within which were deposited large numbers of their glossy globular eggs (plate 134).

For some time before these final goings on in the life history of these insects, the caterpillars had been descending from the foliage meadows high above and using the boxes as daytime shelters, with the inevitable result that large piles of their excrement, in the form of pellets, only added to the foul conditions therein. What happened to the heaps of moist droppings is another part of this interesting story and will be related presently, but the really significant discovery relative to the caterpillars and pupae, to the moths and their egg "cushions," was that the birdhouses were acting as reliable traps for these nuisance insects, and from them, all stages could be collected and destroyed (plate 134).

The static, large-winged, yet flightless females evidently had left within the boxes a lasting sex scent that continued long after the insects and their yellow egg cushions had been removed, for occasional males continued to hover about the entrance holes to the boxes during the entire flight period, lasting from about the second week in July, through the first week in August.

This potent sex attractant secreted by the females has been analyzed and is now being synthesized in the laboratory, and it has been successfully used for trapping the males. The compound is called cis-7, 8-epoxy-2-methyloctadecane. It is hoped that soon it may be available in a cheap and easily useable form so that a simple trap-and-attractant combination package will be sold at most garden-supply establishments.

Gypsy moth caterpillars, when full grown, may measure slightly over two and a quarter inches in length. They are equipped with long protective hairs which probably make them distasteful to most birds. They are basically brownish yellow, with four rows of odd tubercles (four to a segment), some of which are blue, and others red. They are voracious feeders on the leaves of oak, poplar, and many other plants and forest trees. Lutz states in his *Field Book of Insects* (see the list of Recommended Books) that "more than five hundred species of plants, including conifers, are in their dietary."

The male moths are attractive-looking insects, olive-brown, with intricate darker markings, and feathery antennae, while the larger females, as already mentioned, are white, with scattered brown and black spots and other small markings (plate 134).

While caterpillar explosions often panic the uninitiate, it is true, in the case of the gypsy moth, as for many other moths, that, after reaching a zenith year numerically, the species usually declines quite rapidly in that area, or indeed, it may be absent altogether the next year after reaching the apex of its cycle.

In keeping masses of gypsy moth egg cushions for study, the author found minute hymenopterous insects issuing from many single eggs. This minute parasitic wasplike species thus exerted some measure of control over the moth, the latter being an imported species; however, it did become established in this country, where its native parasites were missing, and their absence doubtless hastened its spread as a pest species in America.

If spraying against gypsy moths must be done, the safest pesticide to use is the bacterial preparation *Bacillus thuringiensis*, which only attacks and kills the caterpillars. It is supplied to tree specialists under the trade names of Thuricide HP, Biotrol BTB, and Dipel. It is hoped that it does not injure other forms of animal life.

As mentioned before, the excrement deposited by the caterpillars within the birdhouses during their daytime resting periods accumulated as large piles of moist green pellets composed of partly digested white oak and swamp white oak leaves, digestive fluids, and bacteria.

One such mass had caught the airborne spores of a fungus, probably when the stuff was fresh, and in early October its fruiting (spore-bearing) bodies, in the form of a dozen rather leathery yellowish toadstools, had arisen over the natural manure heap that had nourished them within the box!

By the time these toadstools were discovered (when the

The six largest New England moths, which are growing ever rarer. (A) *Cecropia*. (B) *Polyphemus*. (C) *Promothea* male. (D) *Promothea* female. (E) *Luna*. (F) *Regalis*. (G) *Imperialis* male. (Pages 123, 124.)

Above: The underwing moth, *Catocala amatrix*, form *selecta*, shown with protectively colored wings closed. *Left*: With highly colored underwings exposed. Upper specimen shown natural size. (Page 124.)

Twelve of the fifteen underwing (*Catocala*) moths found in the mini-wood. Stated colors apply to underwings only. *Column 1:* Red with black bands. *Column 2:* Top pair, red with black bands. Next two below, yellow with black bands. Smallest moth, black with white borders. *Column 3:* Top two, black with white borders. Next below, (underside) banded with black and white. Two lower pairs, rich yellow with black markings. All shown one-half natural size. (Pages 124, 125.)

Over four hundred parasites (Hymenoptera) which issued from a single pupa of the common mud wasp, *Sceliphron cementarium,* enlarged about six times life-size. *Inset:* An ichneumon wasp that parasitizes the European cabbage butterfly. $\times$ 4. (Page 125.)

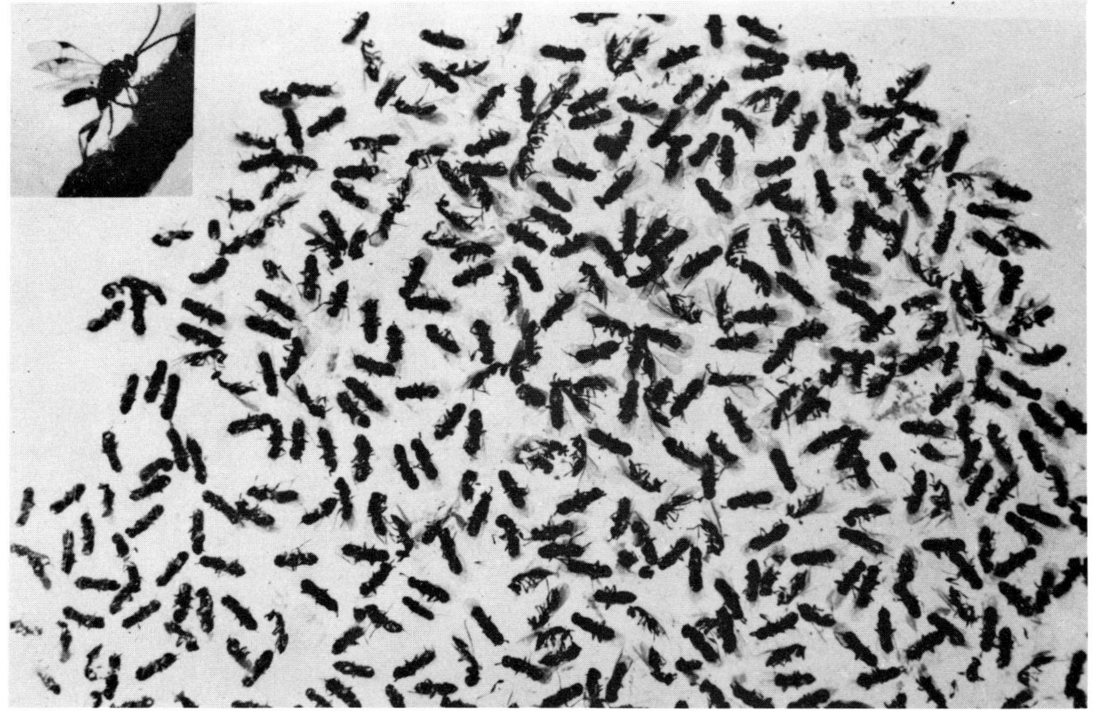

Eastern tailed blue, enlarged × 1 to show the tails, broken off the specimen at J, in picture at left.

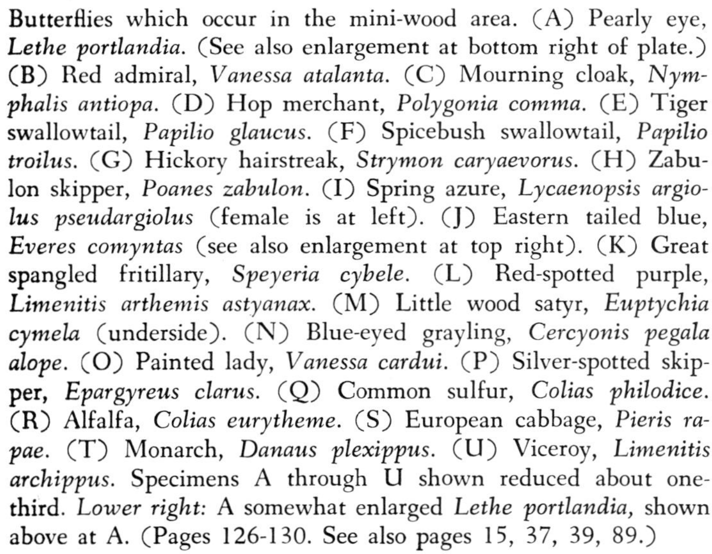

Butterflies which occur in the mini-wood area. (A) Pearly eye, *Lethe portlandia*. (See also enlargement at bottom right of plate.) (B) Red admiral, *Vanessa atalanta*. (C) Mourning cloak, *Nymphalis antiopa*. (D) Hop merchant, *Polygonia comma*. (E) Tiger swallowtail, *Papilio glaucus*. (F) Spicebush swallowtail, *Papilio troilus*. (G) Hickory hairstreak, *Strymon caryaevorus*. (H) Zabulon skipper, *Poanes zabulon*. (I) Spring azure, *Lycaenopsis argiolus pseudargiolus* (female is at left). (J) Eastern tailed blue, *Everes comyntas* (see also enlargement at top right). (K) Great spangled fritillary, *Speyeria cybele*. (L) Red-spotted purple, *Limenitis arthemis astyanax*. (M) Little wood satyr, *Euptychia cymela* (underside). (N) Blue-eyed grayling, *Cercyonis pegala alope*. (O) Painted lady, *Vanessa cardui*. (P) Silver-spotted skipper, *Epargyreus clarus*. (Q) Common sulfur, *Colias philodice*. (R) Alfalfa, *Colias eurytheme*. (S) European cabbage, *Pieris rapae*. (T) Monarch, *Danaus plexippus*. (U) Viceroy, *Limenitis archippus*. Specimens A through U shown reduced about one-third. *Lower right:* A somewhat enlarged *Lethe portlandia*, shown above at A. (Pages 126-130. See also pages 15, 37, 39, 89.)

Top left: A female gypsy moth placing her cushionlike hairy egg covering. × 2. *Top center:* The naked egg mass, the very dissimilar male, and an opened egg cushion. Male about life-size; eggs much enlarged. *Bottom left:* Two veteran birdhouses, and, on roofs, the gypsy moth pupae removed from each. *Above:* Some of the larvae and pupae, shown half life-size. All photographs taken in mid-July. Birdhouses proved to be efficient traps for these insects. (Pages 131, 132.)

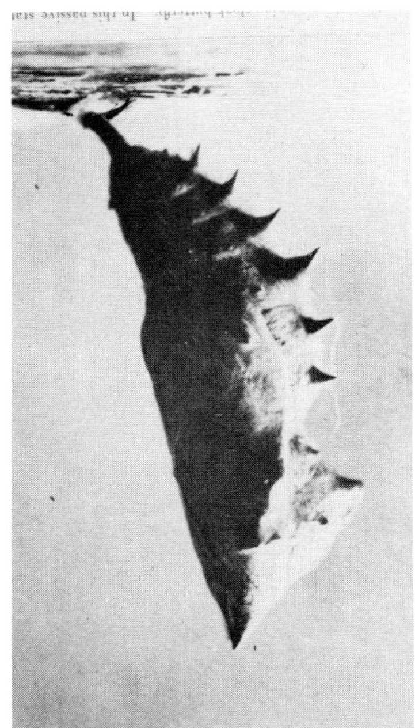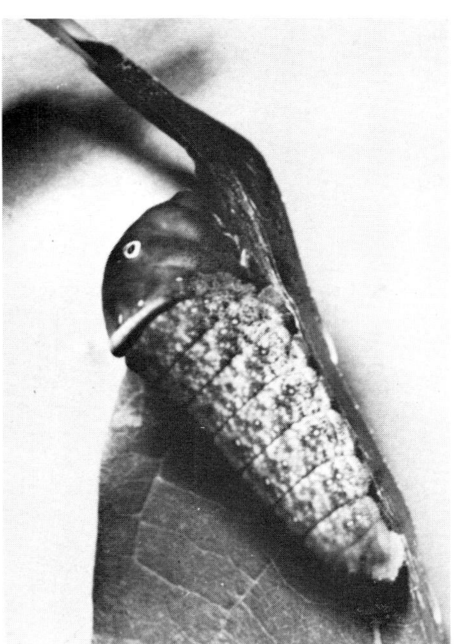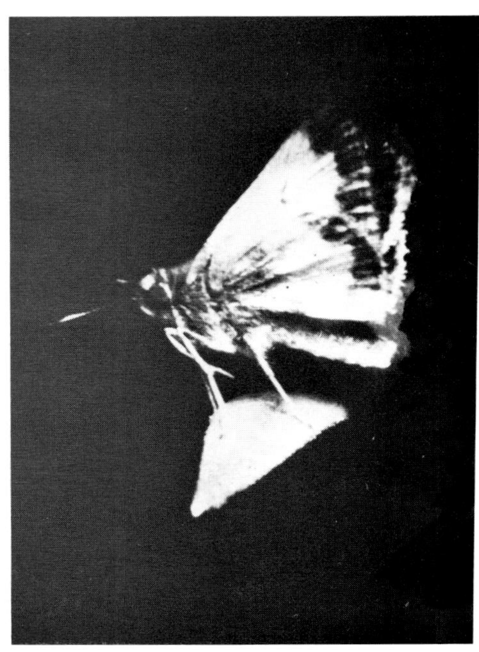

Left: Spiked chrysalis of the mourning cloak butterfly, suspended from button of silk spun by the caterpillar. $\times$ 2½. Date: May 25. (Page 126.) *Center:* False eyes and a ridge suggesting a mouth on the tiger swallowtail's larva may scare predators. $\times$ 2. Date: September 20. (Page 127.) *Right:* Alert "whirlabout" skipper sunning itself on a leaf tip. $\times$ 2½. Date: August 8. (Page 128.)

Left: The fat gypsy moth female looks nothing like the male (see plate 134). Although possessed of large wings, she never flies; the male seeks out her hiding place. $\times$ 2. Date: July 10. (Pages 131, 132.) *Right:* Flashlight photograph of an elm looper moth, *Ennomos subsignarius,* freshly emerged from the pupa—a pest species in its climax years. $\times$ 3. Date: August 11. (Page 134.)

PLATE 135

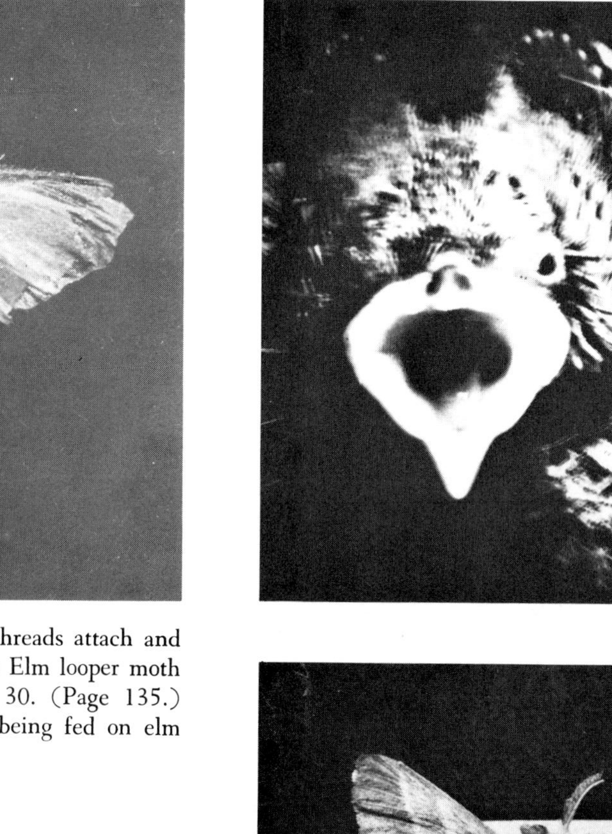

Left: Chrysalis of the spicebush swallowtail butterfly. Silk buttons and threads attach and hold it in position on a leaf. × 2. Date: August 5. (Page 127.) *Center:* Elm looper moth depositing eggs, like miniature bottles, on a twig. × 2. Date: June 30. (Page 135.) *Right:* Eager young Tufted Titmouse just before death, caused by its being fed on elm looper caterpillars. × 2. Date: May 28. (Page 136.)

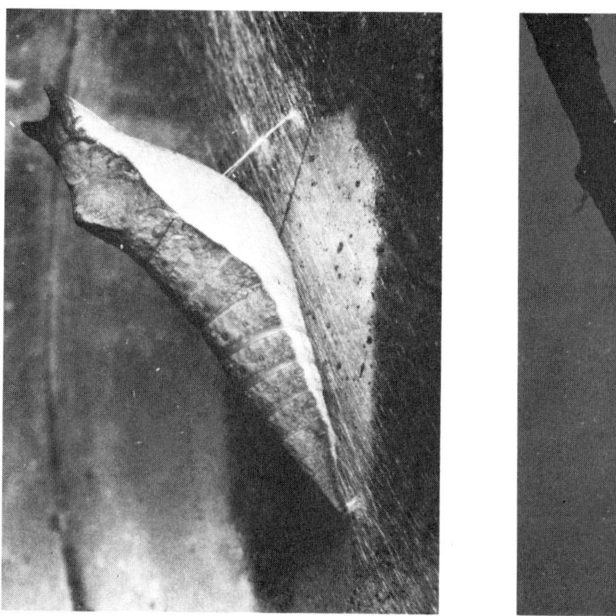

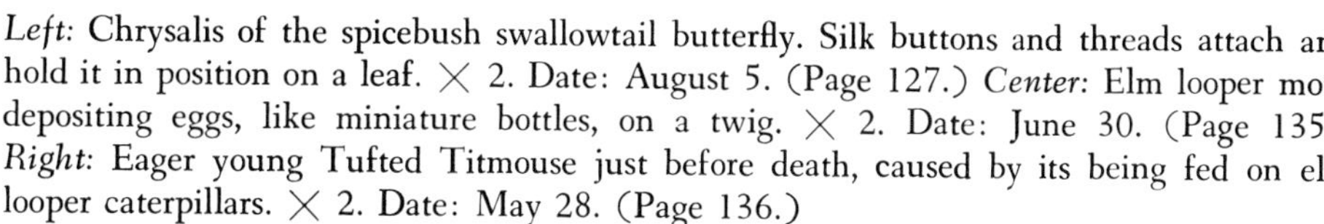

Left: Hairy tent caterpillar. *Above:* The attractive moth which is its parent. Dates: Caterpillar, June 1; moth, June 30. Both about twice life-size. (Pages 136, 137.)

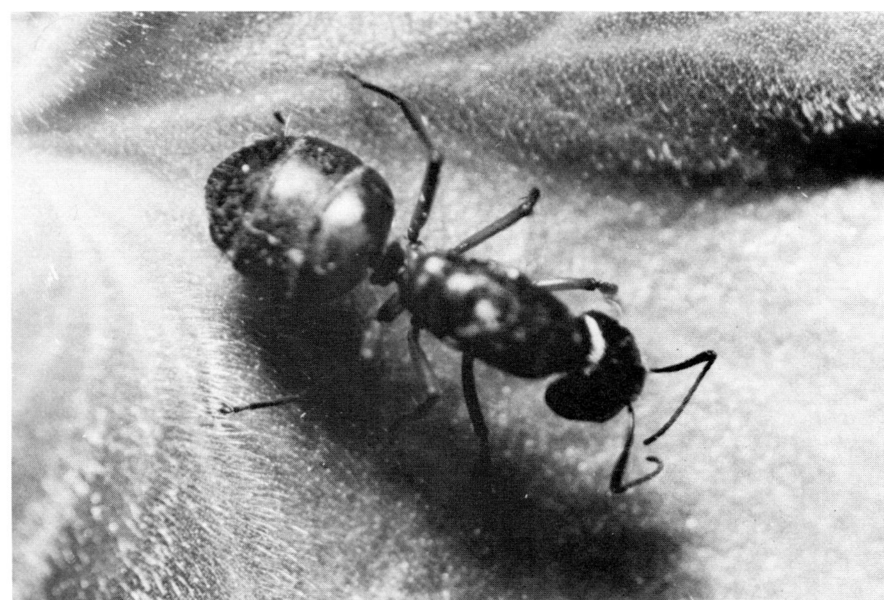

Left: Three examples of the tortricid moth, *Archips argyrespilus* (at top of photo), whose caterpillars are leaf-rollers, occurring some years in great numbers and on many kinds of vegetation. Chrysalis is also shown. Date: June 15. (Page 137.) Large specimen is a male fall cankerworm moth. Date: December 8. (Page 137.) All illustrated natural size. *Right*: Queen ferruginous carpenter ant that outlived her worker ants. Four times life-size. (Pages 139, 140.)

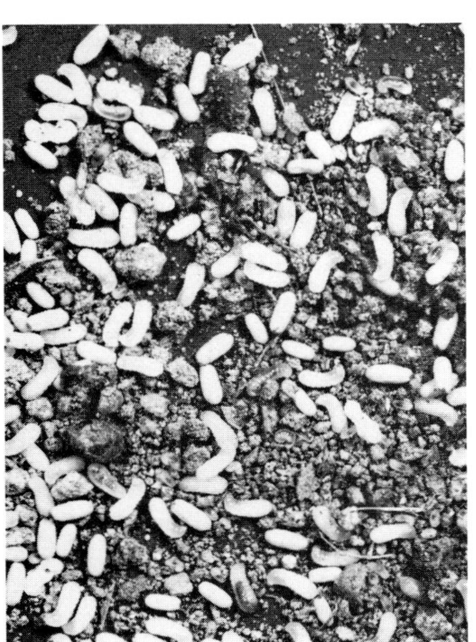

Left, above: White silk cocoons spun by larval *Lasius* ants. *Bottom left*: Some of the larvae (slightly bowed individuals) and newly spun cocoons. Somewhat enlarged. *Above*: *Lasius* ants feeding at night. Flashlight photograph shows how they have placed their "herd" of aphids neatly on ribs of a sugar maple leaf. Somewhat enlarged. Date: November 30. (Pages 11, 142.)

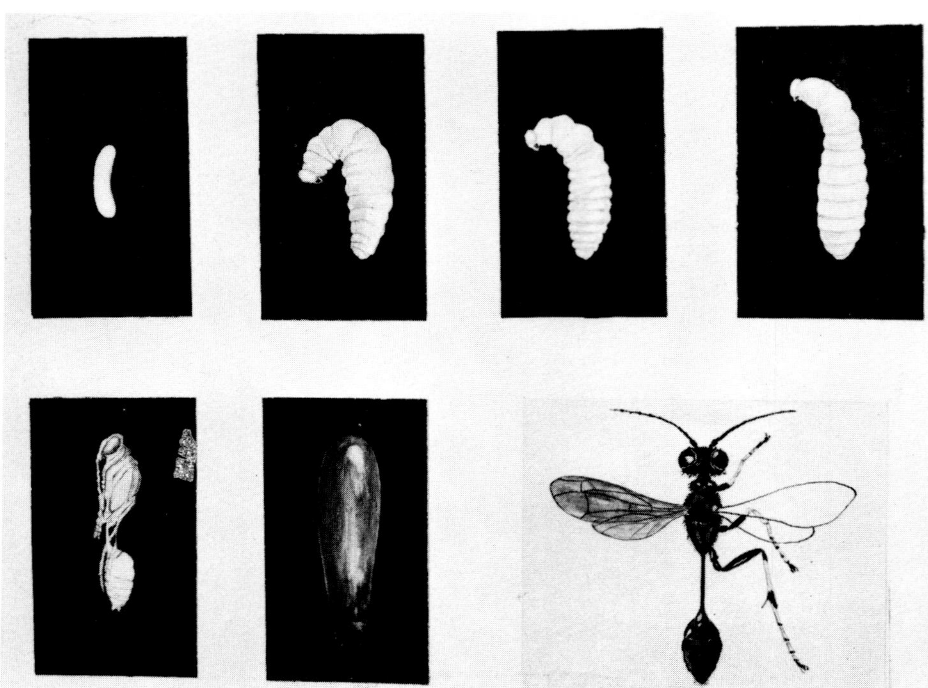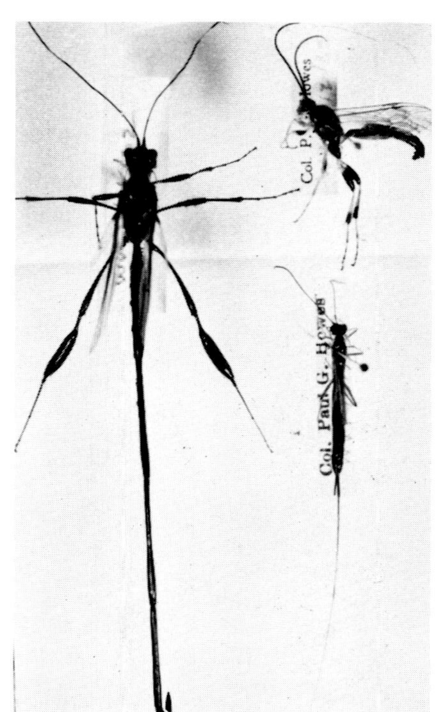

Above: Stages in the life of the mud wasp, *Sceliphron cementarium.* (Top rows, from left to right) egg and three larval stages: pupa, cocoon, and adult wasp. Somewhat enlarged from a painting by the author. (Page 143.) *Right:* Ichneumon wasps with long ovipositors. Natural sizes. (Pages 142, 145.)

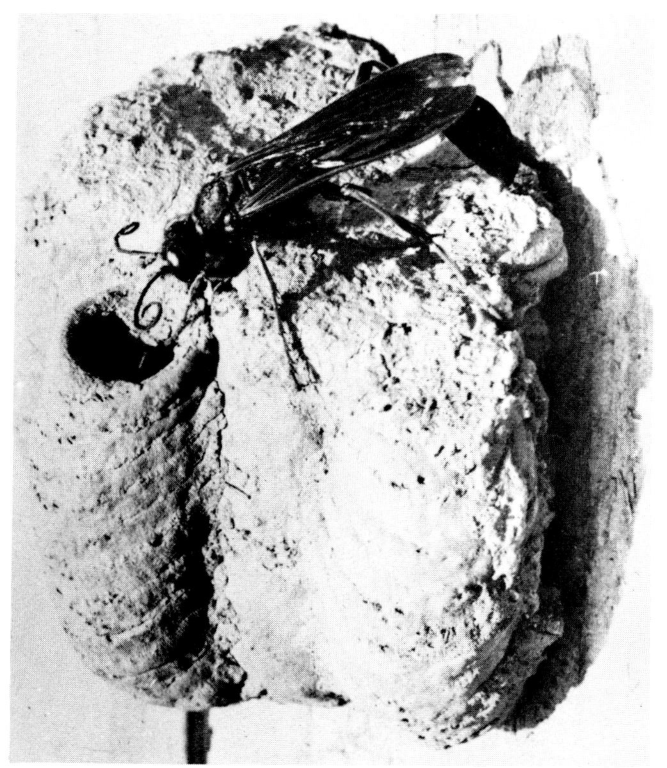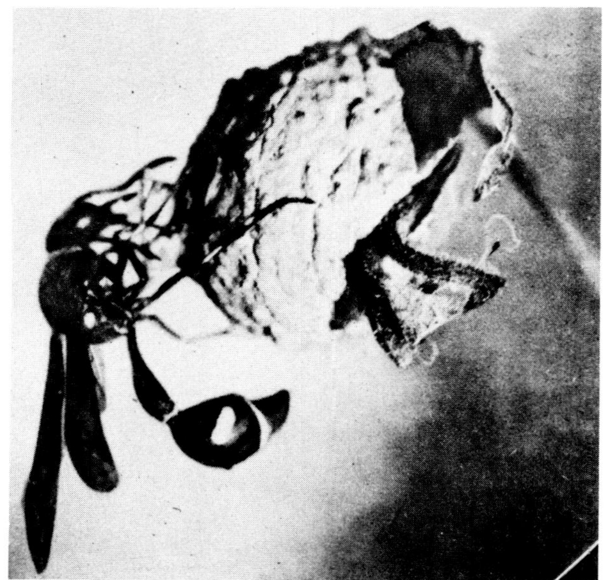

Left: Enlarged photograph of the wasp *Sceliphron cementarium,* ready to fill newly made cell with spiders. Note concentric construction of cell (hidden under additional mud in previously made cells). (Page 43.) *Above:* Potter wasp, *Eumenes fraternus,* on her newly finished clay jug. Four times life-size. (Page 144.)

Top: Row of seven cells fashioned of leaf fragments by the solitary bee *Megachile brevis,* in a nest in rock crevice. Two cells at left were opened to show white grublike larvae. Somewhat enlarged. Date: August 20. (Page 146.) *Center:* A typical cell (top, far left) and the pieces cut by the bee in making a single cell. All shown slightly enlarged. (Page 146.) *Below:* Slits made in small twigs by the seventeen-year cicada for its eggs. $\times$ 2. (Page 147.)

Left: After long periods underground feeding on root sap, nymphs of the common cicada or "hot bug" emerge and leave brittle shells from which, as adults, they have crawled, dried their wings, and flown away. Date: July 25. *Below:* Seventeen-year cicada at top; common cicada under it. Shown somewhat more than twice life-size. (Page 147.)

The three "monster" beetles of the mini-wood. *Left:* The orange-brown longhorn, *Orthosoma brunneum,* whose larvae eat dead wood and timbers. Natural size. Date: August 4. *Center:* The blue, green, gold, and red caterpillar hunter, *Calosoma scrutator. Right:* The glossy, all-black "squeaking" bess beetle, *Popilius disjunctus,* which is sometimes colonial. Last two somewhat enlarged. Dates: June 29 and July 18. (Pages 147, 148.)

Pale green lacewing with golden eyes (genus *Chrysopa*), shown with one of its many curiously stalked eggs. Young and adults feed on aphids. Enlarged four times life-size. Date: July 5. (Page 148.)

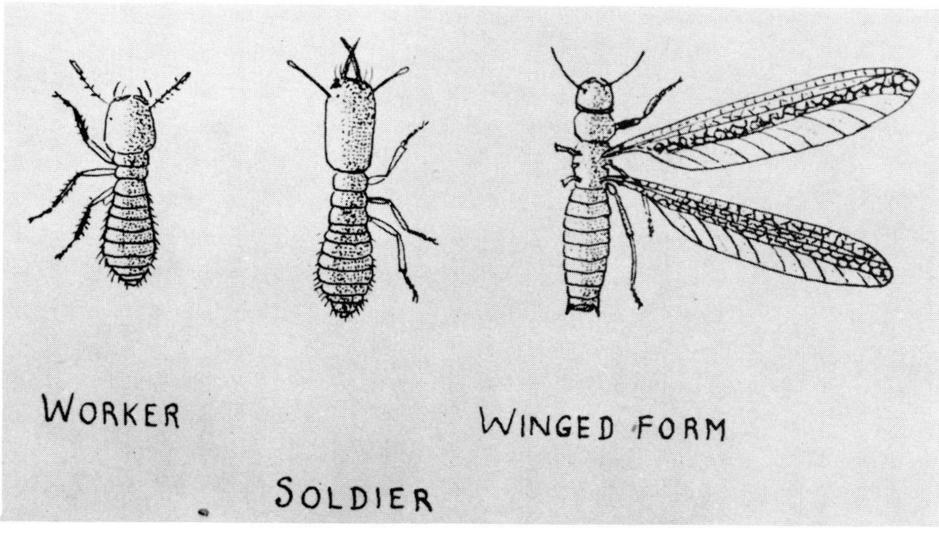

Right: A small portion of a colony of common native white termites, shown natural size. Only the worker caste is illustrated. Flashlight photograph made in the miniwood April 25. (Page 149.) *Above:* Approximate drawings of these insects, highly magnified. Number and arrangement of veins in termite wings are very different from those of ants, for which termites are most often mistaken. (Page 150.)

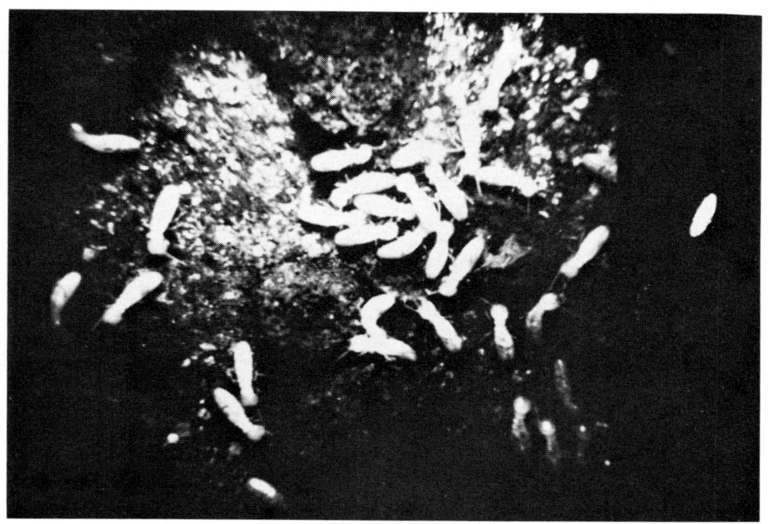

PLATE 141

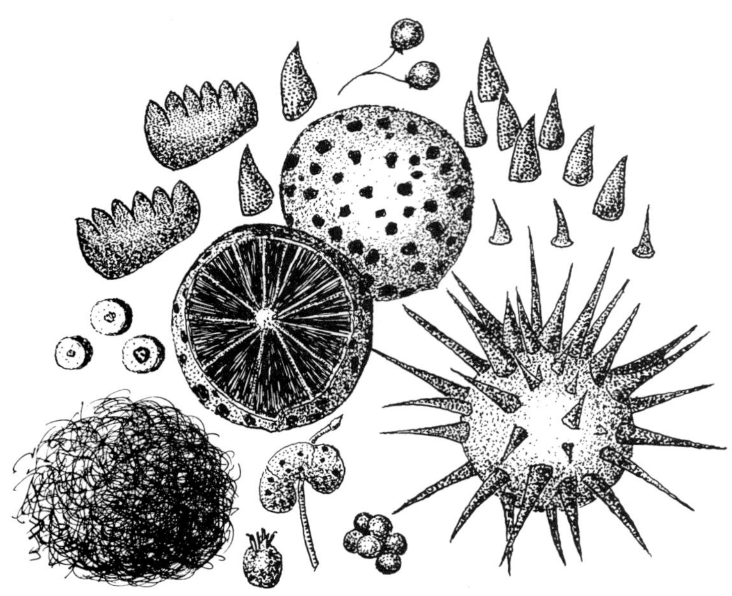

Drawings of common galls. Center two are spotted oak galls, one of which is cut open to show radii supporting central larval cell. At top left, two cockscomb galls from elm leaves; at bottom right, spiny fleshy ball is a rose gall, shown much enlarged. Pea-shaped pair at center top were on jewelweed; cones (to their right and left) were on hickory leaf; furry ball (bottom left) was on wild rose; kidney-shaped gall was on blueberry; cluster of little balls (center bottom) were on northern red oak. Three buttons (left center) were on hackberry. (Pages 150, 151.)

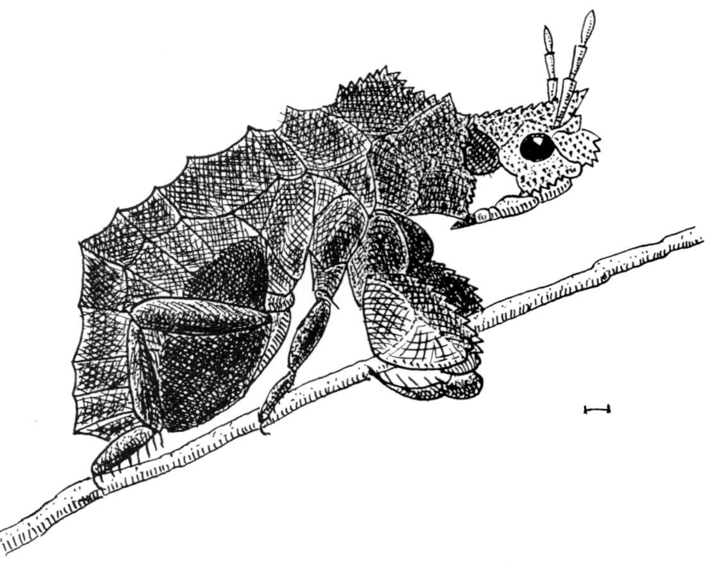

Valuable predators — for man — on other insects are the ambush bugs, with their poisonous (to prey) piercing and sucking beaks. Drawing shows armor of very young specimen. Line at bottom right shows actual length of the bug. (Pages 151-152.)

PLATE 142

Leaves of the white ash almost completely covered with green galls caused by large numbers of mites in various stages of growth within each swelling. One-half life-size. Date: June 15. (Page 151.)

Galls of other forms produced on leaf stems of the white ash, and by mites feeding within. Date: May 30. Shown one-half natural size. (Page 151.)

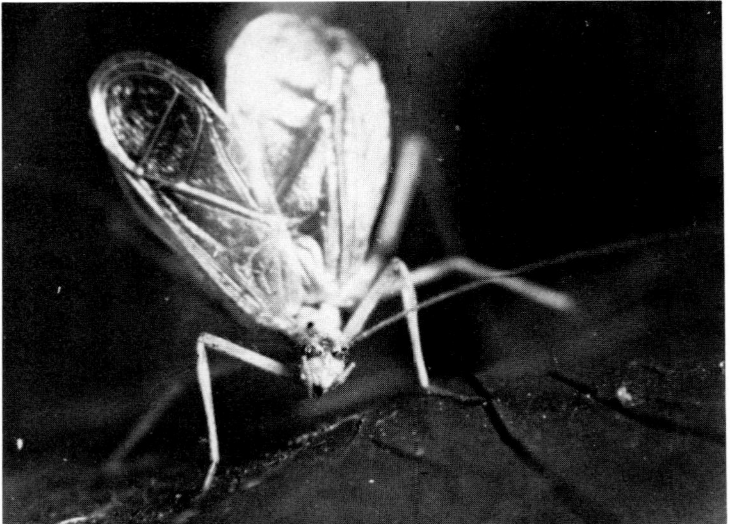

Some insects of the order Orthoptera found in the mini-wood. *Above, left:* A long-horned grasshopper, genus *Neoconcephalus,* with swordlike ovipositor. Date: August 10. *Top right:* A short-horned grasshopper, family Acrididae. Date: July 15. *Right:* With wings expanded, a pale green tree cricket (male) of the genus *Oecanthus* makes its familiar sounds by moving its wings across each other. Date: August 15. *Below:* Oriental mantids hatching from an egg cocoon. *Below, right:* Adult mantid eating a grasshopper. Dates: May 10 (hatching) and September 10. All photographs about twice life-size. (Page 152.)

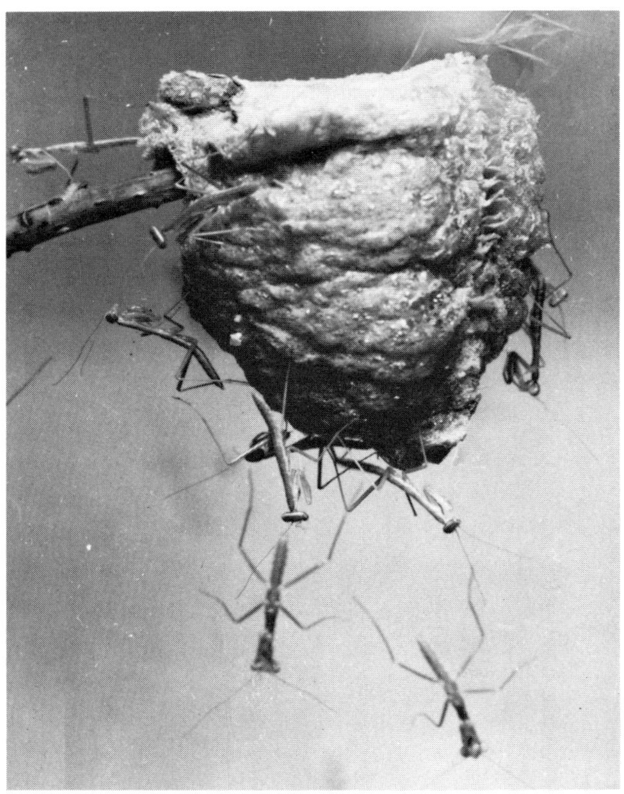

Terrestrial snails *Oxychilus draparnaldi* taken from the mini-wood and maintained as a colony by the author. Shown about one and a half times life-size. (Page 154.)

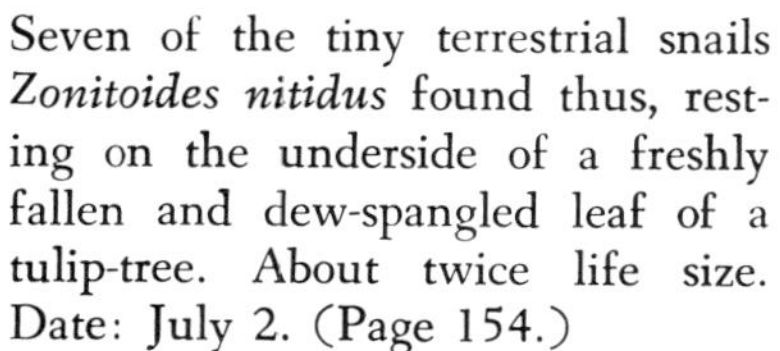

Seven of the tiny terrestrial snails *Zonitoides nitidus* found thus, resting on the underside of a freshly fallen and dew-spangled leaf of a tulip-tree. About twice life size. Date: July 2. (Page 154.)

PLATE 145

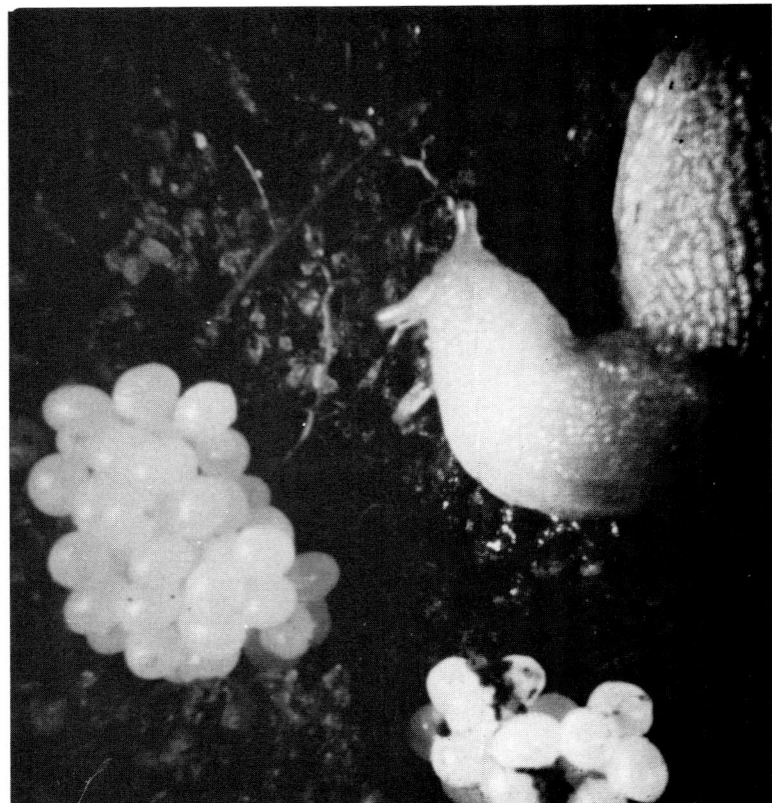

Left, above: Cluster of forty eggs of the slug *Arion subfuscus,* suspended from a twig over a cavity. Somewhat enlarged. Date: October 8. (Page 157.) *Right:* Female *A. subfuscus* with two completed egg clusters, September 27, and (below right) same slug commencing laying, September 26. Both shown much enlarged. (Page 157.) *Left, below:* Three slugs in motion, and three turned over to show their pale-colored undersides.

box was taken down for cleaning), their caps had begun to rot, and now, curious as to what else this birdhouse "complex" might be hiding or might yield, the author removed the whole thing, just as it was, to a covered terrarium, where the droppings and fungus were kept moist by the addition of distilled water.

Within a few days the mass had become very black and partly liquefied, and microscopic examination revealed that it was swarming with nematodes (see chapter XII) resembling so many minute bits of spun glass and pointed at both ends, either free and whipping themselves about in jerky fashion, or half-buried in the gooey stuff with only their "head" ends protruding, swaying from side to side as if moved by some invisible tidal rhythm, and recalling the same phenomenon (probably in the same species of nematodes) found in another kind of dead fungus and described in the earlier chapter.

There were literally millions of these organisms living out their brief life spans in this "waste" matter, brewed of chewed-up leaves, rotting fungi that had already used up some of the medium, bacteria in countless numbers, and subtle chemical compounds, all at the outset begotten from the oak trees (in this case the white and swamp white species).

The nematode consumers were accompanied by larger (but still minute from the human point of view) white grubs, which, after feeding for a time and pupating within the black mass, emerged as delicate little fungus gnats, but with sturdier legs than those of mosquitoes and equipped with very long antennae for their kind. These gnats were members of a very large family technically known as the *Mycetophilidae,* whose still more numerous species are among the standard consumers of fresh and spent toadstools, mushrooms, and other kinds of fungi.

Still other rivals for the slimy food stuff were small brown beetles about two pinheads (three millimeters) in length, while others not much longer but much slenderer, a species of rove beetle (family Staphylinidae), attacked and fed upon some of the grubs of these fungus gnats.

In the past, the author had written of the succession of living forms which had appeared over several years from a single quart of pond muck. The late William Beebe did a fascinating piece on "A Yard of Jungle"—soil—and now here was an entirely unexpected and astonishing series of events revealed by a community of living things concentrated within a single "abandoned" birdhouse!

Small spiders with gray-and-white mottled flattened bodies, resembling diminutive crabs, had also taken up their abodes within this bird box. They rested here, presumably to prey upon the emerging fungus gnats.

Another spider, a tiny green one, constructed a silken shelter *within* an empty pupal shell of a gypsy moth, a convenient layer close by the plentiful food supply. And still this did not write *finis* to the story of this birdhouse complex.

Further scrutiny of its contents revealed tiny winged insects with very much flattened bodies, belonging to the order Psocoptera and popularly known as book "lice" because they are also found among book leaves and in other aging papers (see plate 118).

As we know, the female gypsy moths did not leave the birdhouses in which they emerged. After depositing their eggs, they died, and their spent bodies soon added powdery refuse to the accumulating matter therein.

Another moth, a small gray one, then arrived upon the scene, and laying its eggs upon these gypsy moth remains, begot additional scavengers of "waste" matter in the form of sluggish little caterpillars, which, when full grown, were dull brown, with five lengthwise rows of black dots, and measured half an inch in length.

A few of the little beetles known as "buffalo bugs," *Anthrenus scrophulariae,* whose larvae eat any kind of dead and dried animal matter, including improperly prepared or poisoned museum specimens of all kinds, joined up with the brown caterpillars in helping to dispose of the gypsy moth bodies, and perhaps some of the other accumulated animal refuse.

To recapitulate, and commencing with a nesting pair of White-breasted Nuthatches, these were followed, after their large brood had left the box somewhat fouled, by dozens of gypsy moth caterpillars sheltering therein by day. Not long afterward, these larvae transformed to pupae within the box, and a few weeks later emerged as male and female moths, some of which mated on the spot, males then leaving the premises and the static females remaining to form their egg cushions and then die. Afterward came the toadstools, begotten from the accumulated piles of caterpillar frass, and the fungus gnat larvae and the nematode hordes, all feeding upon the liquefying mass of excrement and dead fungi. Brown beetles and rove beetles preyed upon the emerging adult gnats, and two species of small spiders fed likewise on this abundant game. At length came more scavengers in the forms of small moth larvae, "buffalo bugs," and Psocoptera, gleaners in the dust of a strange succession!

Here, indeed, was one of the more astonishing of the food chains to be revealed in this mini-wood study, and a graphic illustration of natural conservation in full swing.

During the fifteen-year study, the author occasionally collected specimens of a white or (rarely) lightly tinged with green geometrid moth named *Ennomos subsignarius* (plate 136), the so-called elm looper.

Increasing gradually in numbers over a number of years, the species did not reach an explosive stage until after the designated study period for this book had elapsed. But in 1970 this occurred, and with naturalists and tree experts excepted, it seemed to most people that the hordes of brown and black twig-resembling caterpillars had just then burst upon the land as a species. The ugly "things" were then everywhere, consuming the foliage in the trees at an alarming rate, but actually, the species had been building up to an explosive level for some time.

Emerging in early May from rafts of olive-green, mostly unparasitized eggs that had wintered over since the moths' highly successful previous summer, it was not until the minute larvae had grown to an inch or over in length that the explosion manifested itself to the majority of people in its true proportions, the date being about the middle of June.

The *Ennomos* caterpillars, it seemed, would devour most any kind of foliage. Hardest hit in the mini-wood and on the remainder of the author's property, as well as over a wide area of the surrounding country, were the beech trees and their offspring growing in the parents' shade, the red and sugar maples, the tupelo, the dogwood, and the white ash trees, the latter being nearly or completely denuded, according to the size of the victim. Oaks also suffered, but not as seriously as these others in the area under observation.

Many of the caterpillars dropped from the trees and, finding themselves on forsythia, roses, petunias, geraniums, and other cultivated plants, as well as on roadside wildflowers and weeds, proceeded to eat these without hesitation, at least in some cases.

There were so many thousands of the caterpillars, so many on a single tree and even crowding onto single small branches, that in their ravenous haste they cut many leaves into pieces; as a consequence, vast numbers of these green fragments fell to the ground, and though lost to the trees and the caterpillars themselves, they were nevertheless gradually consumed by bacteria and fungi, and earthworms and other animals of the soil, so they were not wasted.

In thousands of cases, upon reaching full growth, these larvae, instead of dropping to the ground to pass the next stage of their lives therein, pupated while still in the trees, drawing partly dried and blackened leaf fragments about themselves with a few silken threads, and transforming under these flimsy networks. Being more or less in plain view, huge numbers of the juicy plump pupae were found and eaten by predators, as we shall see, which had much to do with the decline of the species' cycle to zero by 1972.

Witnessing the *Ennomos* explosion was a truly depressing experience. It burst with such seeming suddenness upon most unsuspecting suburban dwellers that they had little time to take repellent measures before it was too late to save the foliage that season.

Careful and close daily inspection within the half-acre woodland revealed the interesting fact that, even during the worst of the caterpillar damage, the sassafras and spicebush remained untouched by these avid consumers. As we know, the leaves and wood of both species contain aromatic substances pleasing to our human senses but which may have been the deterrent in the case of the *Ennomos* moths at egg-laying time, notwithstanding the fact that another insect, the spicebush swallowtail butterfly, prefers the latter common shrub as the food plant for *its* caterpillars.

We have seen how completely and with what evident ease the wild cherry trees produce new foliage after complete denudation. For many other species, however, the process of recovery is slower and more difficult. Should such pests as the *Ennomos* caterpillars become frequently occurring species, and in such numbers as in 1970, the situation would probably become an insurmountable one for some of these other trees. It is doubtful if such as the beech and white ash could withstand many such complete strippings as we witnessed in 1970 and the somewhat less severe one in 1971. Fortunately, natural controls took over, and the year 1972 was almost caterpillarless in the mini-wood area. How completely the *Ennomos* larvae consumed the leaves of the white ash trees and young specimens of the red maple was truly alarming; the trees prompt comeback was truly astonishing.

The leaflets of the former species grow in opposite pairs from a non-woody leaf stalk, and with a single leaflet at its tip. These stalks, which are also shed after the leaflets have fallen, were all that the caterpillars left uneaten, and they remained thus sticking out in all directions, even up to the time when these struggling ash trees were beginning to send forth small new leaflets. In a few remarkable cases, the new leaflets appeared only two and three weeks after all of the caterpillars had departed.

It would seem to the author that a better common name could be found for *Ennomos subsignarius* than the *elm looper* or *spanworm*, since the animal is neither a worm nor by any means an exclusive consumer of any one tree's foliage.

The effects of defoliation such as we have seen here would doubtless be recorded in the trees' annual growth rings. If, for instance, in 1969 we might have had a cross section taken from any one of the mini-wood's big trees, to examine, and another taken in 1970 (both cut after the period of summer and fall growth) differences in the width of the growth rings would no doubt have been very noticeable. In such a section taken from a tree in 1969, whose leaves had produced a normal amount of sugar and other nutrients, the outer ring of tissue would have been widest, while a corresponding but narrower ring in the 1970 section would have reflected the loss of chlorophyll, and for this the *Ennomos* caterpillars could have been held responsible.

In the case of some of the beeches which suffered complete defoliation in 1970, hardly a new leaf was produced during the hot dry summer that was experienced. Here and there, however, at the tip of a twig, but mostly on the saplings, a fresh tender leaflet or two might be seen moving in the breezes. Within the mini-wood at such times they stood out among the riddled branches, their bright new coloring seeming curiously out of place in late August and early September, with winter not too far away.

Late inspection of even the completely denuded trees revealed that they had, nevertheless, produced good buds for the following spring's foliage, and when that May came around once more, this caterpillar cycle was on the decline, and the worst for the trees was over.

Although hordes of the caterpillars had pupated while still aloft, as already mentioned, thousands of others had dropped to the ground on silken threads to pupate there in the natural debris. Perhaps some others dug into the humus for the purpose, but whatever the case, it seemed extraordinary to the author that these caterpillars selected such varied habitats for this important transformation.

These pupae were brown or reddish brown dorsally, and mostly plain brown on the ventral surface, while a small percentage were inexplicably pale jade green or greenish white, these latter color phases having no significance as far as the moths were concerned, an overwhelming majority of those which emerged being pure white of wing, and with silvery white hairs clothing the body. A moth tinged with pale green was a great, but actually occurring, rarity.

In 1970 the new moths began emerging from the chrysalides around May 10, when an occasional one was seen flying, but the main hordes, delicate fluttering white creatures which were difficult to associate in one's mind with the ugly and disagreeable caterpillar stage, swarmed forth all during June.

Day after day and evening after evening, when the flight was at its height, the foliage of the few untouched trees and the honeysuckle and Virginia creeper vines often appeared to be covered with shimmering white flowers as the insects of both sexes thus congregated in orgies of weakly attempted or less often vigorous and successful copulations. Thousands of the unsuccessful males and virgin females littered the ground under the trees and vines, where they quickly died if they were not already dead upon arriving there.

The sexes were quite evenly divided. From many batches of seventy-five pupae isolated for observation, an average of sixty-three moths emerged, and of these, on an average, thirty-five were females and twenty-eight males. Out of every group of eight males and twelve females, also thus isolated, only two successful matings resulted; the remainder, after much useless fluttering around, died within a day or two without making contact.

Mating seemed to be the sole function of the strongest males, and they died very soon afterward. The females deposited their flat rafts of closely packed olive-green oval eggs on tree trunks and branches or, in their haste before dying, on whatever surface they might happen to find themselves (plate 136, top center).

While the moths appeared to be very numerous when concentrated on certain trees and vines in these mating orgies, so many of them died before succeeding, that the actual number of fertilized eggs deposited was evidently very greatly reduced as compared to the year before.

Contributing to the decline of the cycle were minute egg parasites (even in small numbers) and a parasitic fly whose grubs destroyed some of the caterpillars, but in this stage of their lives, really efficient predators seemed to be absent. As far as the author could observe, the caterpillars were untouched by birds. Evidently there was something in the integument of the caterpillar body distasteful or maybe even dangreous to the usual predators, and it may or may not be significant that an abandoned Tufted Tit-mouse nestling which unwittingly had been fed some of the caterpillars died within a few hours afterward. This little bird had been left in the remains of a nest in a bird box which had been raided the night before, probably by a raccoon. Cold, and nearly dead, it was revived by warmth, then fully recovered. It had been feeding ravenously on green inchworms and other small insects until it was fed the *Ennomos* larvae before the author realized that adult birds were leaving them strictly alone (plate 136).

In view of this sad experience, it came as a satisfying surprise for the author to find that *Ennomos subsignarius* in the pupal stage, that is to say, after the tough brown and black caterpillar skins had been discarded, resulting chrysalides were no longer distasteful to vertebrate preda-tors of certain species.

Most numerous consumers of the pupae were the Com-mon Grackles, lusty birds with powerful appetites which seemed greatly to enjoy the soft and juicy chrysalides lying more or less exposed in the trees. A little later, learning that the dry curled-up or sewn-up leaves still remaining on some of the twigs also contained pupae, the Grackles came in increasing numbers to devour them by the thousand.

Next came the gray squirrels which, the author be-lieves, learned the secret of the dry dead leaves by watch-ing the birds. On one occasion two or more of the squirrels were observed repeatedly visiting the infested trees to eat the pupae. As time went on, other individuals learned to go from branch to branch, methodically smelling out the chrysalides, unraveling the leaves, and evidently relishing the tasty bites within. That these rodents would eat so many insects of any kind seemed an interesting and unu-sual observation to the author, and it would have been satisfying to have heard that these animals had also been feeding on *Ennomos* pupae in other infested localities.

Probably the most familiar moth larva in the eastern United States is the tent caterpillar, *Malacosoma ameri-canum*, the indigenous insect species which constructs those unsightly silken webs or tents for its gregarious larval stage, mostly on our wild cherry trees.

Gradually increasing in numbers over a period of years as its cycle progresses, like the *Ennomos* caterpillars, this species often completely denudes its food trees when the zenith year is reached.

But remember, this moth is a native American species, and it is accordingly controlled to a great extent by native parasites and diseases, with the result that millions of the larvae are killed off naturally when their devastations are at their worst, a time which automatically signals the years of decline which will follow.

One of these killers is a bacterial disease, a sort of "black plague" which gradually mostly liquefies its victims. In the experience of the author, this usually infests the caterpillars in a zenith year of the cycle, when they have become so numerous that all the leaves of their food trees have been eaten before they are mature; thus, in a starving condition and much weakened, they are more susceptible to this apparently contagious and lethal disease. The cat-erpillars may appear and make their tents and eat the foli-age for several years, but sooner or later the natural controls take over.

Rather large ichneumon wasps also exert some control over the tent caterpillars, depositing a single egg to a victim and placing it within the caterpillar's body while it is still well and actively feeding. The grublike ichneumon larva consumes the caterpillar's vital substance, but sometimes without preventing pupation of its host. In such cases, the mature, winged adult ichneumon issues from a neat round hole in the pupal victim, leaving only its shell intact.

Before pupation, the normal procedure of the full-grown tent caterpillar, a two and a quarter inch dark-col-ored, brightly spotted and striped hairy insect, is to spin a compact cocoon of whitish silk. This is long-oval in shape and is often found with the silk intermingled with a curi-ous white to yellowish powder which comes off in little puffs when the cocoons are torn loose or roughly handled.

The moths, which emerge from three to four weeks after the cocoons are spun, are small reddish to yellowish brown chunky insects, with two distinctly lighter-colored bands diagonally crossing each forewing. The full-grown larva and moth are shown in plate 136.

The moths mate soon after emergence, and the eggs, which will not hatch until the following spring, are depos-ited in bands encircling twigs on the food trees. The insects also secrete a tough protective varnish with which the eggs are coated and which may to some extent repel small parasites.

No permanent harm results to the wild cherries from these cyclical explosions of tent caterpillars, which have been recurring every so often, probably for centuries. The viable little trees, having the ability to put forth new leaves very quickly, simply reclothe themselves soon after the caterpillars have eaten the first spring foliage. A specimen which was completely defoliated in May is shown in plate 41 as it looked eight weeks later, densely clothed in all-new foliage. No attacks occurred in the mini-wood during the fifteen-year period of this study, and nearby, wild cherry trees also remained free of these caterpillars, although many of the moths were seen elsewhere during the normal flight period.

Another moth which from near zero in numbers gradually builds up in a few years to the "explosion" status is a small tortricid named *Archips argyrespilus*, a pretty little insect whose forewings, spreading only three-quarters of an inch, vary in different individuals from yellowish brown to reddish brown, but all of them are part checkered with darker brown and glistening gold and silvery scales in the form of squarish spots, details which unfortunately may only be well seen and enjoyed with the aid of a microscope (plate 137).

Indiscriminate feeders, the half-inch greenish caterpillars may be found on oaks and maples, elms and basswood, and many other species of trees and shrubs, and as often on rosebushes, geraniums and petunias, and wildflowers, including loosestrife and jewelweed, while still other larvae may sometimes be found on garden vegetables.

After feeding for many days, the full-grown larvae usually pupate within their rolled-leaf shelters, in which they have successfully hidden themselves from small bird predators.

In their zenith years, the little moths seem to be everywhere, rising in numbers or small "clouds," sometimes as one brushes against shrubbery.

Despite their numbers, at such times their offspring larvae do no permanent damage to their food plants, although they are exasperating, occasionally making holes in the buds of roses and other cultivated flowers.

Plain white little moths of another species sometimes occur in large numbers about June 15, when the *Archips* moths are also on the wing. Also occurring in cycles, they could be mistaken by the casual observer for the latter species.

Two other species whose avid foliage-eating larvae typify these caterpillar explosions as their cycle reaches its zenith are the spring and fall so-called cankerworms, or inchworms, the latter name bestowed upon the caterpillars because of their amusing manner of progressing by alternately stretching and hunching up their bodies, as if measuring the surface beneath them.

The two insects referred to are rather insignificant-looking gray moths, both of whose female forms are *wingless,* so they might therefore easily be mistaken for some species of true bug belonging to the order Hemiptera. Though tolerant of cold, the males are weak fliers.

These males of the fall cankerworm, *Alsophila pometaria,* emerge from their chrysalides in the ground in November and December, mostly during milder periods in these months. Their *apterous* (wingless) females, also emerging from the ground, crawl up the tree trunks, where the males soon find them. Mating follows, and the females then deposit their eggs in compact masses on the bark. These are shaped like miniature gray flowerpots all resting in neat rows. A fall cankerworm male is illustrated in plate 137.

The tiny caterpillars do not emerge until the following spring, then, as the leaves rapidly expand, we begin to notice, in years of severe infestation, that they are punctured with small holes which grow in area from day to day as the foliage nears full growth. Now the caterpillars, nearing the time to pupate and measuring about an inch and a half in length, are black, with one yellow and three lighter-colored stripes on each side.

Presently we see them on all sides, dangling, or slowly letting themselves down to earth on silken threads spun from their mouths. Digging into the ground at once, they pupate there; the moths, as already stated, do not emerge until late fall or early winter.

With similar habits in most ways, the spring cankerworm, *Paleacrita vernata,* is easily mistaken for the fall form, although the moth, *A. pometaria,* belongs to a different genus. The eggs of *P. vernata* are very different from those of the fall species, being nearer oval in shape, indistinctly ridged, and of a purplish color.

The caterpillars of *P. vernata* are frequently seen feeding almost side by side with those of the fall moth and are distinguished by their unstriped bodies, although they vary in color, from black, like the fall worm, through gray to green and yellowish green.

Doubtless these two moths and their caterpillars have been inhabitants of the country for centuries, but the trees have apparently survived them with ease, even through the zenith years of these cycles, when such quantities of their frass (or droppings) were falling in the mini-wood and all about the surrounding area that the sound was like that of light raindrops falling day and night, without letup.

On the credit side, we must recognize the important ecological niche into which these "inchworms" fit. Their frass, for instance, is not waste matter lost to the woodland but actually a light blanket of organic material, possibly containing more nutritious materials than the *dead* leaves which drop in the fall, which so many animals of the soil nevertheless depend upon, year round, for part of their sustenance.

It should also be pointed out here for those who do not know the facts, that these cankerworms are among the most palatable and nourishing insect food items annually sought after by millions of small birds, for themselves and for their potential offspring, at a critical time in their lives. For when the hordes of transient or migrant birds are arriving in or passing quickly through our trees in spring, they are nourished to a great extent by these insects, and the summer resident species which normally remain to raise their young in our woodlands, shade trees, and shrubbery near our houses, continue to consume millions more of such "pest" caterpillars whenever, and as long as, they are available. In the years of their abundance, one sees many bird species, even Common Grackles, which many people do not like, gathering great beakfuls of these small succulent caterpillars that have fallen or purposely descended to the ground, carrying them off to their ravenous broods.

That this advantageous meeting between the cankerworms and the vernal influx of hungry birds was thus so nicely timed by nature, offers perhaps the most convincing argument against incautious spraying with *anything* permanently harmful to these insects. It is the author's opinion that even those bacterial sprays called completely "safe" may one day result in too many consecutive caterpillarless years by killing so many that there may remain too few, or there may be no female moths to lay eggs for the new broods of these beneficial species.

In years of abnormal scarcity, or during May and June of other years when there are practically no caterpillars where suppression previously had been carried out with bacterial sprays, it becomes obvious to anyone paying strict attention that fewer or maybe *no* migrant Warblers and other small transients will be observed working their way northward through the trees in that particular locality.

This was the case in 1972 and 1973, in and about the author's small patch of woods and other property, when the first all but warblerless years ever experienced became a sad truth. Instead of the several kinds of other songbirds customarily raising their young there, only a single pair of Tufted Tits nested within the area. These conditions followed spraying with material thought to be safe, but there were scarcely any food caterpillars in the trees to help sustain these birds in 1973, and very few the year before.

Let us hope that birds are not to suffer a second ordeal as so-called new safe sprays come into general use. The above unhappy observations were written in retrospect, after the fifteen-year study and the manuscript had been completed.

The author is praying that the absence of the caterpillars in question and their small bird consumers in the mini-wood area will prove to be but temporary.

CHAPTER XV

INSECTS III: SOCIAL AND OTHERWISE

The term *social,* as used by entomologists and in this chapter, separates those species which live in organized colonies from those which do not.

As applied here, insects in the strictly social category will include certain ants, bees, wasps, and hornets, all members of the order Hymenoptera, which also includes non-social bees and wasps, and the parasitic ichneumons, chalcids, and sawflies, some of which will again be mentioned in this chapter. Other social insects found in the mini-wood were the native white termites or white "ants," belonging to the order Isoptera, which are *not* ants, but animals more closely related to roaches, and about which more will be said presently.

Generally speaking, colonies of social species (not including certain social beetles) consist of a queen; numerous members of a worker caste or castes; soldiers, which in ant and termite colonies protect the nest to some extent; and males, which are developed at certain periods for the purpose of mating with young queens which appear at the same time. The young insects of both sexes leave the nest in what are known as "marriage" flights, recognized by crowds of wingless workers around the nest entrance energetically seeing to it that the members of this new brood of queens and males get away, all hands urging them to fly off at once in ones and twos for their aerial matings, soon after which, on the ground, they lose their wings for good.

As we have seen in another chapter, the tiniest ants were residents of the mini-wood humus, where they found an easy living feeding on dead animal and plant matter, or, in some cases, preying upon collembolans and other minu-

tiae. These were species whose elongate body forms might, as already stated, cause a hasty observer to mistake them for the equally small, similarly shaped rove beetles which roamed the same habitat, a species only five millimeters (about three-sixteenths of an inch) in length, named *Gyrohypnus hamatus.*

Easiest to observe in the mini-wood were carpenter ants, *Camponotus pennsylvanicus,* those large rugged black insects which sometimes get into our houses in springtime, greatly annoying the owners, but which, upon finding little or no food to their liking (dead insects, preferably) soon depart of their own free will. Carpenter ants dwelling in woodlands make their nest tunnels and galleries in woodpiles or in decaying fallen tree trunks and their stumps, or sometimes in the rotting woodwork of old and leaky buildings, where little piles of sawdust reveal their whereabouts and activities.

Camponotus pennsylvanicus ferruginea, a handsome subspecies of this ant, formed large and very active colonies in the mini-wood, usually excavating deep subterranean passages or tunnels from a surface nest beneath a log or some flatter protective object. Unlike the typical all-black carpenter ants, this subspecies is a more attractive-looking insect whose fat abdomen, a ferruginous orange, is liberally sprinkled with golden yellow hairs. (The queen is pictured in plate 137.)

Observed colonies within the mini-wood emerged from their subterranean galleries where they had wintered as early as April 4. Large broods of young were then hatched from eggs deposited by the queen deep within the nest, these white, grublike larvae being fed and cared for by the worker caste—that is to say, *after* the first brood of workers themselves had been reared solely by the queen, just after the founding of the colony.

When the larval ants reached full growth, they spun neat silken cocoons about themselves in which to transform to pupae, and in this form, on warm days, they were brought by the hundreds to the surface, where the higher temperature (beneath whatever sheltering cover protected the nest entrances) doubtless hastened the final transformation of the pupae into "finished" additional worker ants. (Plate 137 also shows typical larvae and pupae of a smaller ant species, *Lasius.*)

The first *winged* generation (males and queens) appeared in the surface portion of one nest by May 15, and after much nervous milling about, finally took off on their marriage flights, a few at a time. It was assumed, of course, that young males and queens from other colonies not too far away were also initiating sexual flights and that satisfactory cross-matings would thus be assured.

In mid-November, the author found a semi-torpid queen, ten workers, and a few white young of this sub-species under loose bark on a rotting log. The group represented a colony founded late in the fall by this young queen, male ants dying shortly after mating.

The insects were removed to the laboratory, where they were placed in a glassine-covered observation box (two and a half inches in diameter and an inch in depth) which had been previously prepared with a layer of humus and bits of the rotting wood and bark from the spot where the little colony had been found.

There is probably no other sight in nature more interesting or more absorbing to observe than a queen ant and an early brood of her loyal workers carrying out their normal duties and activities—as if they had never been disturbed—under the superb lenses of a modern stereo microscope. Aided by such an instrument, it was as if the observer were being carried right down into the colony, to remain there in intimate three-dimensional, brilliantly lighted association with the insects for as many hours at a stretch as might be desired.

Thus the author watched the endless nuzzling, grooming, feeding, and other manifestations of instinctive but seemingly loving attention lavished on the comparatively huge queen by her loyal little workers. With delicate, often quivering antennae, some examined every part of her big body, while others licked her, or performed *trophallaxis,*

the transfer of food from worker mouth to royal mouth, and at times from worker mouth to worker mouth.

Their care of the few helpless larval ants was as continual and as elaborate. Often the larvae were gently transported by a worker, in little masses adhering to each other, from one location to another, and when the queen occasionally took over, she would hold a ball of the tiny larvae (those which were already out of the eggs when the colony was obtained) in her big mandibles for long periods, as though loath to be away from them for long.

On December 11, about twenty-six days after the colony was confined, the queen laid her first eggs in the observation box, continuing at short intervals until there were ten, long oval in shape, and yellowish white in color. If the colony was disturbed, as when the cover of the box was lifted for the purpose of introducing food, she or a worker would pick up the eggs (as described above, all sticking to each other, as very young larvae did) and hold them securely until the disturbance was over.

By December 20 the eggs numbered about forty, and most of those that had been laid first had hatched. The unhatched eggs were now gathered into groups of from five to twenty-five, and the tiny, newly emerged larvae, hardly distinguishable from the eggs at first, were also segregated, fed, and tended by the worker ants. At this date, some of the larvae which had been on hand in the original colony, when it was found, were as large as the ants caring for them, and some had already spun cocoons about themselves. By the first week in March, new adult ants were emerging from these cocoons; thus the queen and her tiny original group had now been augmented by a small new generation, and, had the colony been living wild and normally, it would have been well launched toward being a populous formicarian community such as those found in the mini-wood by late spring or early summer. No soldier ants had been raised, however, and some of the new workers were not overly strong, probably because of inadequate nourishment supplied by the workers, or, on another level, by the author inadequately supplying the adult ants. Indeed, the question before the author was how to feed the colony at all in winter, when dead insects could not always be supplied.

After some thought, however, it was postulated that if such other highly successful members of the order Hymenoptera as the honeybees could live through the coldest winters relying on nothing but honey for nourishment, on which same wonderful substance their larvae were also reared, why could the adult ants and *their* larvae not be fed the same thing?

A fifty-fifty solution of honey and distilled water was forthwith supplied to the author's captive insects. Using the cap from a vial as a feeding dish, the experiment was instantly successful, although probably the food was not altogether right for ants chemically thus for exclusive use.

Watching the ants line up around the little dish, avidly partaking of the fluid, and knowing that they would feed their queen to repletion as well, was a happy thing for the author.

For many months the colony seemed to flourish, and observing how the workers lapped up the honey-water and how, by regurgitation (after probably altering it somewhat in their stomachs for the purpose), they performed trophallaxis with the queen and the larvae (all seen highly magnified through the stereo microscope) could only be described by the word "fascinating."

But there is always the sad time of parting with pets.

Death occurred from time to time among the workers. Their bodies, pushed aside by the living, simply dried up, and were forthwith ignored. Lucky are these little animals who know not the awful meaning or even the sensation of grief!

Sad from the author's point of view was the fact that the queen was the last to go. She outlived the last worker and larvae by less than three weeks. Nervous and evidently bewildered when at length there was no one left to nuzzle and polish and feed her, she simply gave up and succumbed, although right up until the last looked strong, leaving the author wondering about this curious system which makes the queen unable to care for herself.

Small brown ants belonging to the genus *Lasius,* living in colonies which grew to various proportions, were common within and outside of the mini-wood.

There are many species of *Lasius,* but some of them are so much alike that even myrmecologists (ant specialists) have hard work separating them, and the problems are multiplied by the fact that other almost identical or identical ants from Europe have arrived here, probably in foods and plants, and have now become naturalized.

Lasius ants in the mini-wood made their nests under logs and stones, particularly under flat stones which became sun warmed by day and remained warm for at least part of the night. Others, maybe of another or the same species, revealed their presence in the open by their "hills," or by the presence of Flickers that squatted alongside and inserted their long sticky tongues into the nest holes to capture and then enjoy the ants adhering to them.

Year after year *Lasius* ants persisted in establishing their huge colonies between the roofing-paper cover and the brickwork of the author's outdoor grille. Each time the steak, or chickens, the lobsters, or Block Island swordfish were brought out for broiling, thousands of the little insects would scamper wildly hither and yon, gather up their thousands of eggs and larvae, cocoons and food stores, and remove them to safety somewhere in the dead-leaf cover behind the grill, a tedious chore after the cover was lifted, and would eventually settle down, then carry them all back again when the cover was replaced, even before the bricks had completely cooled.

These topmost bricks, with convenient hollows formed by the molds, were exactly suited to the ants' requirements, with compartments for the eggs, larvae, cocoons, and food stores, and it was so in demand that sometimes larger species of ants took over and appropriated the grille top for themselves.

Lasius ants are the ones so often seen tending their herds of aphids or plant lice. In spring and summer, when the sap flow is strong, these tiny soft insects insert their beaks into the plant veins and tissues, and it now seems that the liquid is moved into their bodies *mechanically*. The sugar-rich aphis feces, as already described in this book, which are periodically excreted in the form of tiny sticky drops, are avidly lapped up by the ants and also by many other insects of the woodland community and its canopy, while millions of the fat sweet aphids themselves are devoured by a dozen species of warblers and other transient or resident birds. The warbler-aphid relationship is an important one during the fall migration; in September and October, the oaks and other large trees are visited daily by numbers of these small birds as they hurriedly work southward. If aphids are absent, so may be the warblers.

In and about the mini-wood, in years of heavy aphid infestation (and they came in cycles on different species of trees), the warblers of one or several species were present nearly every day during the migration period, and the opposite was true when these insects were scarce, or absent.

The swamp white oak and the tulip-tree, for instance, might be cited as examples of species that periodically illustrate the statement regarding aphid cycles. In some years, during the summer and fall, these insects multiplied to such an extent on the leaves of both of these trees that much of the foliage became completely coated with the aphids' sugary feces.

Such aphid abundance meant sweet meals for the migrant warblers and other small birds, but at the same time, the coatings on the leaves (some of which chrystallized) afforded a suitable nutrient medium for the growth of mildew, which once or twice caused the leaves of the swamp white oaks in the mini-wood area to appear brown instead of green, while those on one tulip-tree turned completely black.

Under the microscope, this apparently solid black blanket became a vast network of living threads, but whether the hyphae of an unidentified species of the genus *Microsphaera,* or whether a stage in the life of the oak mildew, *M. alphitoides,* was not determined. The important fact here is that these interrelationships, linking trees, warblers, aphids, and mildews, existed in the mini-wood.

Getting back to the little brown *Lasius* ants: some of them place aphids on roots well below ground level, thus assuring themselves of a supply of honeydew during the winter and giving protection to the aphids in return for their sweet exudations. Such mutually beneficial relationships, as we know, are known as cases of symbiosis.

There was one species of *Lasius* in the mini-wood that was hardier than the rest and which was found tending its small "herds" of aphids in November, after several frosts had knocked out all tender leaves and most of them had fallen. A few, however, still partly green, remained on one or two young sugar maples not yet frostbitten, and these the ants had found for their purpose. Most interesting was the fact that the aphids had been placed neatly in single file along the midribs and some of the veins of these leaves, through which there was evidently still enough sap movement to "pump" some small amount of the nutrient into the insects (plate 137).

Social species of Hymenoptera such as the honeybees and bumblebees, the yellow jackets, the white-faced and *Crabro* (European) hornets, and the paper wasps, are not very numerous (in species) here in northeastern North America, nor are there a great many species of ants. In tropical America, however, the situation is very different, and hardly a day passes, as one tramps the forests and open spaces, that additional species, new to oneself, are not encountered.

But there are plentiful species of bees and wasps here in the north that are *not* social, many of them so small that they are overlooked entirely by most people, but all of them have extremely interesting lives and habits. These are the *solitary* bees and wasps (Hymenoptera) in which, as in the social species, the metamorphosis or life history consists of four stages—the egg, a grublike larval stage, the pupa (often enclosed in a silken cocoon spun by the larva), and the final winged stage or imago. In these species, after mating, the female almost invariably lives singly, building and provisioning the nest herself, and after sealing in batches of stores each lot bearing a single egg in separate cells, she abandons the whole thing, leaving the young to look after themselves.

Plate 138 shows the stages in the life-history of a typical wasp (whether solitary or social), except for the cocoon and the winged stage, which of course differ from species to species. The solitary and social bees have similar egg and larval stages also, most of them not too much unlike those shown in the painting.

It is an interesting and curious fact that adult solitary wasps feed by preference on nectar or other plant juices, while their tender grublike young are supplied with much coarser animal provisions, usually other insects, or spiders stung into an immobilized but still living state. The social wasps, such as the hornets, feed "meat" to their offspring in the form of minced flies and caterpillars, or other animal matter, having on occasion even taken bits of lobster and crab meat from a dish of salad on which the author and his wife were working! The solitary bees and the bumblebees feed their offspring more delicately, using nectar and pollen paste, while the honeybees of course employ the most delectable foodstuff of all, as well as pollen.

As destroyers of house flies and some larger species, the white-faced hornets (seen as sap-drinkers in chapter IX) have no superiors, and it is interesting to see them dive hawklike upon flies, mince them in their mandibles, roll the meat into neat balls, and fly off to their big paper nests with this food for their larvae.

When a solitary wasp has completed the first cell of her nest (constructed by many species of cement composed of mud mixed with saliva), she proceeds to the hunt at once, finding and stinging, but not killing, the game, which thus remains fresh within the cell with the delicate egg until the young wasp emerges and begins feeding itself. This provision of nature was evolved with the result of eliminating fungus (mold), which would soon grow on dead animal matter in a *sealed* cell, destroying it as food for the larva and in effect, killing the larva.

Other solitary species of wasps eliminate the elaborate nest altogether, selecting some convenient hole such as a boring beetle's abandoned tunnel, a hollow plant stem, or a cranny or hole of most any kind in which to place their stores in mud cells or between partitions.

The prey of solitary wasps varies greatly from species to species, and according to habitat and geographic range. It may consist of large insects such as cicadas or katydids and other grasshoppers. Some wasps take crickets, flies of various sizes, or small bees; still others of the tiny forms employ such minute game as aphids and collembolans. Caterpillars, if soft, hairless and spineless, are greatly favored by many wasps, as are spiders of all sizes.

Plate 138 illustrates the stages in the life of the common solitary mud "dauber" wasp, *Sceliphron cementarium*, and its nest. While this species now seldom builds its neat cells (consisting of concentric rings of mud later camouflaged with rough lumps of the same material) on rocks or in hollow trees, as it probably did originally, it was occasionally seen as a transient, traveling into or through the mini-wood on its journeys to and from a building or other adopted shelter in which the nest was under construction. These were trips for gathering mud, or to hunt out the proper stores with which to provision the cells (plate 138).

The author's long study of this wasp, carried out over many years in an area thirteen miles in diameter and of which, for some of these years, the mini-wood was the hub, revealed many interesting things about *S. cementarium*.

It was found, for instance, in the vast majority of cases, that this wasp employed only six species of spiders as food for its larvae, and that fifteen of these arachnids was the average number in a stored cell. Packing the paralyzed victims in firmly, the insect deposited a single egg on each mass, and then sealed the cell entrance with mud.

Careful examination of sixty-seven provisioned cells taken as a sample from many widely separated nests directly after they had been sealed, revealed that 44.55 percent of the victims were immature banded garden spiders, *Metargiope trifasciata*, mostly young females, which at this stage were silvery white. Next preferred were small whitish or yellow spiders named *Misumena aleatoria*, flat bodied and crablike in appearance and making up 19.67 percent of those in the cells. Hairy, dark-colored jumping spiders, *Phiddipus purpuratus*, accounted for 11.28 percent. Young specimens of the familiar big black and yellow orb-building garden spider, *Miranda aurantia*, made up 9.64 percent; and the plump *Epeira trivitata* and *E. trifolium*, 7.15 and 6.55 percent respectively.* The two last named species are common large (when full grown) orb weavers, whose strikingly symmetrical webs have long been admired and written about.

Earlier research during this project also revealed that a diet consisting of these spiders supplied the larval wasps with a high percentage of nitrogen, analyses to determine this being carried out by tedious methods at that time. One was by means of a combustion furnace and elaborate procedures known as the Dumas method; while the simpler technique, known as the Kjeldahl method, will be but sketchily outlined here for those who might be interested.

A very great deal of time had to be devoted to this project, once the material necessary had been gathered during the limited weeks each summer when the wasps were building and provisioning the nests. This, of necessity, had to be prepared for safe storage with great care, and the analyses made whenever time permitted, each such analysis requiring several hours attention.

Employing the Kjeldahl method, the very accurately weighed, completely dried and powdered sample was digested with nitrogen-free sulphuric acid in a specially long-necked glass flask. After completing the digestion (at this point the boiling acid became colorless), a saturated solution of sodium hydroxide was introduced (after certain other additions), and the flask connected by means of tightly fitting stoppers and glass tubing to a glass condenser or "still." The contents were then brought to a boil again, and under this treatment the nitrogen in the sample was released by the sodium hydroxide in the form of ammonia (NH_4). The distillate containing the ammonia and dripping from the end of the condenser was collected in a flask containing a measured amount of what is known as 1/10 normal standard sulphuric acid solution. From all data so far on hand, and by means of a formula, the percentage of nitrogen in the sample could be calculated after performing one final operation to ascertain amount of excess acid remaining in the flask after all the liberated ammonia had been saturated, this manipulation being known as a *titration*.

Each analysis required several hours of meticulous attention to details not included in these descriptions, and collecting sufficient materials for analyses covered many

---

months during several years. Two hundred and thirteen analyses of the banded garden spider were made, and 200 of the crab spider, *Misumena aleatoria*. Far fewer were made of the other four species regularly used by this wasp, because so few were found in the cells, while an occasional additional species found singly on one or two occasions was not considered in the study.

As a result of this work, it was found that both of the preferred and most frequently stored species supplied 4.56 (slightly plus or minus) percent nitrogen to the consuming larvae, a figure which of course also indicated the high protein* value of these foods—richer, indeed, than prime rib roast!

Whether significant or not, it was also interesting to note that the four less-often selected species supplied less and less nitrogen as their percentages in the cells decreased.

The reader might also be interested to know that, left alone and under normal circumstances, the eggs of this wasp hatched in two to three days, and the larva consumed all of the stored spiders within two weeks. It then spun a torpedo-shaped silken cocoon which soon turned brown and became as brittle as the inner skin of a peanut. Next, it discharged all remaining waste matter from its body in a single lump, thus plugging the lower end of the cocoon, as shown in the illustration (plate 138). Wintering in this form, pupation did not take place until the following spring, the winged and powerful spider-hunting imagoes not emerging until June and July. Old barns and sheds were the best places in which to find the nests.

As the author stood one May day observing the migrating warblers along the edge of the mini-wood, a small black and white solitary female wasp alighted on a strand of the wire garden fence near by.

In her mandibles she clutched a slender greenish caterpillar about a half inch in length. Resting for a few seconds after her burdened journey, the wasp then walked along the wire to a perfect little earthen jug of her making, into the small round mouth of which the caterpillar was unceremoniously shoved.

Taking off after grooming her antennae, she was back again with another caterpillar in twenty minutes, this time a green and pink one. For the rest of this day (July 15), this wasp continued with her project, and by dusk of the

following day she had filled the jug, laid her egg on the stores, and had sealed the opening with mud (plate 138).

This was the diminutive potter wasp, *Eumenes fraternus*. She had packed this earthen cell, a beautifully made more or less globular container, very solidly with caterpillars —an energetic little arthropod control organism in the destruction of foliage consumers thus helping to maintain balance in the woodland economy.

How many of these little caterpillar-filled pots each *fraternus* wasp is responsible for in a summer someone else will have to say, but at times two or three of the jugs have been placed close together, and one from such a group opened for study revealed twenty-five caterpillars, fresh-looking, but immobilized, probably by the insect's sting.

The single egg hatched in two days; the larva fed for ten days, almost completely consuming the stores, after which pupation soon followed; and the newly emerged imago was released by the author three weeks later.

Examining the nests of the two species of wasps just described, we wonder how it happened that one—the "mud dauber," *Sceliphron cementarium*—evolved nests consisting of *concentric* rings of mortar, while *Eumenes fraternus* came up with these little storage pots whose globular forms hold so much potential food for the larvae in the amount of space allotted in their construction? The "mud dauber" piles horizontal cell upon cell and plasters them over with pellets of mud until they become completely camouflaged before the maker's working days are over.

A number of small black wasps belonging to the genus *Odynerus,* most of them marked with yellow on the thorax and narrowly ringed with the same color on the abdomen, also played a part in the control of certain foliage-consuming larvae. Although small solitary insects themselves, collectively the members of this genus were doubtless of much value to the woodland vegetation, every normal female being responsible during her short life-span for the destruction of perhaps several dozen small caterpillars.

Utilizing abandoned burrows of boring beetles, crevices in rocks and buildings, hollow plant stems, even keyholes and other odd situations, these little wasps were often observed carrying small limp caterpillars into such spaces and passages. Several of the victims were stored for each potential wasp larva. Upon each such immobilized batch of caterpillars, a single egg was deposited, and each such lot was then separated from the next, in whatever space, by a partition of mud.

---

*Nitrogenous organic compounds yielding amino acids, which are required for all life processes.

The number of such cells thus prepared depended only on the amount of space available within the "nest," and in one instance, when a summer screen was taken down, twenty-four cells containing wintering pupae were found in a groove in the woodwork.

A truly tiny species of all-black solitary wasp was seen using old beetle borings of suitable small diameter, storing these passages with immobilized *aphids* as food for its footless, grublike larvae. While such a minute predator could not make much of a dent in the enormous aphid populations individually, collectively, as with *Odynerus,* the species was actually of value as another of the many controls seldom noticed by most of us, and therefore, of course, generally unrecognized for what they are doing in our favor.

Small as these wasps were, each one accounted for many aphids that would otherwise have *given birth* to large numbers of young, and whose offspring, thus liberated, and *their* young as well, would have done likewise during the same summer season.

Of still greater economic importance within any woodland, and wherever there is vegetation, are the large numbers of wasplike insects belonging to this great order (Hymenoptera) and the superfamily, Ichneumonoidea, and which are parasitic on other insects. Popularly called *ichneumons* and, sometimes, *ichneumon flies,* there are a great many known and named species, and they are much more numerous in and out of the woodland than most people realize.

We have already seen how they destroyed many of the tent caterpillars (chapter XIV), and not infrequently they have surprised and disappointed the lepidopterist who has collected moth cocoons or butterfly chrysalides by suddenly emerging from his prizes in place of the splendid expected insects.

Ichneumon species of greatly varying sizes, from those much smaller than a gnat up to insects (including their ovipositors) several inches in length, were found in the mini-wood. Species were observed throughout the warm months during every year of this study; some of them are shown in plate 138.

These insects were most often seen hunting over dead tree trunks and stumps, or nervously hunting over and under leaves, their antennae aquiver as if from the excitement in the quest for their special victims, and their quick movements and whole attitude spelling alertness.

While many ichneumons look so much like typical wasps that the layman would have difficulty in telling them apart, a great many others, especially the larger, and largest species, may be recognized instantly by their posteriorly protruding, often hairlike, and sometimes very long ovipositors (plate 138). Most interesting to watch are the females which are somehow capable of boring into dead wood with these flexible hairlike organic "drills," and through which an egg is eventually passed and deposited in a grub somewhere within the wood, far from the surface. How the ichneumon in such cases senses the location of its host species is astonishing, and as surprising as its ability to reach the grub with its remarkable appendage.

A great many smaller ichneumon species deposit a single egg or group of eggs by alighting directly on the victim and sometimes puncturing its skin and tissues without apparently greatly disturbing it. A caterpillar or other insect thus parasitized often lives on for many days before the grub eating it from within grows large enough to kill it.

The abandoned burrows from which boring beetles had emerged from dead wood of various kinds within and on the borders of the mini-wood proved attractive to small solitary bees as well as wasps. Of particular beauty, when examined with the stereo microscope, were two species of solitary bees measuring about six millimeters (less than a quarter of an inch) in length, whose chitinous outer armor or exoskeleton was a glistening metallic emerald green.

Although these bees were equipped with closely set hairs on their femurs and some on the tibiae, they did not seem to be carrying adhering pollen at any time during three summer months, and it may be that they were *inquilines,* entering the burrows to deposit their eggs on the food stored up by other species of small bees.

Unlike most wasps, which supply their larvae with *animal* food, the solitary bees not only live on nectar and pollen themselves, but also store a pollen-and-nectar paste or perhaps, sometimes, unmixed pollen in their nest cells as standard rations for their offspring.

One of the more interesting, not to say astonishing, cases among all Hymenoptera pertains to the nest-building habits of the leaf-cutter bees.

The species found by the author was *Megachile brevis.* Not strictly a forest insect, it was, nevertheless, observed at its nesting activities, once in a crevice in a rock, and once beneath a bit of loose cement within a few yards of the mini-wood proper, into which the females often flew on their journeys for pollen and building materials. They resembled small, somewhat flat honeybees.

Once the site had been chosen, the bee lost no time in commencing her labors. Seeking a particular species of plant (which in these cases was a rosebush) she proceeded to cut especially shaped slabs from the leaves, at times roughly long oval in shape; at times smaller, almost round fragments, according to the stage of her work within the nesting space.

Making an average of nineteen trips to and from the rosebush, each time cutting and carrying back one of these flexible green strips, she quickly formed them into an expertly worked cell averaging seven-eighths of an inch in length, and tapering from five-sixteenths of an inch in diameter at one end to a quarter of an inch at the other (plate 139).

When these pieces had been fashioned into this tube to her satisfaction, she proceeded to cut out the nearly circular pieces already mentioned, which measured about a quarter of an inch in diameter. With these the bee closed the smaller end of the cell, tamping one fragment on top of the other, using a dozen or more altogether, and here, because they were of somewhat greater diameter than the rest of the tube, tension held them until they dried, when they formed a solid plug.

Now, with a cell in readiness, she turned to the simpler task of collecting the nectar and pollen, from which she prepared a wad of yellow paste made up of many small loads well tamped down, one on top of the other, and filling about a third of the cell from the small end.

On these stores she laid a single white, slightly bowed egg, after which the opening in the cell was closed very neatly with many more freshly cut nearly circular leaf fragments.

But this was not the end of the small bee's work by any means, for now she proceeded to construct, provision, and lay her single eggs, in numerous additional cells, placed in parallel rows of three or more, and with from five to seven cells in a row. Then, with her labor at an end and her lifework attended to, she suddenly disappeared from the nest area and was never seen again thereafter. In July, a leaf-cutter bee was seen collecting pollen from winterberry flowers.

As for the young wasps within the cells, which were pudgy little grublike insects at this stage, feeding upon the pollen and nectar paste, they would pupate within their leafen bowers soon after consuming these stores, and later transform and emerge as winged adults themselves, with no further parental care. As for the "mother," she probably lived but a short time after the nesting activity, for this seems to be the sole purpose of these short-lived insects as adults.

Nests observed by the author near the mini-wood, and elsewhere, at times, were all found in August between the first and twentieth of the month. The larvae may therefore pass the winter as dormant full-grown grubs, pupating in the spring, when the imagoes emerge early enough to make other nests. In other words, there may have been two broods annually, the bee whose activities have just been described representing one of the offspring of much earlier (spring) emerging individuals.

When the hottest days of July and early August arrive, we hear the cicadas' or "hot bugs'" long, permeating "songs" issuing from the trees; and clinging to the trunks and branches we see the empty brown nymphal shells from which these robust winged insects have emerged (plates 139 and 140).

Having spent a long period underground, feeding in the early stages on sap extracted from roots, the winged adults are subject to the attacks of our largest native solitary wasp, *Sphecius speciosus*, whose emergence upon the scene corresponds annually with the "singing" season of the cicadas (plate 140), which, by the way, many people improperly refer to as the "locusts." The shrill, monotonous "singing" of these insects (messages serving to bring the sexes together) is not produced orally, but by the vibrations of membranes stretched over sound chambers located on the cicada's abdomen.

Often, however, instead of attracting a mate, the sound lures this large predatory wasp. Then a drama accompanied by sounds of violence ensues as the wasp pounces upon its victim, stinging it in its nerve centers and thus immobilizing it, but not before a few short agonized sounds, what some might interpret as screams, issue from the doomed insect. Sometimes the cicada's steady calling notes change into a few painful bursts of sound followed by silence as the wasp, now clutching the victim securely between its spiked legs, falls to the ground. Usually in such cases the wasp is then in trouble, for its victim is so heavy that it is almost impossible for the predator to take wing again from such a position.

Those wasps which can successfully launch themselves

from the trees with their heavy burdens fly to previously excavated deep burrows in open ground in which they have stored one or more cicadas in earthen cells as food for the young wasps. The predator is single brooded; adults of the new generation do not issue from the ground until the following July or August, when once again their cicada prey is awaiting them.

The two best-known cicadas are the common annual form (that described above), which is *Tibicen sayi*, and the "seventeen-year" cicada, *Magicicada septendecim*, a smaller species which actually may appear aboveground oftener than every seventeen years (plate 140).

Cicada females somehow make slits in twigs in which the eggs are deposited (plate 139). The newborn insects soon drop to the ground, where they burrow in to remain during the long early stages of their metamorphosis. As their sap-sucking and twig-slitting activities do not exactly benefit the trees of the woodland and elsewhere, it was fitting that nature interrelated this common cicada with a powerful predator—the wasp—as a control. Birds also help to reduce the cicada's numbers, and the author has observed House Sparrows, Catbirds, and Starlings feasting on these fat cicadas, sometimes on those left on the ground by predator wasps.

One wonders at what period this wasp-cicada association began—how many centuries ago it may have been. While the two now emerge from the ground as adults at the same time of year, the question arises: Did they interrelate from the beginning, or did this wasp once use some other insect for its young to feed upon? Many other species of wasps use several kinds of spiders, or caterpillars, or locusts (grasshoppers), or still other kinds of insects in their nests, but not so *Specious speciosus*. Cicadas, and only cicadas of this one species, will serve for them. We wonder also, had indiscriminate spraying not been banned and had all the *T. sayi* cicadas been killed off, would this magnificent wasp, measuring an inch and a half in length, and with a wingspread of over two and a half inches, have also disappeared from our fauna?

No nest burrows of this species were found within the mini-wood, as the wasps choose open ground for this purpose, but they were seen there, taking toll of the cicada population on many occasions, and their life history was studied elsewhere.

There were three "monster" species of beetles that occurred in the mini-wood—that is, monsters when compared with the other species of Coleoptera found there. These three appeared off and on about the same dates in each of these years.

One, a member of the long-horn family Cerambycidae, was an orange-brown insect with a slim body with parallel sides. Nearly an inch and a quarter in length, it had strong wings neatly folded under hard wing covers or *elytra*. Protected by a hard exoskeleton, its thorax featured three lateral spikes on each side, and its head had large compound eyes and long, eleven-jointed antennae. The beetle is known to entomologists as *Orthosoma brunneum* (plate 140).

The large grublike larvae of *Orthosoma* were borers in rotting and also more or less sound logs and tree trunks, and they are known to attack lumber as well. Although never observed emerging from their burrows by the author, it was assumed that they were probably the cause of large holes occasionally observed in old trees within the mini-wood.

The adult beetles were active at night, and were sometimes seen clinging to porch screens, to which they had been attracted by lights within.

They were equipped with powerful, incurving, pointed black mandibles, probably for digging their way out of wood, and perhaps for predation, although they were never induced to feed on other insects. The beetle's large larvae are also fitted with powerful wood cutting mandibles for burrowing in and feeding upon ligneous materials.

Another of the "monster" beetles was a very hard-shelled glossy black insect with a quadrate thorax (viewed dorsally) and deeply striated elytra. Unlike *Orthosoma*, it was equipped with much smaller compound eyes and short, club-ended antennae. The species was *Popilius disjunctus*, the only member of this genus found in the United States, and what it was doing within the mini-wood (plate 140) was not determined by the author. It is well known, however, and interesting from the biologist's point of view, because it appears to be one of the few species of beetles that is somewhat colonial in its manner of life, and is said to congregate at times, along with its larvae, in rotting wood.

A very strong insect, it can push its way out of a closed cardboard box with comparative ease, and it may sometimes startle the uninitiate who picks one up, feeling, or perhaps faintly *hearing*, its stridulations, which are produced in some way not entirely clear to the author, who calls it the "squeaking" beetle.

* * * * *

An important predator beetle also of a large size and already mentioned in another chapter, which was found in and outside of the mini-wood, was *Calosoma scrutator,* a brilliantly iridescent animal (plate 140), an avid feeder on caterpillars and thus one of the important natural controls for which we should be thankful; unfortunately, many of these were no doubt poisoned when DDT-sprayed caterpillars were generally available.

Many other insect predators are constantly at work within a woodland and elsewhere. Some of these are often seen but not generally recognized for what they are doing toward the control of other species. The little red lady beetles with two black spots on their wing covers are exceptions, known to all as tremendously valuable predators, feeding, in both larval and adult stages, upon plant lice, but other useful predators remain unnoticed by most persons.

Among the good species are the shimmering, pale green, golden-eyed lacewing "flies," which on summer nights so often flutter about and into our houses, attracted by the lights. Their larvae are predators upon several species of aphids.

These insects are not "flies," as their name would imply; they are members of a much lower group of insects more nearly related to such things as ant lions and dragonflies than to true two-winged flies. The common ones referred to here, which were numerous in the mini-wood and elsewhere nearby, belonged to the genus *Chrysopa,* and were insects which deposited their curious eggs (each one fastened to the top of a hairlike stalk) near or in among the colonies of aphids (plate 141).

The larvae, known as *aphis lions,* are quite unlike the adult parents, being black or nearly so, having tapered wingless bodies and being fitted with strong pincer mandibles for grasping and chewing up their prey. They are like minute dragons in the insect world, doing away with large numbers of the little sap-sucking nuisance bugs, and one such larva will eat many aphids in a single day. The larvae of both the lacewings and the lady beetles also feed upon the woolly white scale insects or "mealybugs," which often grow to large colonies on white pines and which formed colonies on the trunks of the mini-wood beeches.

The adult lacewings give off a rank protective odor if touched.

* * * * *

Things quite new to the author continued to show up occasionally in the mini-wood, but a number of them appeared only once during the fifteen-year study period, like the minute armored ambush bug shown in plate 142 (the slime mold that climbed the wire of a fence [see chapter VI] was another such novel example), while a tiny second insect should be mentioned as another representative from the animal world.

A member of the order Hymenoptera and the genus *Inostemma,* it was unlike any other insect parasite the author had seen before. About two millimeters (slightly over a sixteenth of an inch) in length, glossy black, with rather a large head and thorax for the rest of its body, it would not have been extraordinary except for its abdomen, from the forward end of which a curious object resembling a black sausage arched backward and high over the insect, reaching in front almost to the center of the head.

Examination and dissection under the stereo microscope revealed that this curious extension was a tube or "scabbard," for the owner's long ovipositor, which could be drawn backward when not in use and thus be neatly housed within this sheath.

It would be hard to imagine why such a protective covering had become necessary in the case of this minute species, when, as we know, the majority of the parasitic Hymenoptera, including the largest native species, with ovipositors three and four inches in length, carry theirs uncovered and often sticking straight out behind, even though they may be only hairlike in diameter.

The little insect in question proved to be a parasite of the larvae of certain gall insects, of which numerous species, belonging to several orders, cause a variety of curious growths found on plants, a subject which will be treated in more detail presently. The single example of this "scabbard carrier" insect found in the mini-wood was visiting the multiple little flowers of the wild white aster, and presumably feeding on nectar.

Returning once more to the social insects observed within the mini-wood, we now come to those soft-bodied and sluggish little colonists, the termites of the order Isoptera, popularly but erroneously called "white ants."

On a much lower rung in the ladder of insect classification, termites are not ants in any sense of the word. More closely related to such insects as roaches and the stone flies (which make good trout bait) than to the lordly ant fam-

Above: The glossy black *Milax gagates,* often called the "greenhouse slug," here feeds on toadstools. *Right:* A smaller species, pale gray in color, feeding on a single toadstool growing from a log. It consumed half the cap in one night. Both shown life-size. (Page 158.)

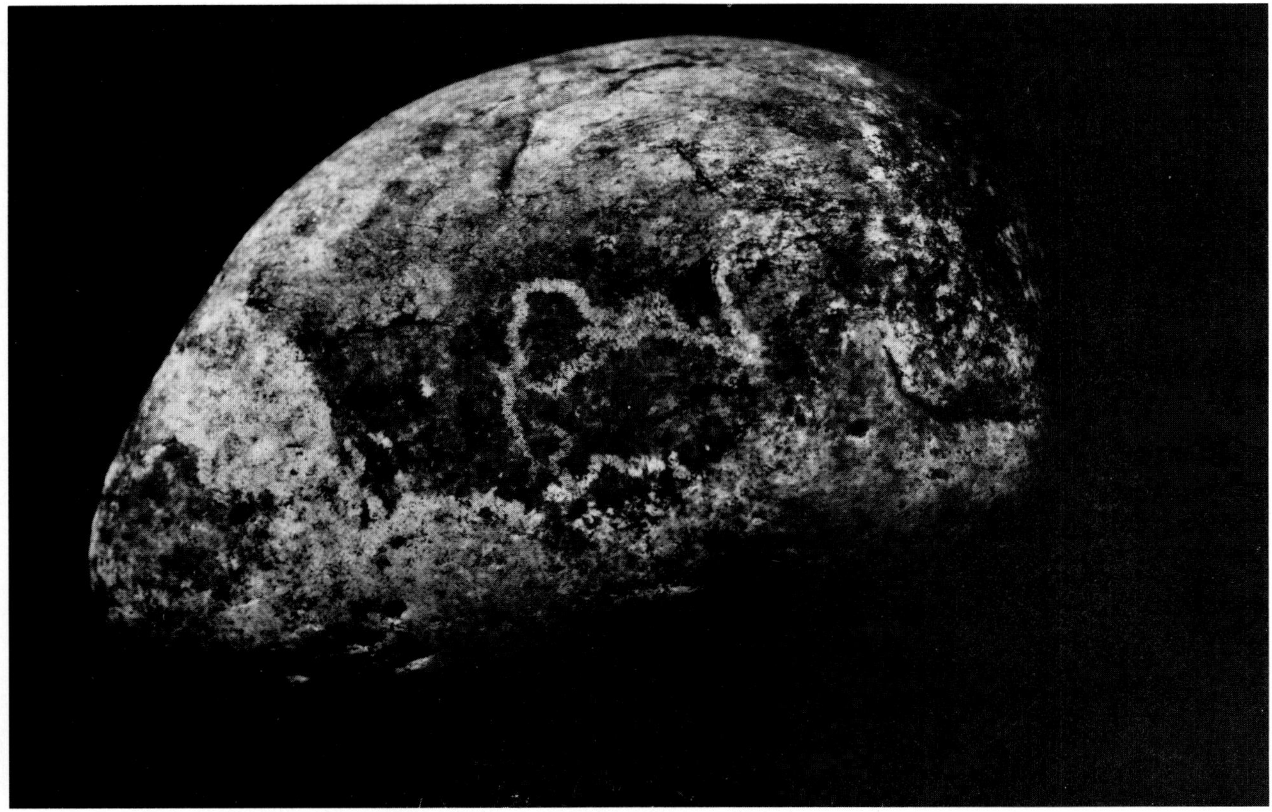

A coating of algae grew on this glacially shaped and smoothed quartz rock when it was exposed to the weather. An orange-colored slug, *Arion subfuscus,* feeding on the algae, left this record as it scraped off the plant with the back-and-forth motions of its radula. Shown natural size. Date: October 12. (Page 156.) Same rock as it looked when first dug up is shown in plate 3.

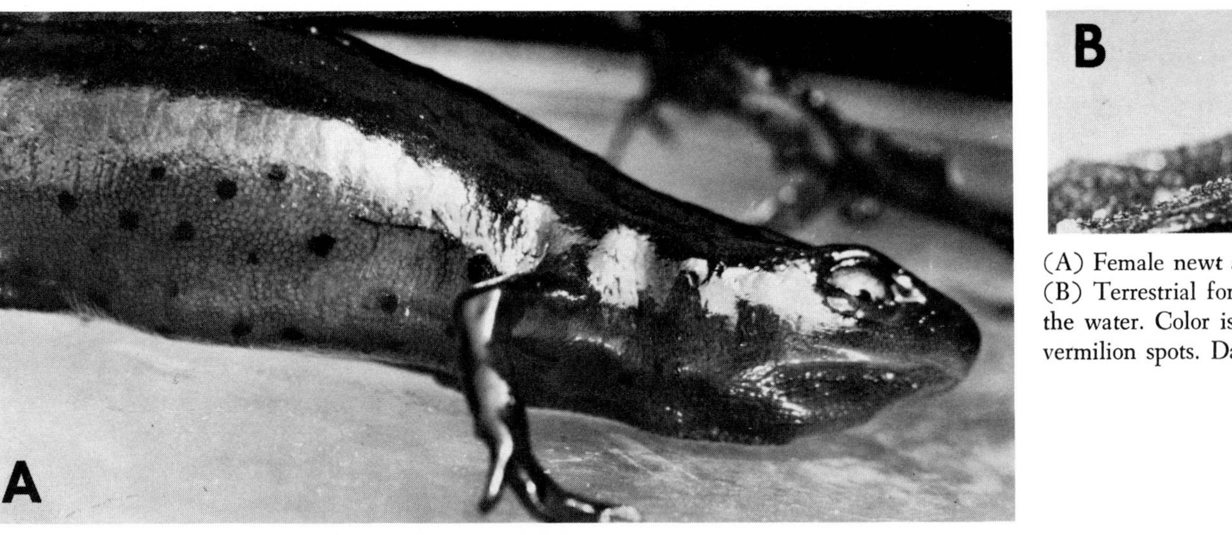

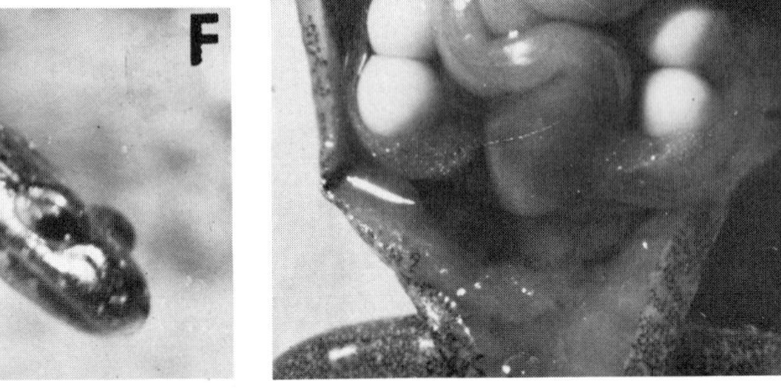

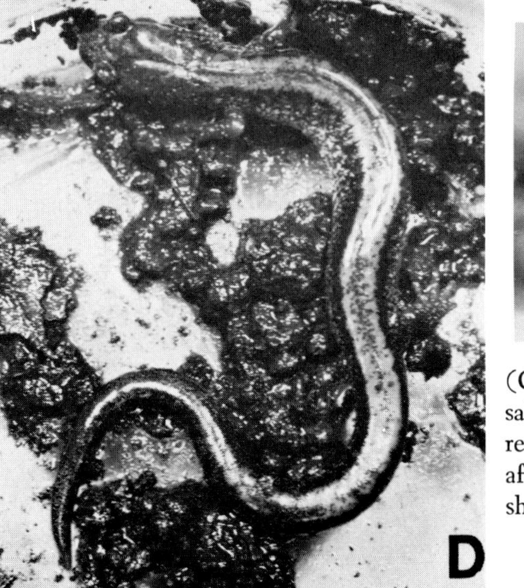

(A) Female newt at adult aquatic stage, about four times life-size.
(B) Terrestrial form of the newt, natural size, at time of leaving
the water. Color is then orange-red or orange-brown with brillant
vermilion spots. Date: July 25. (Page 162.)

(C) Young red-backed salamander, natural size. Date: June 6. (D) An adult red-backed
salamander, somewhat enlarged. Date: November 11. (Still active.) (E) Hibernating
red-backed salamander. Date: March 6. (F) The same individual (shown in E), just
after emerging from hibernation, March 12. (G) Dissected female red-backed salamander,
showing eggs about ready to be laid. Much enlarged. Date: March 18. (Page 164.)

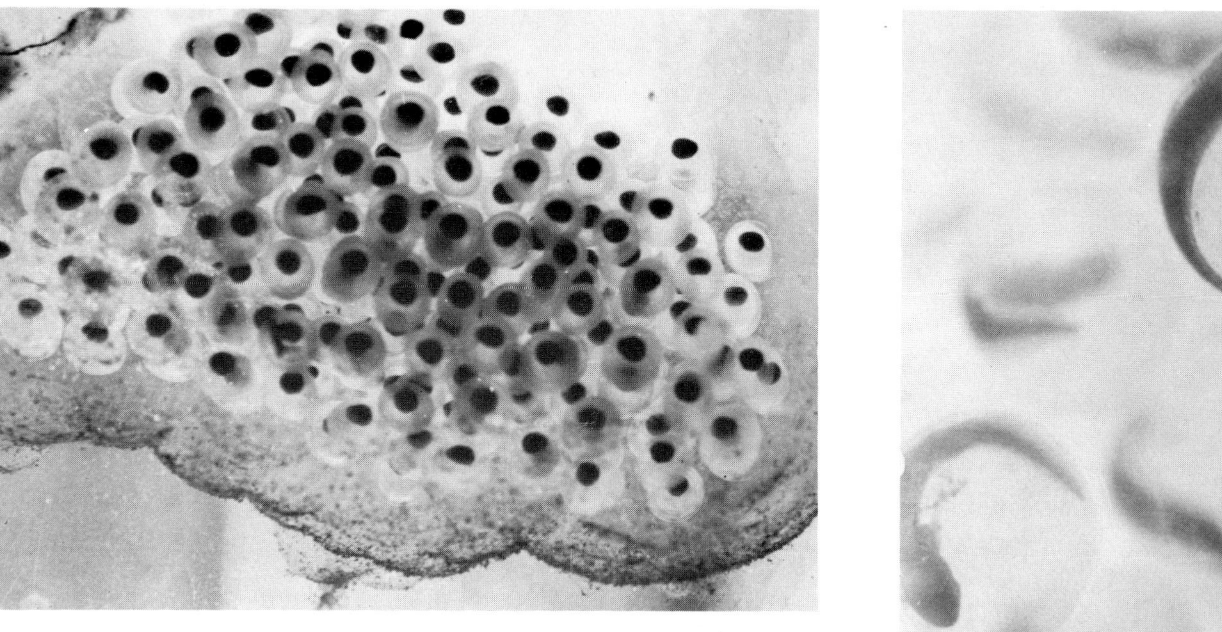

Spotted salamander. (A) The clear eggs embedded in a milky jelly envelope. Date: April 12. Natural size. (B) Greatly enlarged eggs showing salamanders in very fishlike gilled stage. Date: April 28. (C) The adult female, a seven-inch specimen, that laid these eggs in icy water, April 10. (Page 165.)

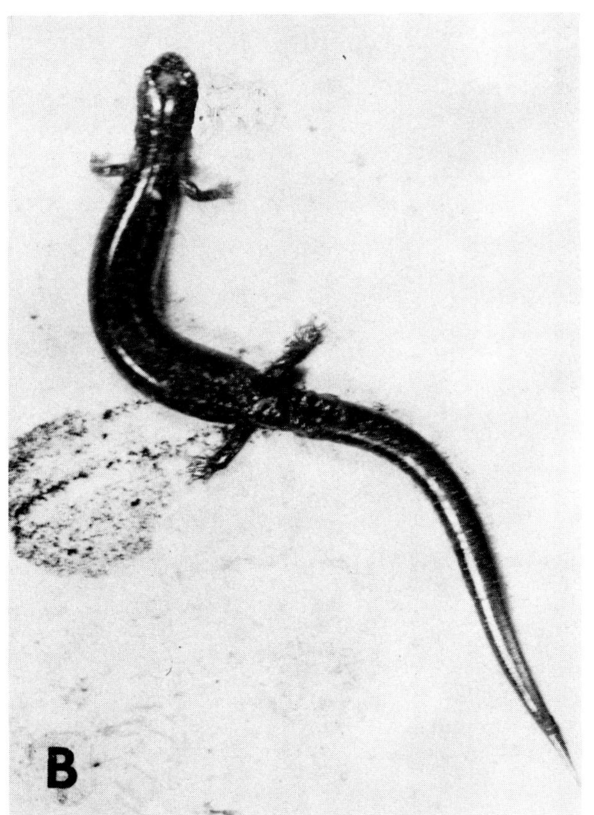

Mini-wood salamanders. (A) Blunt profile of a full-grown spotted salamander, only dorsally blue-back species, with two rows of round yellow spots. (See also plate 149C.) Note fleshy toes. Twice life-size. Date: April 1. (Page 165.) (B) A six-inch slimy salamander. Note cylindrical form, minutely spotted with white on black ground. Date: April 25. (Page 165.) (C) Two "communicating" salamanders are in lead-colored phase of the red-backed species. Natural size. Date: April 28. (Page 165.) Other two are typical red-backed salamanders. Same date. (Page 165.)

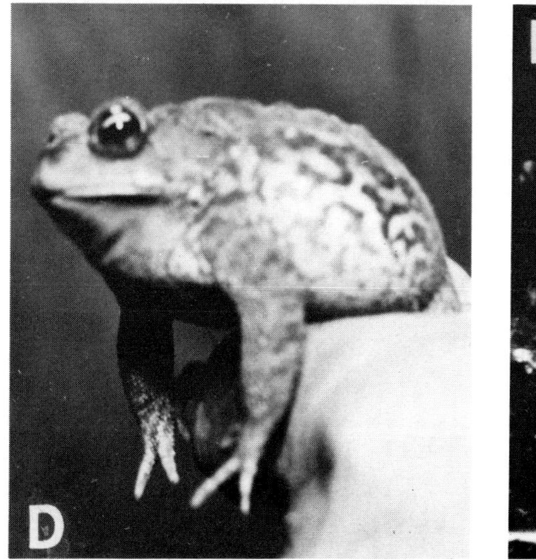

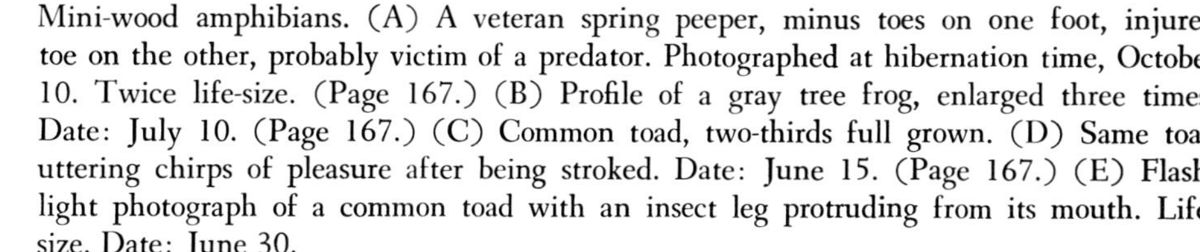

Mini-wood amphibians. (A) A veteran spring peeper, minus toes on one foot, injured toe on the other, probably victim of a predator. Photographed at hibernation time, October 10. Twice life-size. (Page 167.) (B) Profile of a gray tree frog, enlarged three times. Date: July 10. (Page 167.) (C) Common toad, two-thirds full grown. (D) Same toad uttering chirps of pleasure after being stroked. Date: June 15. (Page 167.) (E) Flashlight photograph of a common toad with an insect leg protruding from its mouth. Life-size. Date: June 30.

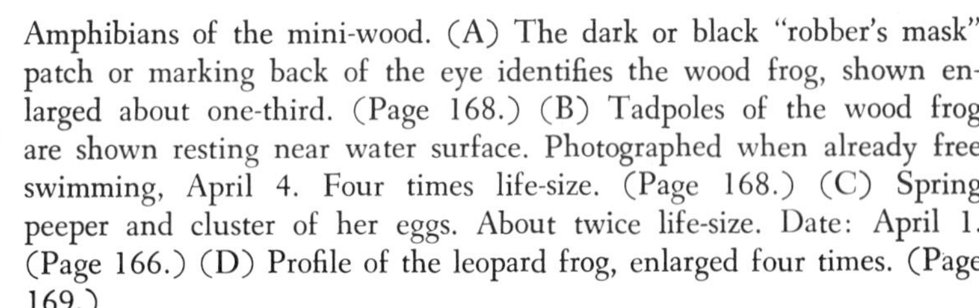

Amphibians of the mini-wood. (A) The dark or black "robber's mask" patch or marking back of the eye identifies the wood frog, shown enlarged about one-third. (Page 168.) (B) Tadpoles of the wood frog are shown resting near water surface. Photographed when already free swimming, April 4. Four times life-size. (Page 168.) (C) Spring peeper and cluster of her eggs. About twice life-size. Date: April 1. (Page 166.) (D) Profile of the leopard frog, enlarged four times. (Page 169.)

Mini-wood reptiles. (A) Box turtles eating a banana. (B) Box turtle with its hinged plastron tightly closed. (C) Wood turtle (at left) and box turtle about to dine on lettuce and banana. (Pages 171, 172.)

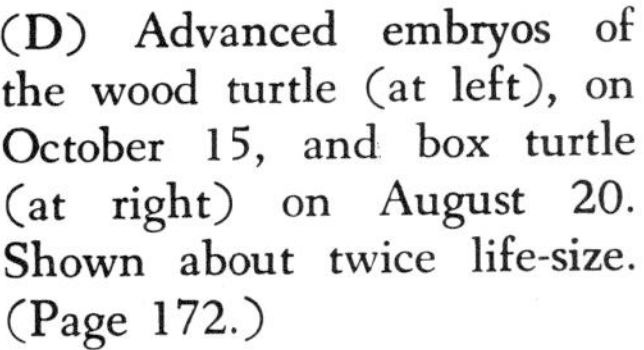

(D) Advanced embryos of the wood turtle (at left), on October 15, and box turtle (at right) on August 20. Shown about twice life-size. (Page 172.)

(E) Sculptured quality of the scutes and orange skin on neck and limbs identify the wood turtle, shown here about half life-size. Note the powerful posterior limbs and claws. (Page 172.)

Left: Blunt profile and head markings of the common northern water snake. Two-thirds life-size. (Page 173.)

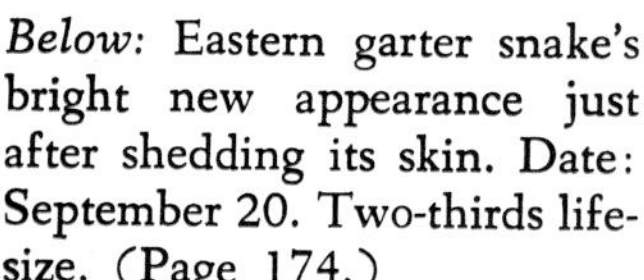

Below: Eastern garter snake's bright new appearance just after shedding its skin. Date: September 20. Two-thirds life-size. (Page 174.)

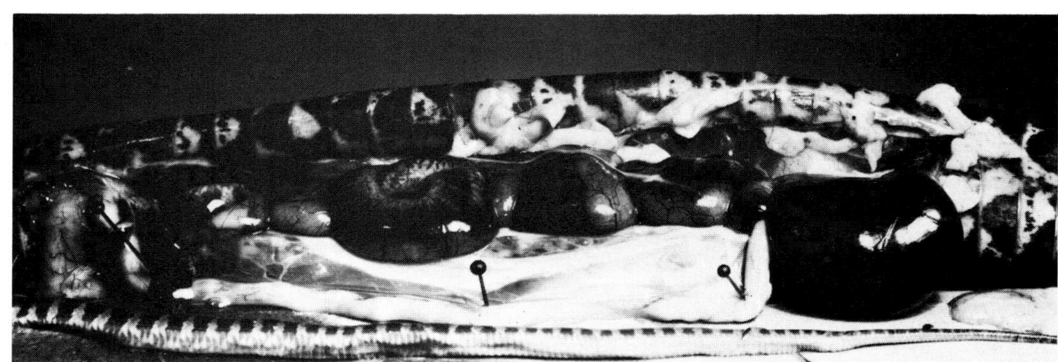

Above: Dissected water snake, showing six of the many young that were about to be born, still in sacs except for one in center. Date: July 25. About two-thirds life-size. (Page 174.)

Right: Portrait of an eastern garter snake sunning itself and alert. A life-size photograph. Date: September 26. (Page 174.)

ilies, they nevertheless have evolved elaborate systems of colonization which have caused many people to regard them as members of the highest insect order.

The termites found within the mini-wood were, technically, *Reticulitermes flavipes,* family Rhinotermitidae, and known as the "subterranean" or "damp-wood termites," the only species native to the northeastern United States (plate 141).

Often confused with the much more detrimental southern species which seem to be spreading north, our common white termites also do damage to man's buildings and lumber at times; but these insects are rightful members of the woodland association and important animals whose function in the forest is the reduction of dead wood to materials useful to other organisms of the complex.

Before briefly describing the life history and other facts about their occurrence in the mini-wood, it might be mentioned here that termites have often been attracted to the foundations of houses by the numerous fragments of waste wood buried there during the grading of the land after new construction. As it is often but a step for possible termite consumers from such hidden decaying wood to the building itself, where access to its timbers is facilitated by the foundation stones or cement, the danger is obvious, and the fault usually the builder's, if trouble results.

Baffling indeed is the fact that two orders of insects whose members are so far apart structurally became similarly organized into colonies of great solidarity. Really living in a state of successful socialism, each nest of ants or termites has its queen at the head; often has fabulously numerous individuals of a worker caste, with consequent division of labor; and in many cases has considerable numbers equipped with huge mandibles, the "soldiers," or "protectors" of the colony. Among all the other orders of insects there is really nothing just like these organizations. Even the colonies of social bees and wasps (which belong to the same order as the ants) are not quite the same.

The queens among ants and termites, are truly regal individuals upon whose presence and good health (as we have seen in another chapter) the successful continuance of the colony depends.

For a time, however, after leaving the old nest in her marriage flight, a newly impregnated young queen does a certain amount of work herself, but only during the period required to establish a new nest: hatching, feeding, and bringing the first brood of new workers to maturity.

After the mating flights, the expendable male ants soon die off, but in the case of the termites, the young and newly mated pairs may remain together for some time. All of these insects lose their wings soon after mating.

Termites, ants, and social bees and wasps raise these special broods of queens and males periodically, and see that they leave the old nest successfully for the good and continuance of the species.

The worker insects, whether ants, bees, wasps, or termites, and which make up the vast majority of individuals in every colony, are sexually arrested (non-functional) members, upon whom falls the labor of foraging, the responsibility for the feeding and care of the queen and her eggs and the feeding and care of the larvae, and the extension and construction of the nest, whenever necessary. Needless to say, the workers wear themselves out in such a system, and new broods are constantly needed. The queen's sole duties in an established nest are egg laying, and inspiring all other members of the colony.

Except in the winged form, at the time of the marriage flights, termites cannot tolerate much daylight. When uncovered in the mini-wood, those caught unawares in the surface portion of the nests moved as fast as their sluggish muscles could carry them into the elaborate passages below ground, the soldier termites among them, and which the author did not find in the least aggressive in this northern species. As was expected, the queen and the bulk of the nest inhabitants remained out of sight underground, and could only be obtained for observation by excavation.

Such soldiers in the tropics were more aggressive than their northern relatives. Those from some of the huge nests found in Guyana in the rain-forests were more determined to use their saberlike equipment when disturbed. Soldiers of the genus *Globitermes* have been reported by other observers to possess glands secreting a repellent which they spray upon intruders.

The food of termites consists of cellulose, a harsh woody diet, the bulk of which is ejected in the form of fecal matter which eventually becomes partly intermixed with the humus and is partly used by other woodland organisms.

Cellulose passes through the bodies of caterpillars, as well as through termites, and is then, in one form or another, taken over by several kinds of soil microbes, which by very involved processes break it down into substances that may be utilized by plants and animals. Bacteria, in breaking down these organic compounds of nitrogen which occur in animal waste matter, convert them to nitrates; these, as we know, are important to plants. Similar processes are carried out with the nitrogenous matter at first locked in the dead bodies of the woodland animals, whether they be insects or still smaller organisms or the largest mammals. The recycling processes thus go on and on, not only in the case of nitrogen but in carbon, phosphorus, sulfur, and other elements essential to life.

Cellulose being the tough and difficult-to-digest material that it is, it would be impossible for termites to retain enough of it to nourish their bodies, were it not for certain microscopic *internal* partners which the various species of these insects acquired ages ago. These tiny internal organisms are flagellate protozoans of the order Hypermastigida, curious "tailed" animal forms which are able in some way to make cellulose digestible as long as they live within their host insects. They are able, in other words, to work on and process cellulose already injested by the termites, and to convert enough of it into sugar to in part nourish themselves as well as their fat insect partners. Here we have a classic example of symbiosis.

Functionally sexed termites—queens and males raised for the marriage flights—possess compound eyes at this crucial time in their lives. These winged individuals are also dark colored instead of white like the workers and soldiers of this species.

Once they have established the nests, the young queens soon grow much larger than before—larger than all other members of the colony. They become, in fact, simply egg-producing machines. As they become swollen with the eggs, their abdomens lengthen greatly, and they are so fat at times that all their six legs become useless.

Unlike true ants, young termites reach adult size by a series of molts interspersed with periods of quiescence, and it is believed that they acquire their quotas of internal symbionts (protozoan partners) by oral contact with the anus of an adult!

In addition to the males, workers, and soldiers, there may also be members of a caste in the colonies known as *nasuti*. In these insects the head is drawn out into a snout-like shape. These individuals are believed to be in some way functional in the defense of the nest.

In winter, the termites of the mini-wood descended to a depth safely below the frost line, disappearing from the surface portions of the sheltered nests well before cold weather set in. Colonies that became established under pieces of wood, building paper, and other purposely placed objects returned to the surface as soon as the sun warmed the ground in April. From such locations or nests, a few individuals tunneled or traveled overland to some of these other fragments of woody material used in the experiments. Small colonies thus became established under such objects in a single instance in the same summer, while other fragments were not utilized until very much later or not at all.

When termite or ant marriage flights occur, as they frequently do in summer and early fall, the myriad winged individuals scurrying about and intermingled with wingless workers helping the young insects on their way often worry and convince people that such gatherings must all be outpourings of the dreaded termites which have probably been eating up their nearby homes.

At such times of "crisis," if a few of the winged individuals can be captured and viewed with a powerful magnifying glass, the question can be quickly settled.

Termites have four rather elongated narrow wings of nearly equal lengths, which are thickly reticulated with a *great many* small cross veins connecting with *very few* lengthwise veins in each wing. Winged ants, on the other hand, may be easily recognized (if taken during these gatherings) by their two, moderately long forewings with *very few* veins, and two *much shorter* underwings with even fewer veins. Identification is as simple as that (plate 141).

Nearly every species of shrub and tree, and some wild-flowers, are subject to the attacks of certain insects and mites known as *gall-makers*. Galls, it hardly need be said, are those very interesting though often puzzling growths in the form of small globes, cones, furry mounds, lenticular objects, cockscombs, stem swellings and a great variety of other abnormalities so often observed upon and often disfiguring leaves and stems, and in a few cases even flowers.

Although they may be harmful to vegetation when they happen to occur in great numbers on one species of host, most people completely overlook their interesting side, for these often remarkably shaped and constructed objects are composed of the plants' *own tissues*, which are thus influenced into producing abnormal growths through the agencies of many small species of arthropods (plates 142 and 143).

The majority of galls are produced by sawflies (Hymenoptera), two-winged flies (Diptera), a lesser number of small moths (Lepidoptera), and by many species of eight-legged mites (Acarina).

Gall-makers deposit their eggs on or within the tissues of the plant's leaves, stems, or flowers; but whether the growths are initiated by the adult forms or their offspring (feeding on the tissues), no one seems to know for certain. In the opinion of the author, based on long observation in the mini-wood and elsewhere, it is a stimulus set in motion by a specific exudation of the feeding inmates which brings about each kind of gall.

Perhaps the most interesting fact concerning them is just that each different type of gall—its shape, construction, and adornment, if present—can only be produced by one species of mite or insect.

When galls are kept for observation, the adult specimens which finally issue from them may be very confusing, for they may not be the rightful makers of such a gall. All kinds of things go on within these strange growths; thus the uninitiated student will no doubt be baffled upon seeing two or more species of insects emerge from the same gall, but this occurrence is simply explained by the fact that the rightful inmates of these growths are frequently parasitized, and in some cases, the parasite is itself found infested or eaten by a hyperparasite!

In addition to these abundant galls on leaves and stems, there are the very important galls or nodules which occur on the roots of clover and beans and other legumes, and which are caused by nitrogen fixing *bacteria*, hence their importance to agriculture, hence the reason for ploughing-in leguminous crops before planting others, or for growing another crop alongside a legume, where it may receive the benefit of the nitrogen gathered by these bacteria.

Within and on the borders of the mini-wood, species of oak trees seemed to be the favorite hosts of many kinds of gall-makers.

Most remarkable, in the author's opinion, were two globular species. One, about three quarters of an inch in diameter, named *Amphibolips confluentus,* was a sphere filled with a spongy material enclosing the larvae and whose outer skin when fresh was green and liberally sprinkled with purplish red spots. The other globular gall species, *A. inanis* (thin-skinned, and brown when dry), contained a single central cell for the larva supported by dozens of delicate radii extending to attachments on the inner walls of the gall. What baffling chemistry, we might well won-

der, could produce such an organic object! Here was an object which, should *we* try to reproduce it in any size or material, would indeed call for more than engineering skill. (Both of these galls are shown in plate 142, top drawing, center.)

Galls seem to be specific to one species of plant or tree. Northern red oaks of the mini-wood border were infested from time to time (but never two years in succession), with rounded, ocher-colored leaf galls with surfaces like fine wool. On the wild blackberry vines, hard green and reddish galls, like clusters of sausages encircling the stem, were often found, the growths caused by a cynipid, *Diastrophus nebulosus.* Red maple leaves were often found thickly sprinkled with lenticular little galls, each with a nipple in its center, and probably caused by a two-winged fly of the genus *Cecidomyia.* Curious white ash galls caused by a mite were often numerous around the end of May. Occurring on the leaves and their stalks, some of them, when opened, were found to contain (in a large hollow space) one apparent full-grown mite and a whole "herd" of younger ones of all sizes, from pinhead dimensions on upward, and all the inmates about the color of the greenish yellow interior of the gall. The species, shown in plate 143, was not identified.

A few other galls found not far beyond the borders of the area covered by this book are also shown in plate 142. Those who become enthused and wish to know more of these things are referred to the well-known *Field Book of Insects,* by the late Dr. Frank E. Lutz (see the list of recommended books at the back), an authoritative little volume illustrating and identifying a large number of the species.

The author learned long ago that the tiniest animated object will bear close observation, and if feasible, examination with the stereo microscope. Animated brown or green, or more or less brightly colored moving specks may prove to be monsters, or living things of unimagined details and perfection.

Thus it came as a revelation upon such microscopic examination of some active greenish objects to find that they were newly hatched ambush bugs of the order Hemiptera and the family Phymatidae, insects even at this early stage of growth strongly suggesting minute organic "tanks," and truly amazingly constructed bits of living machinery.

With their vital organs well protected under an armor-like exoskeleton, and with extra armor resembling chaps covering their forelegs; possessed of poisonous beaks that one might imagine to be "guns" of a sort, and with what through the microscope even looked like rivets, the illusion of viewing miniature living "tanks" was hard to dismiss (plate 142, lower drawing).

Lying concealed in flowers or motionless on leaves and ready to pounce upon, kill, and suck dry much bigger insects than themselves, these bugs were consistently beneficial organisms; when about a half an inch in length at maturity, they no doubt accounted in a single summer for numerous detrimental insects of other kinds.

Before reluctantly terminating these insect chapters (for there is still a great deal more that might be recounted about even those which were found in the area of the mini-wood), a few words will be added about the familiar katydids, grasshoppers, tree crickets and others, to which we owe the pleasant sounds, in chorus, in the evenings of late summer and early fall.

During the earlier summer weeks, these insects have been growing and molting, many of them producing four good wings or wings and wing covers, and each molt bringing them closer to sexual maturity and their season of "song" by stridulation, which collectively produces that medley of sound which replaces the bird chorus by then already terminated until the arrival of another spring.

These are the males we hear "singing," although some of the females are capable of making lesser sounds. The sound-making apparatus varies in appearance according to the species, but, fundamentally, construction consists of a "pick" or scraper on one wing which is rubbed across a series of ridges on the other, these stridulations in some species made louder by an amplifying device below.

The above facts pertain to the long-horned cone-nosed and meadow grasshoppers, the katydids, and the crickets. Some of the short stubby grasshoppers stridulate by rubbing the large hind legs against a thickened portion of their forewings, and many of the large locusts produce a clacking series of sounds in flying.

Many of these orthopterous sounds are ventriloquial to a remarkable degree. Some of the most incessant (yet weakest in volume) trills of the summer nights are all but impossible to trace to their sources. To this day, the author remains uncertain what species produces the ubiquitous faintest little trills that seem to be all about us on warm summer nights but always seem to transfer to where the searcher is not.

All of the insects mentioned above are vegetarians, and most of them feed at night. The katydids are usually too high in the trees to be easily observed when "singing," but persistence may sometimes reward the investigator with the sight of a long-horned grasshopper "fiddling" in the flashlight beam. Again, if one is lucky, the pale green snowy tree cricket, genus *Oecanthus,* may also be located producing its short, more musical sounds by rhythmic, long-continued wing scrapings (plate 144).

The reader may have seen it stated that one may calculate the temperature by the number of "chirps" per minute made by this insect. In testing this supposition within and near the mini-wood, it did not hold true, and as an example negating the idea, on one night when in late summer the temperature was 61° F., tree crickets were stridulating ninety-four times per minute.

Eggs of the orthopterans so far mentioned are deposited in the ground, or on twigs and other vegetation. There are not a great many eggs to a batch, and they are protected with tough "shells," for these must remain viable through the winter. The author has found rather flat-looking grasshopper eggs inserted in the ground, but where the snowy tree cricket deposits hers safely for the long cold Connecticut winters, someone else will have to inform us.

Belonging to the same order (Orthoptera) as the katydids, grasshoppers and crickets that were met in the mini-wood area were rare, small, *native* mantises, about two inches in length, and green or brownish in color. The much larger *introduced* oriental mantis, *Tenodera sinensis,* also showed up now and then, leaving its oddly shaped, well-varnished egg mass or *ootheca* attached to a twig or weed stalk. Wintering in this form, the young hatch from these eggs and pour forth around May 10. From their time of hatching and throughout their lives, they are ravenous predators feeding on whatever other insects they are able to capture and hold in their sharply spined forelegs, munching the prey as we eat corn-on-the-cob, and with as much evident relish. Hatching mantises, like performing acrobats in their early motions, and a large adult female at her meal are illustrated in plate 144.

CHAPTER XVI

MINI-WOOD MOLLUSCS

There are perhaps eighty thousand known species of molluscs, constituting a phylum which is divided into numerous classes and orders. Familiar animals included are the *bivalves,* such as the clams and oysters and other double-shelled species; the snails and other *univalves;* the snaillike *slugs,* most of which have no external shell; and the *cephalopods,* or squids, octopi, and nautili, the latter being the highest members of this second largest phylum in the animal kingdom. Only the phylum Arthropoda, which includes the crustaceans, insects and spiders, myriapods, and numerous others, outnumbers the phylum Mollusca in known species.

Technically, the bivalves belong to the class Pelecypoda, while the marine and terrestrial snails, the limpets, abalones, and slugs are grouped in the class Gastropoda. The squids, octopi, and nautili, which are comparatively large-brained, very active predatory animals of the sea, constitute the class Cephalopoda.

Molluscs which inhabited the mini-wood were few in species and varied greatly from year to year numerically, but, generally speaking, were commoner than might be supposed by a casual observer.

The smallest of these were gastropods, terrestrial snails seldom visible for the reason that they were chiefly dwellers of the humus and damp refuse, while the much larger slugs, one species of which was at times abundant, were mostly nocturnal feeders, hiding away during the day and appearing in numbers, about dusk, to feed avidly, too often on cultivated plants and flowers, petunias in particular.

After a good rain had soaked the mini-wood, these terrestrial molluscs, both the slugs and the mini-snails, were more active than at any other time. The latter tiny univalves were usually feeders upon black detritus in the humus, which sometimes showed through the animals' paper-thin, translucent shells. Rains, however, seemed to stir them to move out and upward at times; as mentioned in an earlier chapter, they were then found browsing on the algae growing on tree trunks, perhaps enjoying this change in diet.

Some slugs also ascended tree trunks after a rain, sometimes, if venturing onto bark that was partly dry, leaving a silvery trail of mucus on which to travel back.

Despite the long terrestrial history of these well-adapted snails and slugs, neither can tolerate dryness for long, the ancestry of the sea being still strong in their genes. While they no longer require the chemistry of seawater, plenty of moisture in their habitats is an absolute necessity or they quickly die, if shelless; otherwise they must seal themselves temporarily in their shells, if they have such housing. The tiny snails of the mini-wood pulled themselves far back into their shells if their surroundings became too dry, and then sealed the entrances with a secretion which dried into a tough protective wall behind which the inmate was safe until the return of moisture.

In the tree of animal life, molluscs place somewhat below the crustaceans, the insects and spiders and many other arthropods, and even on a lower branch than the annelid worms. They differ radically in body plans from these other invertebrates, being equipped, for instance, with a muscular "foot," varying in structure according to the mode of life of the animal. In the snails and slugs it affords a strong, flat-surfaced plane especially adapted to their smooth, gliding mode of locomotion.

There are no rings encircling gastropod bodies as we saw them in the annelid worms, no rounded or flattened somites as we found them in the millipedes and centipedes, no head-thorax-abdomen combinations set apart by distinct constrictions as in insects, no jointed appendages of any kind. In these molluscan organisms, the head, foot, and body become one smoothly integrated whole.

In the snails, most of the vital organs remain within the protective shell and only the head and foot are projected; as the animal grows, it enlarges the shell by secreting calcium carbonate, depositing it in layers around the opening of its "house." For escape from unsuitable conditions and some enemies, snails withdraw into their shells, sealing them, if necessary, while the slugs (with only degenerate caplike shells in some species, or useless *internal* ones in others) depend for protection upon heavy secretions of alkaline slime, which will even save them for a considerable time if they are completely submerged in kerosene.* Under adverse conditions, the snails collected in the mini-wood sealed their doorways as mentioned, sheathing being no doubt mostly of the calcareous material secreted for the shells.

A few brief additional notes on gastropod anatomy may be of interest to the reader.

Covering the soft visceral mass within the shell there is a protective mantle which extends to the shell aperture. The calcareous material (carbonate of lime) of which the shell is composed is secreted by this mantle.

Terrestrial snails and slugs, being air breathers, are equipped with a lung and respiratory pore. On one side of a slug, this entrance may be plainly seen opening and closing as the animal takes in and expels the air.

From the heart, a branching pulmonary artery carries the blood. A pedal artery traverses the length of the fleshy foot, and a branch of it enters the head. The animal also has a kidney and ureter, and of course many glands. The slime glands serve to keep the mollusc's skin moist and to protect it when attacked.

No other animals are equipped for eating as are these molluscs. Within the mouth there is a "ribbon" bearing rows of strong, chitinous teeth. As with a file, when this is

*Slime from one species tested pH 8.0 to 9.0.

protruded and moved back and forth by powerful muscles, the food substance over which the animal is gliding is rasped off in small fragments (see plate 147, bottom). This organ is called the *radula*, and as the form of the teeth varies in different species, it is important in identifications.

Behind the mouth and radula there are a pharynx and jaw, then the alimentary canal, crop, and stomach, and a long, coiled intestine terminating in a *side* anus. Two pair of fleshy tentacles, the longest pair bearing the eyes, surmount the head. There is also, of course, a well-developed nervous system.

In North America there are a great many species of terrestrial molluscs, and in the rest of the world, especially in the tropics, there are hundreds more. But out of all this great assemblage, in this book we are concerned with only three species of snails (all of them small forms with discoidal shells) and with three or four molluscs minus visible shells, which we know as the slugs.

The commonest snails found living in the mini-wood humus—and also under refuse and discarded lumber, if wet —carried shells averaging six millimeters (about a quarter of an inch) in width. Extended, the animals sometimes stretched out to fourteen millimeters, but when thus drawn out to the full, their bodies were only about a sixteenth of an inch (1.6 mm.) in width. These were of the species *Zonitoides nitidus*, in which the shells in living specimens looked dark brown, but when empty, were semi-transparent, amber colored, and glossy (plate 145, lower photograph).

Species other than Z. *nitidus* were usually found less often, but in one extraordinary instance, on December 21, the author removed a sheet of plywood that had been lying on the moist dead-leaf carpet in the mini-wood for several months. On its underside, where they had gone into hibernation, were over fifty land snails and some empty shells. These averaged eleven millimeters (about seven-sixteenths of an inch) in width, and looked very dark brown with the animals sealed within. The empty shells were semi-transparent, vitreous, and pale amber in color. These were paper thin. The snails being a bluish lead color, produced the brown hue of the occupied parts of the shells.

Tentatively identified as *Oxychilus draparnaldi*, thirty of the torpid wintering snails were removed to a terrarium, the atmosphere within kept humid by a deep layer of black humus from their normal habitat. In the warmth of the laboratory, these snails returned quickly to normal active life, apparently having eaten the seals which they had placed in their doorways, and it was then seen that their bodies were of a distinct bluish lead color, as mentioned before, with the foot pale grayish yellow on the underside (plate 145, top photograph).

In addition to their natural food, available in the humus layer, these thirty individuals were supplied with occasional leaves of lettuce, which they ate with apparent relish, sensing the presence of the food as soon as it was dropped in the cage.

These thin-shelled snails could do a remarkable thing.

As they explored their cage, or glided in their slow smooth manner toward newly introduced food, their longest tentacles were alternately retracted and thrust forward, eyes and all, giving an erroneous but nevertheless startling illusion that they were being sucked in in some way, and then blown out again. Viewing these actions through the stereo microscope was a most interesting experience, but not all gastropods, and few, if any, other animals do this. Why those who can, do so, is a mystery, for the actions take place whether or not the eye tentacles encounter obstructions.

While the head and foot are directly joined in these animals, as already stated, a distinct head *seemed* evident as they held this end erect with their four tentacles directed forward, the longer pair with the bright little eyes well displayed. Indeed, the sympathetic observer and lover of all these small living things might even see in them a characteristic sort of gastropod "face."

Evidently the snails received welcome messages through their tentacles almost as soon as fresh food was placed in the terrarium, and as they headed slowly in the right direction, it reminded the author of a snake's way of approaching its prey, with its sensitive tongue (in place of tentacles) thrusting forward and backward until the food is reached.

After two months in captivity, these snails began to deposit tiny round white eggs, dropping them singly or in twos and threes, on the surface of the humus and wherever they happened to be, no further parental incubation or care being required. The young snails emerged twenty-five to thirty days later. Housed in pale yellowish shells already two whorls wide even at this early age, the youngsters themselves were pale grayish yellow.

Their first meals consisted of the shells of the eggs from which they had just liberated themselves. Very soon thereafter, they too were able to sense fragments of lettuce placed some distance away, and then, like their elders, become eager consumers. As most of the eggs had been removed from the terrarium into mini-petri dishes containing sifted layers of humus, hatching and other details could all be observed through the microscope.

The adult colony remained in apparent good health for five months, but then, for no discernible reason, commenced to die off one or two at a time. Perhaps these deaths were natural, the animals having reached their time to go, once a new generation was coming on. Having become attached to the group, it was sad for the author to observe its decline and see its population dwindling. Within seven months after they were found, these snails were all dead, and the author, not wishing further to reduce their numbers living naturally in the mini-wood, returned most of the young snails to its rich acid humus in the spring.

A third species was rare within the woodland proper, but on one occasion several were found huddled together under mulching paper in the author's vegetable garden not far away.

This species, which up to this writing had not been identified, had a glistening black body, including the four tentacles, and the shells, which were semi-transparent, accordingly looked black. Actually an amber color, after several weeks these shells mysteriously became lighter and lighter, until about five months after being collected and maintained in a petri dish with the usual layer of sifted humus, they had lost their partial transparency and had become a light cream color, except within the deep depression in the center of the ventral side which is called the *umbilicus*.

As the animals themselves remained black, this color change in the shells was possibly due to a deposit related to the snails' new or "artificial" diet.

The shells of this species were very distinctly marked by delicately raised growth lines, each such encircling ridge indicating fresh activity of the mantle and its secretion of calcium carbonate.

In the acid humus of the mini-wood, one wonders what *natural* foods found there supplied these animals with the required amount of the calcium element. The thinness of these shells may indicate its scarcity, while at the same time it should be remembered that the heavy shells of marine forms come from water very rich in calcium. The little gastropod shells which gradually turned to a cream color retained their vitreous gloss, and in certain lights were slightly iridescent.

Although close watch had been kept on the few snails of this species that were obtained, somehow eggs which must have been deposited in the petri-dish were not observed, but five and a half months after the adults had been confined, young appeared.

Very neat, dusky little creatures, they carried almost transparent, pale yellowish shells already of two and a half whorls which measured one and a half millimeters (approximately one sixteenth of an inch) in width.

As in the case of the adults, these youngsters turned eagerly to the offered bits of lettuce, and with their microscopic radulae, soon rasped holes in the leaf fragments bigger than themselves. They grew very slowly, but when returned to the mini-wood in April, all were in good health.

What enemies the three small species described had to contend with was not revealed to the author. While many snails are known to be infested with nematodes or trematodes, none such were found in individuals that died and were dissected under the microscope. Although Common Grackles, Starlings, Towhees and White-throated Sparrows often scratched for animal food in the dead-leaf carpet, no birds seemed to actually dig into the humus where these small species of molluscs are found. Wood frogs and salamanders like the large yellow-spotted black species may occasionally eat these molluscs, and the woodland form of the water newt (see chapter XVII) may take young and adult snails if it happens to find them.

When themselves feeding in captivity, all of these snails were observed leaving large amounts of fecal material behind as they progressed over the lettuce leaves and other foods. As a good proportion of each such deposit was pale green in color, it was easily seen against the black humus layer in the cage. While the snails were very small individually, it was evident that, under natural conditions in a woodland, collectively they would eject a considerable amount of the vegetable and detritus foods ingested, after due dimunition during its passage through their systems, processed and of value to other small animal members of the community. Microscopic examination revealed, for instance, that at least two kinds of soil animals—collembolans and acarines (mites)—were feeding on the freshly excreted "waste."

* * * * *

We come now to those molluscs which we usually and disdainfully call the slugs but which naturalists look upon as strangely interesting organisms—not exactly snails as we usually think of them, but closely related gastropods nevertheless.

A few slugs have small, thin, calcareous objects resembling very flat shells which cover limited dorsal areas. Many common species have *internal* similar shells; and all carry a fleshy "saddle" of sorts, either covering much of the back, or a smaller area anteriorly. This is also called a *mantle,* and near its lower border on the right-hand side, the breathing pore is located.

Like the land snails, the slugs have one short pair of plain fleshy tentacles, and a longer upper pair, each bearing a single eye. The curious habit of slowly inverting this longest pair, eyes and all, and then thrusting them out again as they move about, is also common in these molluscs.

They glide along on a muscular foot as snails do, and they are equally slow movers. For protection, when disturbed, they contract the whole body and sometimes roll over, or if feeding on leaves or flowers, they drop to the ground, and if further annoyed, may exude a copious protective slime.

Smooth rocks on which has grown green or grayish algae will sometimes be observed bearing interesting chainlike patterns cut by a slug's zig-zagging radula (the toothed organ in its mouth) when feeding on the plant cells (plate 147, bottom photograph). Some species, when gliding over dead leaves or rocks, or even when feeding on living leaves which are not sufficiently moist, leave behind trails of mucus which turn silvery as they dry, thus revealing the animals' nocturnal routes.

Slugs are hermaphrodites, and sperm is exchanged from one to the other at the same time upon mating. In some species this is "packaged" with glandular secretions in spermatophores, somewhat as we have seen in other animals (arthropods of certain species) in earlier chapters. Individual sperms are not released until the jellylike substance of the spermatophore dissolves within the genital orifice of the recipient. In these animals the procedures leading to and involving the mating act may be intricate, and for details the reader is referred to the list of works for supplementary reading at the end of the book.

Our small to medium-sized *native* slugs, which have inhabited our land for no one knows how many centuries, have never in the author's experience become numerous enough to cause serious damage, as have some of the introduced species.

The native species are members of long-established and stabilized associations, fitting into their particular niches in the habitat, taking their natural intended portion of the available food, supplying food in the form of their own bodies to natural predators, and supplying fecal matter usable in various ways by still other organisms, as we have seen to be true of so many other woodland animals. Native predators probably hold native slugs as well as the native snails in balance; thus they do not increase in numbers to any great extent as long as the natural community of which they are a part is left undisturbed.

Introduce a foreign species, however, and (as we have learned again and again, to our sorrow), since the alien's controlling foreign predators and parasites are missing, there is often nothing standing in the way of its increasing enormously in numbers.

While the native species are also vegetarians, they live their lives singly. Seldom are more than two or three found in close proximity under the same shelter. They never seem to be gregarious like at least one of the ravenous introduced European species, a conspicuous orange-yellow to brownish orange animal named *Arion subfuscus,* some of which measure over fifty millimeters (more than two inches) in length (plate 146). Slugs of the family Arionidae, to which this animal belongs, have the saddlelike mantle situated far forward, where it covers only about one-third of the upper dorsal surface, and in which the breathing pore may be seen low on the right-hand side, somewhat anterior of the center. In these fat slugs the opening and closing of this respiratory "valve" may be easily observed as the animal slowly breathes. Below this orifice the reproductive pore is situated.

During the eighth year of this study, this species suddenly appeared in the author's area, including the mini-wood. For the next few years, the animals came from hiding each spring, but in numbers that were in no way alarming. These were young individuals, and no full-grown slugs were seen until well into the summers. The author believes that the pregnant, well-fattened individuals die off in late fall after depositing their eggs in from one to three layings, the ova being dropped singly or in small clusters under flat stones, discarded lumber, rotting logs, or refuse under which the ground remains damp. Clusters were also found suspended from twigs or rootlets and held thus by strings or stems of hardened whitish mucus. These suspended clusters resembled miniature bunches of white or yellowish white grapes (plate 146).

No adults of this species were found at any time burrowed into the humus, and none were found after they had disappeared, following the egg-laying period. None were found hibernating under rocks, logs, refuse, or other possible winter shelters, notwithstanding the fact that these slugs became gregarious at times, numbers having been found clustering together toward fall.

A cluster of milky white eggs, found dangling from a rootlet over a hole dug in the mini-wood by the author in which to keep a thermometer, was removed to the laboratory on October 8. The hole had been kept covered with a square of roofing paper, and the slug, finding the dark damp interior to its liking, had crawled out on the rootlet and there suspended its egg cluster over the cool hole by a band of silvery white mucus. The eggs were jellylike in consistency, with thin skins, and measured three millimeters (less than an eighth of an inch) in diameter (plate 146).

In the warmth of the laboratory, these eggs commenced hatching thirty-nine days later. The newborn slugs measured two and a half millimeters in length when contracted, and up to four millimeters when on the move. They were pale gray in some cases; more yellowish gray in others.

At first maintained only on what they could find in damp humus taken from the woodland where the eggs were located, they increased in size very slowly. Their rate of growth was somewhat accelerated when lettuce was added to their diet, but these youngsters were not a vigorous lot, and all of them died within a few weeks after hatching, no doubt for the reason that the eggs had not been allowed to winter over normally in a cold temperature.

Certain vegetation on the mini-wood borders and in the author's garden, including flowers and tomatoes, proved irresistible to these robust slugs, and after a time they became so numerous and proved to be such ravenous feeders that drastic measures against them had to be taken, particularly when they began to destroy not only beds of petunias and —strange to say—the rank-smelling foliage of marigolds, but also the author's prized tomatoes.

Slugs are without doubt among the most difficult garden pests to control, once a foreign species multiplies the way this one did in the author's area. The difficulty lies in the fact that mild sprays have little or no effect upon these animals, which are capable of secreting large amounts of protective slime that will even exclude kerosene for some time when slugs are dropped into this oil.

The only sure way to rid the flowers of these animals is to hunt them at night with a flashlight, when they are avidly feeding, and quickly snip them in two with long-bladed sharp scissors. It is a disagreeable method, to be sure; but dozens, or hundreds of potential egg layers may thus be instantly executed during a single summer, and the damage to one's flowers greatly reduced. Also, dishes of beer placed below the plants will attract some slugs, which die after drinking this fluid. To protect tomatoes from slug infestations, the author found that clear plastic bags tied over the growing fruit clusters saved them, while acting as individual "greenhouses" at the same time. A few holes punched in the bags for air circulation, and others at the bottom corners to let out accumulating moisture, were of course necessary when this method was followed.

Getting back to the smaller, harmless native slugs, there were two species inhabiting the mini-wood, both averaging twenty-five millimeters, or slightly less than an inch in length. One was *Pallifera dorsalis,* a mostly bluish gray animal, and the other a similar, but almost black slug, apparently of the same genus, but unidentified as to species. In both, the mantle covered the whole dorsal surface, the head arising from under the anterior margin.

With their skins moist with glandular secretions, these little slugs glistened in the flashlight beam. On rainy nights they ascended the tree trunks to feed on algae, by day hiding in litter, where occasionally small clusters of jelly-like eggs were deposited, although more often they were buried in the humus.

Pallifera dorsalis and the other slugs of this genus found in the mini-wood or elsewhere belong to the family Philomycidae, a group of molluscs which may be distinguished from other northeastern species by the mantle, which, as already stated, covers the entire dorsal surface. There are several other species in addition to the two described from the mini-wood.

The largest slug that occurred in the study area, a consistent nocturnal feeder upon the many species of fungi which grew there (see chapter VII), was another introduced European species named *Milax gagates,* a member of the family Limacidae, and commonly referred to as the "greenhouse slug" for the simple reason that it is often

found in these glass houses with their year-round succulent food supply and congenially moist atmosphere. And it must be stated that this slug has at times caused considerable damage to greenhouse plants.

Those individuals found inhabiting the mini-wood had, however, fitted harmlessly into the scheme of things along with many other kinds of animals, eating a share of the various toadstools as they appeared during the summer and fall (plate 147).

Some of these slugs measured up to two and a half inches in length. Their color was very dark gray to glossy black, with the mantle situated far forward and only covering about a third of the dorsal surface, and with the respiratory pore located on the lower margin of this saddle slightly back of center.

When extended, the posterior end of the foot in both *Milax* and *Pallifera* was distinctly pointed, whereas in slugs of the genus *Arion* it was notably and rather plumply rounded.

Milax gagates proved to be a very heavy feeder. The specimen shown at its nocturnal meal in plate 147 ate the rounded cap (entire) of each of the toadstools tackled, and consumed two or three in a single night. Stems of the plants were not attacked. With the animal's filelike radula rasping away, and with only occasional rests, a single cap was consumed in about one hour.

Watching *Milax* effortlessly consuming a toadstool cap, a substance nearly as soft as butter, recalled how another mollusc, a marine snail named *Urosalpinx cinereus,* actually drills with its radula through the extremely hard shells of oysters, its chief food item, and through the small round hole thus made, extracts the soft helpless mollusc within.

Microscopic examination of their feces and of food taken from the crops of slugs has revealed (to other investigators) the remains of aphids or plant lice therein, indicating that animal food may be regularly included in their diet. But in such cases, in the opinion of the author, it would seem more likely that the insects were unwittingly, rather than purposely, eaten as the slugs were consuming leaves infested with these minute bugs whose sucking beaks are often very firmly planted in the leaf or stem tissues of the plant.

In regard to the introduced species in general, one may never be certain which plants these slugs will choose to attack. On one occasion late one summer, after *Arion* slugs had eaten up most of the foliage of the author's marigolds, they turned their attention to a heavy stand of wild jewel-weeds (touch-me-nots; see plate 75) growing nearby, which were thriving in their customary place on the lower wet border of the mini-wood.

In the leaves of these plants, sometimes five feet above the ground, these slugs ate large rounded holes, browsing here and there like cattle slowly moving across a green meadow. Feeding thus in the ample foliage of these massed jewelweeds, they did no important damage, but each morning the silvery trails of dried mucus could be seen all through the foliage where the slugs had fed and where they had ascended the plants. Would that *Arion* slugs would confine their feeding to these massive annuals, rather than preferring the author's cultivated flowers!

CHAPTER XVII

AMPHIBIANS OF THE MINI-WOOD

Having delved for some time into the realms and lives of invertebrates—dominant animal forms of the mini-wood community—we come now in these last four chapters to the phylum Chordata, of which four classes of the subphylum Vertebrata were more or less well represented in the study area, namely, members of the Amphibia, Reptilia, Aves (birds), and Mammalia.

While in number of species vertebrates were comparatively few in this fragmentary mini-wood, they nevertheless constituted a no-less-interesting and important group in the compact interrelated whole of the little community.

The class Amphibia, as most of us know, includes the toads, the frogs, including the tree frogs, and the less familiar salamanders, which, like the frogs, are moist- and smooth-skinned, and have *clawless* toes, but have long tails in the adult stages. Eastern North America is rich in salamander species, some of which live under stones along stream banks, while others inhabit caves, ponds, and springs, but the woodland-inhabiting forms are the most numerous and most often observed. They are frequently mistaken for lizards, which to be sure they resemble in general form; the latter, however, are dry-skinned, scaly, sharp-clawed, sun-loving reptiles, which seldom if ever occur as far north as the mini-wood area.

Close by the mini-wood, the author's spring-fed pool attracted many creatures. Among these—and doubtless having wandered away from nearby larger pools or ponds—a salamander commonly known as the newt, has bred occasionally. This was the species *Diemictylus viridescens*, an amphibian which, as we shall presently see, has a curious life history, for it spends a year or more of its life as a land animal after leaving the water as a young adult; then, after having grown considerably larger, it returns to the aquatic habitat for the purpose of breeding and remains there permanently. (Plate 148.)

Here is its interesting story in detail.

The acquatic breeding adult may be four inches or a little longer in length. It is then greenish brown or yellowish green above, with many red spots, but yellow, sprinkled with black below. The male's tail is strongly developed and bears a wavy keel of thin skin. On the head there is a dark line running from the nostril and passing the eye to the neck. The hind legs are much thickened, and like the belly, sprinkled with black dots.

Pairing occurs in the water as early as April and continues into May, quiet ponds with plenty of aquatic vegetation being the favorite habitats. In courting the somewhat smaller female, the male approaches stealthily, then caresses her suggestively, nudging her gently, or rubbing her with his nose in a preliminary and gentlemanly sort of procedure. If she starts to swim away after such attentions, the vigorous male, with sudden new energy, darts after her, and clasping her at once between his powerful legs, holds her fast, squeezing in her sides so deeply that sometimes it might seem as if about to pinch her in two.

Bred in aquaria by the author, mating pairs, with the male often underneath, remained clasped for only short periods, sometimes only for minutes. Whether or not there was actual intercourse could not be determined, but it appeared more likely that the sperm was being shed into the water or the aquatic plants, or perhaps a spermatophore was being transferred.

Most eggs that had already been deposited had been partly rolled up in the delicate leaflets of the water weed *Elodia,* but *Vallisneria* and watercress, the common aquarium plants, were also used by the newts in the leaves of which to "glue" their transparent oval eggs, within which the yolks, when fresh, were greenish black on one side and yellowish on the other.

The developing embryos were fascinating to observe under the stereo microscope, the eggs being easily handled in a water-cell on the microscope stage where they could be held thus in a bright light for considerable periods without damage if kept cool.

Measuring from three to four millimeters (about three-sixteenths of an inch) lengthwise, and distinctly oval in shape, the eggs were splendid objects for study. Removed from the aquaria on May 25, still rolled up singly or in twos and threes in the leaflets and placed in finger bowls of pond water with *Elodia* for an oxygen liberator, the developing embryos resembled kidney beans in outline by June 1. By June 4, the larvae were moving about within their envelopes of glasslike jelly. By June 7 there was still more activity, and the tails of the rapidly developing newts were distinctly visible through the albumen. By June 11 hatching commenced, and it continued through the fifteenth of that month.

The newly emerged newts were curious little things, six millimeters (about a quarter of an inch) in length, with large feathery tufts extending at each side just back of the head. These were, of course, the gill tufts (retained, according to the species, by young salamanders that are hatched in water) for varying periods in their early lives. Such gills in larval amphibians reveal something of their early history—their descent from more fishlike ancestors—and, as already stated, some salamanders are permanently aquatic, some species retaining the gills throughout their lives. Indeed, these newly hatched newts reminded the author very strongly of fish hatchlings, except that their rather long black speckled tails were broadly tapering and without the caudal widening or V-shape of their distant finny relations.

When two days out of the eggs, their forelegs were already formed and body skin was yellowish studded with numerous stellate black pigment spots. In a week, the posterior pair of legs had sprouted, and at twelve days after hatching, these newts measured somewhat over ten milli-

meters (over three-eights of an inch) in length.

Their rate of growth thereafter seemed to be governed by the temperature of the water and variations in their diet, composed of minute aquatic animals. In the author's area, newts which had hatched in natural ponds measured from twenty-one to thirty-five millimeters in length by the end of July, and upon reaching these sizes, commenced to assume a more mature look and the truly beautiful, almost glowing orange coloring of the terrestrial stage, some of the newts paler than others after just losing their gills and upon first leaving the water, but darkening within a short time thereafter, so the fine black speckling and rounded vermilion spots then showed up with great brilliancy.

No one seems to know why it is that this species, after leaving the water, remains for an extensive period on land but finally returns to an aquatic existence permanently. It then changes color, becoming dark yellowish green or brownish green with a black-spotted yellow belly, and the rounded spots on its back become more red than brilliant vermilion. The author cannot say if the newts spend one, two, or more years in the attractively colored land phase. Why this terrestrial period is necessary in the life of this species is an ecological puzzle of the first order. (Plate 148.)

Although land-phase newts were found only occasionally in the mini-wood, they occur in large numbers elsewhere, and it may interest the reader to hear about an astonishing encounter experienced with these newts some years back, when camping in the wooded hills above the Mianus River in nearby Stamford, Connecticut.

It was mid-August, and the period had been hot and breathless. No rain had fallen for days, but distant rumblings seemed to indicate relief, which my companion and I prayed would not disappoint us.

Cicadas were "singing." A hawk or two circled high in the hazy blue against a backdrop of cumulus clouds that looked like rippling whipped cream. But no rain came in daylight, and the woodland carpet remained crackling dry. At midnight, flashes on the tent canvas followed by rumblings heartened us once more, for the land was in desperate need of a soaking.

My companion and I had been talking about how we might feed ourselves in the woods or elsewhere in camp, had we no other means of supply. We spoke of the various mushrooms and edible toadstools, were one certain of their identities, and especially of the large bunches of certain species of coral fungi or *Clavaria* which popped up here and there in Connecticut's forested hills like cascading miniature fountains of frozen cream. We mentioned wild grapes and other berries, spring's fern fiddleheads; cattail and milkweed new shoots, cowslip greens and water-lily roots, and the wild asparagus, self-planted at the borders of the salt meadows and thus established from seed brought from the Sound by high tides and deposited around the inlet edges. We spoke of crows' eggs, snapping turtle stew, the leathery lichen sheets called rock tripe if one was ever that hungry.

The storm was coming nearer, the lightning flashes more intense, the rumbling growing louder. After midnight it hit in earnest. It snorted and blew, it rained with all its might, quickly filling our tent trenches, which ran like miniature rivers on each side of the waterproof canvas floor. Then it quickly passed over and petered out, leaving us to fall safely and peacefully to sleep, deeply thankful for the rejuvenated air smelling of freshly bathed foliage and limbs and bark, and a lingering scent of ozone which the author has always believed to be left by lightning.

Day dawned fresh and clear. The forest carpet, long dry and pale, had turned umber and sepia and burnt sienna from its drenching, and the shadow beneath each partly curled dead leaf was now cast black as night by the risen sun.

When the sun had risen enough to send its brilliant beams through gaps in the canopy, like little spotlights, we dressed and wandered leisurely to the spring for water and to recover the can of milk before making our tea and coffee. Then suddenly we came upon the newts, one or two at first, then in tens, dozens, and all at once we realized that the woods were literally swarming with them.

Orange-red to brownish orange above, with two rows of vermilion spots along the back, and warm yellowish below, hundreds of these leisurely land-form salamanders were crawling over the leaves and rocks, logs and other fallen debris, and in the wood paths. Wherever we looked, efts (which we then called them) were meandering from one place to another, their bodies almost a luminous orange color when passing through little pools of sunlight sifting into the woodland.

Here indeed was one of the most astonishing sights I had ever experienced in Connecticut. In our daily hunts before, we had found them occasionally, often along with other common species hiding under logs or fallen bark or large flat stones, if the ground was damp enough. But where had these great numbers of newts come from so suddenly?

They must have been all around our camp site some-where, perhaps far down under the leaves burrowed into the cool, still moist humus, thus spending the days of drought, for salamanders must have moisture; and if their skins dry up they quickly die.

There were no ponds anywhere near our camping area from which hundreds of newts at this stage in their lives might have all emerged at once to explain the exodus we had witnessed. As for the river in the valley of the Mianus not far below us, we had never found newts breeding in its waters at any season, and the available food supply for their larvae there would never have been suitable.

The simple truth was that we did not know, and still do not know many things about the lives of the common creatures which live all around us. For reasons unknown, this thunderstorm breaking over the Mianus woodlands had stirred the whole terrestrial-phase population of newts into action all at once, producing a sight never to be forgotten.

Certain animals, in widely separated phyla, have the odd habit of "freezing" when suddenly disturbed, which may, in some cases, be a protective action, as when their predators are those which depend upon movements of their prey for success. Certain myriapods and insects practice this device and sometimes thus escape being eaten. But with the 'possum and the land form of the newt, which also sometimes "play dead," it would seem that lying still and helpless might only aid whatever more intelligent animals might be hunting them. The author believes that it is fright and shock which cause a form of fainting in these two vertebrate animals, rather than a deliberate act of feigning death.

Newts and other salamanders are endowed with the power of regenerating lost limbs and tails to a remarkable degree. A newt was once discovered in the author's mini-wood which had suffered the loss of both forelegs, one above the wrist, the other well above the elbow, yet the little creature had not in this instance feigned death when picked up. Taken into the laboratory and carefully coaxed to feed upon white termites taken from its own woodland habitat, it made a fairly rapid recovery, and, most astonishingly, both injuries were repaired, the miraculously and exactly coded genes in the cells at the borders of the wounds creating a new arm, new wrists, and beautifully delicate tiny new fingers, marvelous processes to observe, and feats no *human* cells or other anatomical setup may duplicate.

* * * * *

A secretive but not rare little amphibian of the mini-wood during the years of this study was *Plethodon cinereus,* the red-backed salamander. Four to four and a half inches in length, and occurring in two color forms, the commonest was brown, gray, or almost black, showing a wide, distinctly red lengthwise dorsal stripe. In the other color phase as the lead-backed salamander, the ground color was dark gray dorsally, lighter on the head and limbs, and with the underside finely mottled with dark gray on a lighter ground. Both varieties had rather broad heads and prominent bright, wide-awake-looking little eyes (plates 148 and 150).

In the mini-wood they were found under decaying logs and fallen bark, beneath stones and refuse, under dead leaves, and in some instances under thin weed-deterring plywood strips placed between the rows in the author's vegetable garden. Damp soil or humus and darkness seemed to be their basic requirements.

Feeding upon ant larvae, termites, collembolans, very small worms, and possibly young oniscoids and snails, these salamanders occupied their little niches in the mini-wood, and probably in a small way made contributions to the composition of the humus in the moist habitats in which they dwelt.

Actual stomach examinations made for the purpose of this book revealed and verified the statement that both the red-backed and lead-backed salamanders feed to a considerable extent upon those primitive insects known as springtails (Collembola), which have been described in detail in an earlier chapter. Springtails of many species are abundant in the typical habitats of these predators. Many kinds of insects were found in the stomachs, mostly in fragments, but rarely, those that had been swallowed whole. Small beetle larvae were also identified, as well as white termites, the latter a soft food upon which the author twice successfully fed these salamanders in captivity.

All of the stomachs examined contained minute fragments of mineral matter which doubtless acted as grinding "stones" in these little animals, just as fair-sized pebbles perform this function in turkeys and other large birds, and as larger rocks once did in the stomachs of dinosaurs.

If the early spring was mild, these salamanders might come from hibernation as early as March 12 (but more normally not until April), then remaining active until late in the fall (plates 148 E and F).

When the days are warm, the female prepares a brood chamber. This may be under a rotting log, under a flat stone, or excavated in one of the damp sawdust piles often left by the larvae of large boring beetles in woodpiles. In the little cell prepared, about ten eggs are deposited, and about these the female is said sometimes to curl and possibly brood, although she is of course, a cold-blooded little creature. The author has never found a female thus in attendance upon her eggs. In one case a clutch discovered in one of the beetle sawdust piles, like that mentioned, had been suspended from the roof of the cavity, but as they did not hatch, they may have been those of some other species. The salamander's eggs are white or slightly yellowish, and quite large for the size of the salamander's body cavity, as will be seen by examining the anatomical illustration showing eggs in situ (plate 148 G).

The larval development takes place entirely within the eggs, and while it has larval gills, once it is hatched, *Plethodon cinereus* remains a strictly land animal throughout its life, notwithstanding its early fishlike characteristics, respiration taking place through the skin and the roof of the mouth, as a lung does not seem to be developed.

At breeding time the male produces a spermatophore, a capsulelike object containing the sperm, which* is either left for the attending female to pick up with the lips of her vent and push into the internal receptacle called the *spermatheca,* or which the male may himself insert into her *cloaca,* the chamber into which the intestinal, urinary, and generative canals discharge in birds, reptiles, amphibians, and some fishes. The reader may remember here with considerable interest that other organisms have been met in this book—the minute pseudoscorpions, for instance—in which the male also produces a spermatophore.

A mystery surrounds the disappearance of the very young red-backed salamanders once they leave the brood chamber. The author was never fortunate in finding them when they were, say, under an inch in length, although, on the other hand, it was usually fairly easy to locate an adult or two most any time during the summer and early fall. There is evidently a period during the life of the species when it burrows into the humus or other medium, perhaps in quest of some special food required by the young salamanders in this early stage. (Plate 148C shows the smallest one found by the author.)

---

*According to the species.

What the advantages of the spermatophore are to animals employing this method of sperm transfer are hard to imagine. Certainly it would seem no more certain of success than actual intercourse with an intromittent organ, but it is at least an improvement over still other procedures evolved by nature, as we shall see later in the chapter in the case of some of our North American frogs and toads.

It might be well to mention here that salamanders may vary greatly in color; thus, in some instances, it is difficult to be sure of a species. Even in the usually distinctly marked red-backed phase of *P. cinereus,* the broad dorsal stripe may be orange or yellowish as well as some distinct shade of red, or the coloring may intergrade with that of the lead-backed phase and thus be quite puzzling to the uninitiate. A modern guide will be required by anyone making a study of these amphibians, of which there are several dozen species in the United States, more than in any other country.

Occasionally the author found red-backed salamanders curled up and motionless under dead leaves in cold weather and at other times beneath large flat stones or other objects within the mini-wood. While these salamanders were under protective cover, they had nevertheless actually been on the woodland surface of the ground, and it seemed remarkable that they had not been frozen. But salamanders are not alone in possessing natural "antifreeze" in their systems (plate 148 E); readily coming to mind we might mention collembolans that live on bark, many spiders, and the wintering pupae of butterflies and moths; or even flowers like those of the crocus, which may arise to bloom from melting snow after a freezing night.

Rare in the mini-wood was the slimy salamander, *Plethodon g. glutinosus,* the only species that is glossy black above and with numerous small white or silvery spots and dots on its sides and back and the underside deep slaty blue (plate 150). With a cylindrical body, it may reach six or seven inches in length, and its feeding habits are much like those of the red-backed species. In the mini-wood it was found always under very wet debris, and as early in the spring as April 25. (Plate 150.)

The largest species in Connecticut, and found twice within the mini-wood, was the spotted salamander, *Ambystoma maculatum,* a seven-inch or longer very robustly built bluish black creature with a conspicuous row of large round yellow spots on each side dorsally (plate 149). As these salamanders sometimes reach eight inches in length, they have occasionally astonished people who have suddenly come on them when turning over decaying logs or removing old woodpiles. And on rare occasions, when a number of these big plump salamanders have been discovered floating in icy pond water in March or early April, and which is this species' normal breeding season, the human eyes that saw them seemed to have deceived their owners.

Many times in the past, while the author was director-curator of the Bruce Museum in Greenwich, specimens of the spotted salamander were brought in for identification, their finders not only astonished at the fact of the animal's presence in Connecticut, but finding it hard to believe, when so informed, that *Ambystoma maculatum* was as utterly harmless as any native frog or toad.

The animals' unperturbed presence in icy water during the breeding season serves well to illustrate the term *ectotherm;* applied to fishes, amphibians, and reptiles, it means cold-blooded organisms depending only on the surrounding environment as the source of whatever body heat may be required.

Somewhat globular heavy masses of eggs are deposited in ponds, swamps, and ditches where these salamanders occur, and, as stated, early in the spring. The masses may be found resting on grasses or otherwise floated, or attached to partly submerged twigs. Individual eggs are large and very clear, but the main mass in which they are embedded is more or less milky. In bulk, each mass makes a good human handful.

In these eggs, the biologist or interested observer has superb objects in which to study the whole animal development in detail, and at one stage, not long before the larvae leave the jelly, their resemblance to young fishes is striking (plate 149B).

The eggs shown in illustration 149A, taken April 12, hatched in the author's laboratory between April 30 and May 5, the young salamanders remaining for a time clinging to the mass.

In examining fresh batches, it was found if one pressed the jelly envelope very gently, individual eggs might be made to pop out for examination. These eggs measured seven millimeters (around a quarter of an inch) in diameter, and being as clear as glass, when placed in a dish of water under the stereo microscope, every slightest detail might be watched in three-dimensional brilliance. Sometimes remarkable things occurred within these jelly spheres, such as the appearance of swarms of micro organisms that had nothing to do with the embryo salamanders per se.

Very gradually, while the embryos were increasing in size, what appeared to the naked eye to be a fine green powder or haze might be seen accumulating. As the embryos began to move about within the jelly, this green wave moved around with them. Under the microscope it was seen at once to be a vast assemblage of motile algal cells in perpetual motion. By the time the larval salamanders had moved to the *outside* of their egg capsules, the jelly seemed to have turned entirely green with the fabulous numbers of these plant cells.

They apparently did the young salamanders no harm, and possibly they might have supplied some oxygen to them. They *fed and multiplied* upon what might otherwise have been wasted.

In one extreme case, after the gilled amphibians had left the jelly mass for good, the author observed its gradual reduction by these scavenger plants until the whole bottom of the glass container held a green layer a sixteenth of an inch in depth and consisting of countless millions of these organisms!

Here once more, as we have seen again and again in this book, "waste matter" is indeed a poor phrase to apply to anything in the undisturbed natural world. The albumin of which this jelly consisted, if such is its correct designation, served two distinct and important purposes: the safeguarding of the embryo salamanders, and the nourishment of these millions of algal cells, but it was a species which may of course also enter into food chains with other kinds of organisms in the economy of the pool or pond, as well as exploiting the jelly eggs of amphibians in which to multiply in such fabulous numbers.

Upon leaving the eggs, the salamander larvae measured ten millimeters in length. They bore large, triple gill-tufts on each side of the head. They were very active, and as they increased in size, avid feeders, eventually eating aquatic worms, water asels (a form of isopod), small snails, and insects. By July 20, they measured fifty millimeters (about two inches) in length, and the gill "feathers" were beginning to disappear. As they continued to grow and resemble the adults more and more, they turned dark brown to black, the color being due to myriad pigment dots upon the somewhat lighter ground, while the skin on the underside of the animals was bluish white with opaline tints.

All of our salamanders—and there are several other species in New England in addition to those found in the mini-wood—are among the most harmless of creatures. They may be safely handled and studied at any time, but there is danger to the little animals themselves if they are not gently treated and always with moist hands, as their skins must not be allowed to become dry.

As for the frog and toad group of amphibians of the mini-wood, it was limited to five species, and had there not been temporary pools, a small pond, and a swamp within reasonable distances, there would have been none, for all of them require water in which to lay their eggs, and all are hatched with gills and tails, and long coiled intestines to cope with a predominantly vegetable diet suited to small toothless browsing mouths. Lungs of sorts are developed later.

Earliest to appear and to lay their eggs (singly or in tiny bunches) in the icy water of early spring, were the "peepers," *Hyla crucifer,* smallest of our northeastern frogs and known to all country dwellers by their loud calls which often issue (ventriloquilly) from where they are not when one endeavors to locate individuals. At a distance, when the "peepers" are calling in great assemblages in the swamps and ponds, the sounds become a melodious din like "bells across the meadows."

As familiar as their "knee-deep—knee-deep" calls may be, there are comparatively few folks who would recognize a "peeper" as such were it placed before them, for they remain in the water only long enough to breed and are seldom captured; for the remainder of the spring and summer they are animals of the woods and low vegetation and are seen even less often (plate 151 and egg cluster, plate 152).

Delicate, and about an inch and a quarter in length, the frogs may be reddish brown, yellowish brown, or fawn color, always with darker dorsal markings more or less distinctly in the form of an X.

Usually avoiding discovery during the summer, their presence often becomes known again when, during cool fall days, they may call a few times, but with weaker voices, just before burrowing into the woodland carpet of dead leaves to hibernate until the following March or early April, when all are gathering to breed once more.

At this time, the tiny jellylike eggs are deposited on water plants or submerged dead leaves and as stated, singly or, occasionally, in small clusters. The tadpoles develop quickly, and being very small tender morsels, are preyed upon by fishes, and aquatic insects such as the water tiger (the larva of our largest black water beetle) and by the voracious nymphs of dragonflies, while the adult frogs are doubtless taken by water snakes, ribbon snakes, Herons, and very possibly by raccoons, who are very fond of larger frogs. (See peeper minus toes, plate 151, probably predator maimed.)

When the breeding season is at its height, if one is lucky enough to discover a pair, the male will be seen on the female's back and clasping her very firmly in his tiny arms. Strange to say, nature neglected to develop an intromittent organ in male frogs and toads as in most higher vertebrate animals. Sperm is deposited as the ova emerge, the individual sperms somehow swimming to and entering the eggs unassisted; thus we have the curious truth that some animals high in the ladder of vertebrate classification are no better off in this respect than are the ferns and many other primitive plants.

During the actual breeding procedures, the author believes that the "peepers" suspend feeding altogether until leaving the water for the remainder of the year.

At all other times these little amphibians become tree frogs, the minute disks on their toes aiding them greatly in climbing. In wet woodlands they are often found on the dead-leaf carpet, with which their coloring blends almost perfectly.

A formerly common amphibian and always a cheery sounding summer caller from the treetops, a species once familiar to all who lived or still live in New England country areas, was *Hyla versicolor*, the gray tree frog (plate 151), or "rain frog," as it was often known, for the simple reason that its pleasant trills *did* often precede the summer showers.

An almost exclusively insectivorous amphibian (although it also eats spiders) and a generally useful gleaner of foliage-eaters, a species dependent upon an unlimited, uncontaminated arboreal arthropod supply, the author lays the blame for this beautiful frog's almost complete disappearance from many areas squarely upon uncontrolled spraying with DDT and the other long-lasting pesticides that have unintentionally proved to be lethal to so many other forms of wildlife.

Adult gray tree frogs were found on rare occasions within the mini-wood, but only during the first few years of this study. They measured one and a half or two inches in length, had rather rough skin, and in color varied through shades of gray to white (one specimen had a greenish gray hue), with dark to almost black blotches and other markings on the back, black barring on the legs, and with conspicuously *bright orange* skin on the inner (concealed) sides of the thighs.

This species may change its color rather rapidly according to the surface upon which it is resting. A specimen found clinging to an almost white fence rail closely matched its weathered paint, while the dirtier areas and black cracks in the wood blended well with the frog's various blotches and barrings.

Captured individuals climbed with great ease. Their conspicuous toe pads, evolved for arboreal life, felt cool and clammy as the little creatures clung to the author's hand or arm, and their webbed hind feet and legs proved surprisingly powerful when the animals made long leaps toward nearby shrubbery or a tree trunk.

Breeding in swamps and ponds in early May, the development from jellylike eggs through the tadpole stages to tiny perfect toads is quite rapid, and the author found such specimens as early as the first week in June, beautiful little things when first taking up an arboreal existence, those which were obtained being bright green at this stage, and extremely agile.

Where these frogs hibernated remained a puzzle. Possibly they utilized natural cavities in the trees, or maybe they burrowed into the dead-leaf carpet or the humus. Someone else will have to say.

When spring days become really warm, the sustained, almost musical trills which reach us from ponds and swamps, and sometimes from our lily and goldfish pools, issue from curiously inflated throat "balloons" of the common toads, *Bufo terrestris* (plate 151).

These amphibians come from hibernation in rather drab raiment, wearing a coating of the earth into which they burrowed for the winter. Upon reaching the mating location, however, these familiar hop-toads, once bathed by swamp or pond water, become bright with nuptial colors —olive and pink, buff and brown, brick red and gray—almost as if they had been scrubbed and newly painted for the occasion.

Great activity takes place when many toads gather at one breeding spot. The females, already bulging with unfertilized eggs, seem as gaily colored as the males, and the latter now develop hard and dark-colored pads, one on each palm, which, as they tightly clasp the females, prevent them from slipping off or being pushed off by vigorous although clumsy rivals.

As the eggs are ejected, they are fertilized externally by the motile sperms shed by the male, as described in the case of the "peepers." The eggs are laid in long strings of jelly and appear like black beads embedded in this material. The egg strings may extend over several feet of aquatic plants, sticks, and submerged branches and other objects.

Development is rapid, the black tadpoles growing legs, losing most of their tails, and emerging within a few weeks as delightful perfect miniature toads. Frequently they leave the water in large numbers at the same time, and may then be seen hopping about everywhere, crossing roads, appearing in gardens and pathways, and often some distance from where they were hatched. Their rapid metamorphosis is a nice adaptation to the adults' unwitting habit of often depositing the eggs in temporary pools, such as flooded depressions in meadows, and other bits of water which dry up long before summer.

These very young toads are almost black, and sometimes still carry the incompletely absorbed stubs of their tails. Occasionally one of these youngsters or an adult appeared within the mini-wood but remained there for only a short time, preferring its borders or the adjoining vegetable garden, where the soil was warm. Mature toads fed nocturnally, taking a wide variety of insects, spiders, isopods, and even an occasional snail or slug, but their favorite food seemed to be earthworms, always available at night as they came to the surface to deposit their casts (see worm and cast, plate 125).

Adult toads were observed swallowing even large night crawlers, shoving even the most bulky individuals mouthward with their strong forelegs and "hands" once an end of the worm had been reached by the sticky tongue, then *pulling in* their widely bulging eyes until these were almost flush with the sides of their heads as the worms disappeared. One got the comical impression that the eyes were helping to push the victim stomachward. A few nights of following these animals with a flashlight, which did not bother them in the least, convinced the author of how beneficial toads really are, for their appetites are enormous, and any sort of a bug or crawling thing appeared to be palatable. A toad that had been run over on the driveway had its stomach pushed out intact, and on examining the contents with which it was distended, the author calcu-

lated, from whole and fragmentary remains, that this large individual had eaten twenty-two carpenter and red ants, five beetles, several immature crickets, two caterpillars, and two oniscoids or sow "bugs." Multiply this night's menu or catch by the number of summer nights alone, and again by the whole toad population, and one may get some idea of the vast numbers of arthropods destroyed by these amphibians.

On more than one occasion, a clever little toad was observed squatting on the ground directly below the author's outside garage light; there, having eliminated the necessity of hunting, it gobbled down partly toasted insects that hit the hot bulb and tumbled to the ground.

And a final note on toads: Handling them does *not* give one warts. The two elongated warty objects seen back of the eyes, one on each side of the raised median line, are important glands which secrete a protective exudation. The bitter substance will cause an incautious dog's saliva to foam slightly, which is nothing to be alarmed about, although the toad's glandular exudation may be somewhat poisonous to small predators. It has no effect upon garter snakes, however, which sometimes tackle and swallow toads. The common toad, if gently held and stroked during the breeding season, will often puff up and give forth little chirps of pleasure (see plate 151D). At the same time it may urinate on the handler, and it has been this harmless fluid that was credited with resulting in warts!

An inconspicuous dweller, or sometimes temporary visitor on the woodland carpet of dead leaves and litter, but which was only found occasionally in the mini-wood, when the leaves there were not dry, was the big-eyed, attractive wood frog, *Rana sylvatica* (plate 152A). Strictly sylvan, the species enters ponds and leafy pools *within* the wood only for the very brief breeding season early in April, sometimes even before all of the ice has melted. The frogs prefer shallow leaf-lined bodies of water with mud bottoms, around the edges of which grow elders and other low shrubs. Suddenly, on some sunny day, even as early as March 25 some years, crowds of wood frogs that hibernated in the vicinity may join up in joyous choruses at such breeding places. Later, entering the surrounding woods, they remain in evidence only so long as the leaf carpet remains wet, burrowing below or retiring to moist litter until rain moistens the surface again after dry weather.

Eggs are deposited promptly, as the frogs reach the pools. As with those of the other frogs and toads described, the eggs are externally fertilized, but their form differs from the others. At first rounded blobs or masses which become submerged, they soon float to the surface again, where they spread out into more or less flattened rafts apparently buoyed up by gas bubbles.

Hatching in ten days or more, according to the water temperature, the tadpoles, like other species of frogs and toads, remain clinging to the egg jelly for a short time before taking up an active life in search of food. They grow rapidly, and most of them get away as froglets before hot weather sets in and the smaller, sometimes temporary woodland pools dry up.

The author has seen an exodus of these tiny newly transformed wood frogs similar to the sudden emergence of the common toads in large numbers. It was an experience to delight a naturalist, what with the woodland carpet alive with these perfectly formed pale-fawn-colored froglets already showing something of the black-mask pattern, and diagnostic marking of this usually reddish brown to pale brown species.

Their diet includes insects of many species and earthworms, which as we know, constitute such an important animal food for so many other kinds of organisms. What these frogs themselves give to the woodland complex is probably a small amount of nitrogen in their more or less fluid excrement.* Among their enemies are raccoons, and at one breeding gathering (not in the mini-wood) a Red-shouldered Hawk was observed that had been attracted by the clacking chorus. Every now and then this predator swooped down from a nearby tree to grasp a frog from the abundant supply, which it then quickly devoured with apparent relish on its high perch once more. Thus it stuffed itself after the long winter of less-easy hunting.

Some of the wood frogs varied in ground color from those already given, being pinkish brown to dark brown, and fawn color in varying depths. The species differs from all of the more familiar northeastern frogs in having the distinct black patch in back of the eye on each side, and these together form the "robber's mask" effect, although the eyes, of course, remain fully exposed.

* * * * *

____
*Nitrogen in urea, a highly soluble nitrogenous compound secreted by frogs and toads and many other kinds of animals.

One other species of which occasional individuals entered the borders of the mini-wood annually was the leopard frog, *Rana pipiens*. Two and a half to three inches in length when mature, brown or greenish brown, this frog bears a narrow, pale-colored *ridge* down the back on each side (the dorsal-laterals), and between these ridges are rows of more or less rounded large dark spots with paler borders showing clearly against the brown or greenish brown ground color (plate 152D).

The species can be confused with the pickerel frog, *Rana palustris,* whose dorsal spots are squarish against a ground color of varying shades of brown; and distinguishing the species at once from the leopard frog are its orange-colored inner thighs. The pickerel frog was not observed in the mini-wood area.

The croakings and other sounds made by these two frogs issue from vocal sacs, one of which protrudes on either side of the animal between the eye and the arm, while the toads and many other frogs protrude a balloon-like throat bubble from which the sounds issue.

Leopard frogs are early breeders, in mild springs depositing their eggs in water as early as March. Their tadpoles make rapid growth.

At breeding time these frogs are capable of making odd croakings and other interesting vocalizations that are hard to describe. When a number of them are thus "conferring" all at once, a listener at a distance might imagine that a group of incoherent human beings was talking or trying to communicate.

Tireless insect hunters, leopard frogs occasionally entered the author's garden. Otherwise, they were observed mostly at night and were sometimes encountered some distance away from any standing water, or again, in the daytime, hiding in long wet grass near the lily pool. Those which reached the garden remained hidden during the day beneath sheets of plywood laid between the rows of vegetables to eliminate weeds. These plywood strips in fact became hideouts for many other animal forms, including red-backed salamanders, small black ground beetles, hunting spiders, and occasional small toads. In addition, pockets beneath the plywood were used as storerooms for cherry pits by deer mice more than once, and in the late fall, large more or less orange-colored slugs deposited many batches of their eggs under these shelters.

CHAPTER XVIII

REPTILES OF THE MINI-WOOD

Reptiles, the highest class of cold-blooded vertebrates, are distinguished from amphibians in their manner of reproduction. Fertilization is accomplished by copulation, and while these animals may either lay eggs or give birth to their young, according to the species, they are not dependent upon water in which to breed, as are most of the former.

In reptiles there is no larval stage, no flimsy-tailed swimming stage, nor respiration for a time by means of gills. The newly hatched or newborn serpent, like the parent, is already breathing air by means of lungs, and, also like the parent, is clothed in scales and broader ventral plates or scutes. Whereas the lizards and terrestrial and aquatic turtles have strong legs and sharply clawed toes, the limbless snakes must rely upon the ventral scutes, which may be moved in such a way as to give these reptiles their smooth gliding form of locomotion.

Two terrestrial turtle species reached the mini-wood on a number of occasions, having wandered in from other patches of woodland or from nearby swampy areas to take up temporary residence. These were slow, nonchalant members, apparently afraid of no other animal, including human beings.

Admiring its coolness, the author once watched a box turtle, *Terrapene carolina*, as it wandered into the home gateway, and with many pauses, as it traversed the long traprock driveway, progressed past the native aquatic plant pool, and finally reached the wood, over an hour having been consumed by the journey. It did not eat anything on the way, although tempting red-capped *Russula* toad-stools, often nibbled by squirrels, were passed close by. At each pause it raised its head, surveyed the territory for a bit, and after thus giving the impression that it possessed very good eyesight, moved on toward the shelter of the mini-wood. (Plate 153.)

This is the six-inch land turtle with the domelike carapace (upper portion of the shell) with a low keel sometimes indicated in its center, and possessing a remarkably *hinged* plastron (lower shell) behind which the turtle may withdraw and completely cover its head and forelimbs (plate 153B).

In color the carapace varies in individuals through shades of brown to almost black, with yellow markings in large patches. There may also be black spots and lines interspersed with yellow on the plastron, but here again, the coloring varies greatly in individual specimens.

The turtle that wandered up the driveway and into the mini-wood was a male with *red* eyes, although most of these turtles seem to have eyes of brown.

That all of them have great viability and recuperative powers has been proved many times by completely healed shells of individuals that appear to have been run over by automobiles or otherwise badly crushed. A much deformed but thoroughly healed and still useful carapace on an otherwise healthy and apparently happy box turtle was examined on more than one occasion by the author, cases in which suffering for the little animals must have been intense during the period of recovery, aggravated by the dirt which must have been ground into the bloody fissures between the carapace scutes.

Turtles are long lived, truly tough creatures, as we

know, and sometimes authentically attested to as being of *great* age by specimens found with dated shells, such as the one found at Gettysburg battlefield a few years back with a long deep (bullet?) crease through its long-healed shell. Turtles with holes bored through the edges of the shell (for tethering) have also often been found.

These terrestrial turtles seem to be natural wanderers or tramps, moving from place to place in search of their food, which includes fungi, leaves, fallen berries, insects, and earthworms—the prey relished by so many kinds of animals—and on rare occasions this species probably devours a slug or two. In captivity at the Bruce Museum, they readily accepted bananas, lettuce, and tomatoes (plate 153A and C).

In summer the female box turtle digs a hole in soft ground into which she drops a few somewhat oval, tough-shelled eggs, which are often dug up and eaten by skunks and raccoons—one possible reason why so few newly hatched specimens are ever found.

The eggs hatch before winter, but average dates cannot be stated by the author. In two cases, however, equally far advanced embryos of both species described above were obtained, that of the box turtles on August 20, and of the wood turtles on October 15, indicating that both species normally emerge during the fall months (plate 153D).

Both turtles seem to hold their own against DDT, probably because large tracts of the natural woodland escaped spraying of any kind, as such areas of natural habitat still may.

Aquatic animals, on the other hand, especially swamp and still-pond inhabitants, where mosquito control has been practiced without proper knowledge of its dangers, have suffered greatly, frogs especially; the spring peeper has been all but wiped out in many parts of suburbia. This happened in the author's mini-wood area, but, happily, once the lethal sprays were widely banned or voluntarily abandoned, the comparatively few survivors initiated a slow but definite comeback in numbers.

The wood turtle, *Clemmys insculpta,* the other mostly terrestrial species in New England, appeared in the mini-wood, as already mentioned, now and then during the fifteen-year study period. Like the box turtle, it just wandered in, stayed for a time, and then crawled off to pastures new. This species always seemed to prefer damper situations than the former, occasionally even going into water deliberately, perhaps in search of aquatic vegetation or in-sects. One wood turtle wandered into the mini-wood three years in succession, being recognized by two holes bored in its shell.

About the same length or a little longer than the box turtle, the shell of the wood turtle is lower and not so globular in form, and the scutes, which are dark brown on the carapace with some black and yellow, are concentrically grooved, giving the shell a sculptured appearance, and there is yellow and black in some specimens on the plastron. Most striking is the orange color of the neck and on the legs, especially the forelegs, which are in sharp contrast to the dark-colored head (plate 153E).

A specimen kept within a large wired-off observation area within the author's half-acre wood was fed on lettuce, fungi, and berries to supplement what earthworms and insects it might pick up at night and in rainy weather, when so many of the former come to the surface while their burrows are filled with water.

In early July, this turtle was seen to burrow to the bottom of the natural leaf carpet. There it dug a hole well down through the humus, and after backing into it, deposited eight white eggs with tough shells, not brittle with calcium, like those of birds.

In mid-October, three of these eggs gave up their young turtles, which were characterized by extraordinarily long tails for the size of their bodies, a puzzling observation, it would seem, as the tail in the adult is only moderately long. When dug up, the remaining five eggs were found to be shriveled. Four contained very small embryos in various stages of disintegration, and the fifth held the single normal but dead one shown in plate 153D.

Probably this species digs a deep hole in which to hibernate. It possesses very powerful limbs and good claws, longest and sharpest on the hind toes. Some have said that it crawls into natural holes and even muskrat burrows in stream banks. The author has never found a hibernating individual of either this species or the box turtle.

Holding one of these small modern reptiles in the hands, one suddenly remembers that it is a representative of one of the oldest classes of land vertebrates to evolve on the earth. Now believed by scientists to have endured as a class for 130 million years, there are still many hundreds of healthy species walking, slithering, or swimming on this planet. Despite evolution's extinction of such giant experiments as the dinosaurs, flying pterodactyls, great marine ichthyosaurs, and other huge reptiles eliminated millions of years ago, we still have such healthy representatives as

A ribbon snake that was dropped by a small hawk. Date: April 22. About twice life-size. (Page 174.)

Right: A ringneck snake recovering after being rescued from mauling by a feral house cat. Date: September 4. About three times life-size. (Page 175.) *Below:* Belly pattern of the milk snake. (See page 174.)

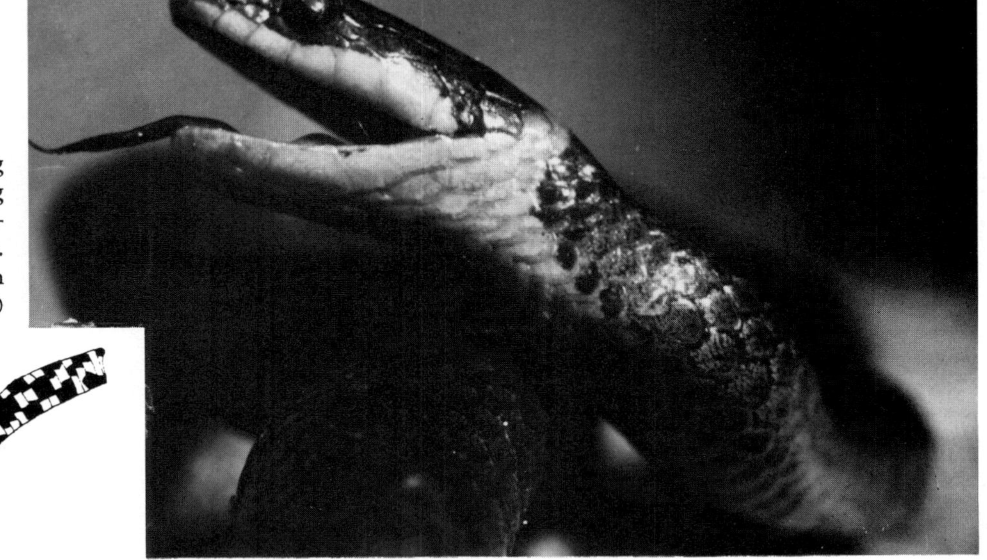

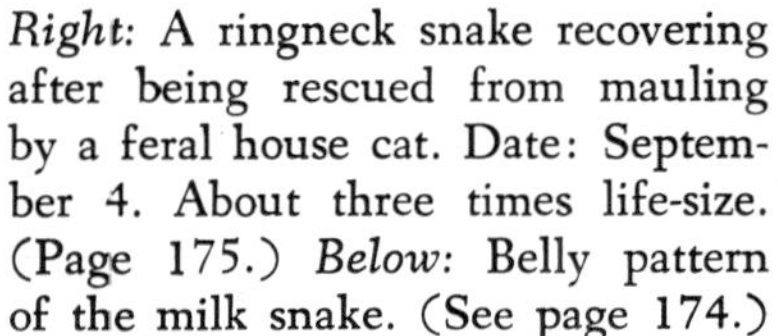

The beautiful and docile smooth green snake, which climbs into the shrubbery with great ease and is there well camouflaged among the leaves. Shown about twice life-size. Date: June 7. (Page 175.)

P LATE 155

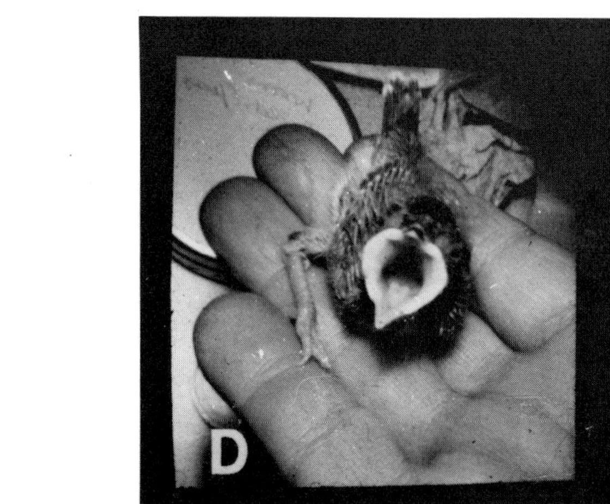

(A) White-breasted Nuthatch at a double-entrance nest box. (Page 179.) (B) Same box with lid lifted to show set of ten eggs. Date: April 18. (Page 179.) (C) Nest and eggs of the Tufted Titmouse that were broken up by some predator. Date: May 5. Natural size. (Page 180.) (D) A week-old Titmouse in the author's hand. (E) A close-up of its gape. This was the nestling that died after eating elm looper caterpillars May 28. See plate 136. (Page 181.)

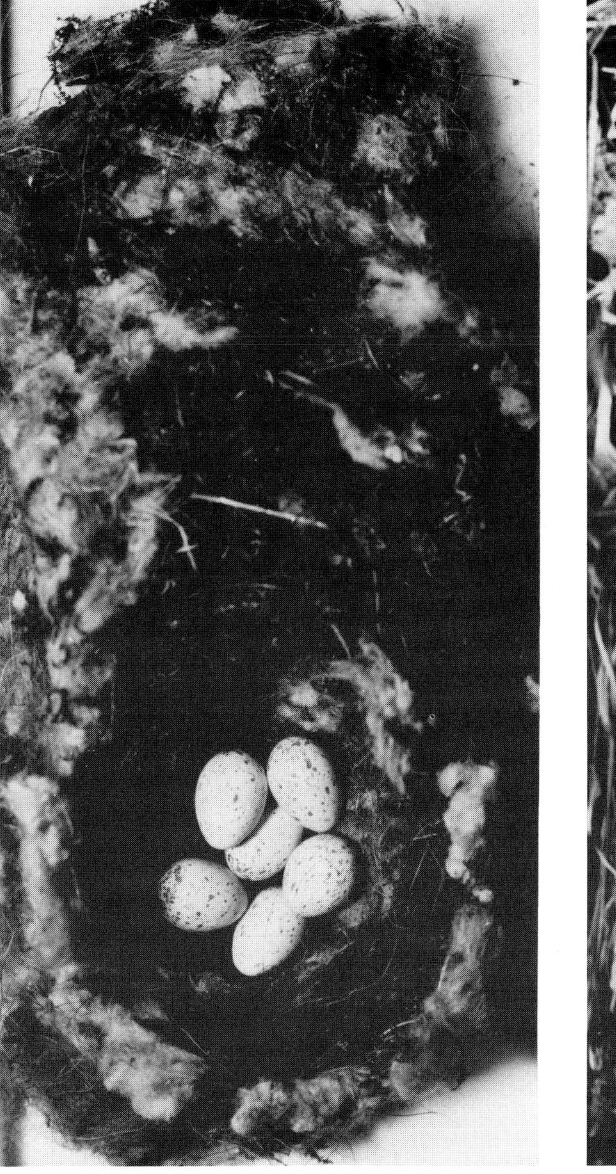

Left: Chickadees excavated a nesting cavity in this decayed stub only ten inches above the ground, but broke through the back. Shown one-third natural size. Date: May 10. (Page 181.) *Center:* Typical eggs of the Chickadee in a nest built in a long narrow birdhouse. Date: June 10. (Page 182.) *Right:* A bird box nest of the Crested Flycatcher, showing the curiously marked eggs. Date: June 15. (Page 182.) Both nests shown natural size.

PLATE 157

Left: White-breasted Nuthatch performing the "fan dance" threat against a strange bird on its box (from a motion picture sequence.) Date: April 30. (Page 179.) *Right*: A Starling's pale blue eggs and rough grass nest in a birdhouse. Date: May 3. (Page 186.)

Left: House Wrens' nest and seven eggs abandoned during period when spraying with DDT was general. Shown natural size. Date: May 20. (Page 184.) *Right*: Skeletons of nestling Wrens, found in a box during the same spraying period. The probable parent Wrens were picked up dead below the nesting site. Date: July 20. (Page 185.)

PLATE 158

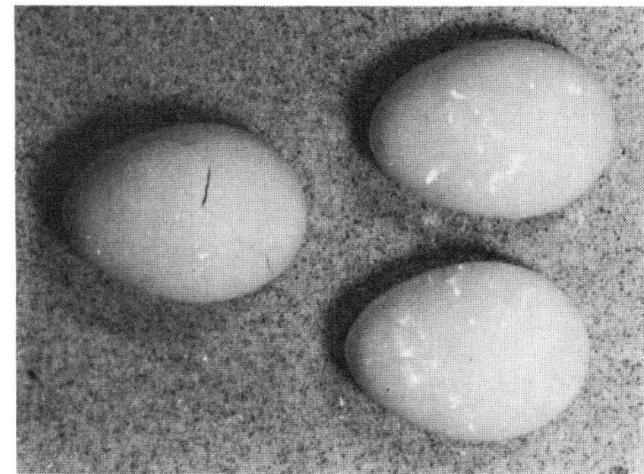

Above: Pure white eggs of the Flicker, about two-thirds life-size. Nests may contain as many as nine eggs. Date: May 25. (Page 185.)

Above: Typical nesting stub of the Downey Woodpecker, with entrance hole drilled by the birds. About two-thirds natural size. *Right:* Rear view of the same stub, sectioned to show cavity and white eggs. One-half natural size. Date: June 10. (Page 185.) *Below:* Eggs of the Brown Thrasher. Colors were bluish or grayish white, speckled with cinnamon brown. Shown about two-thirds natural size. Date: June 1. (Page 194.)

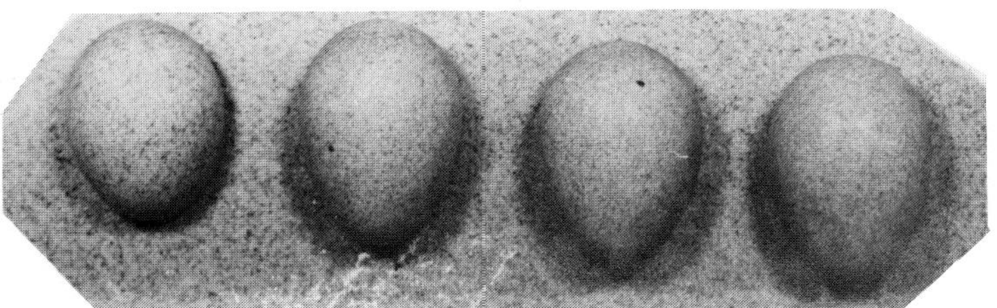

Right: The firmly woven cup of the Red-eyed Vireo, suspended between basswood twigs. Date: June 14. (Page 187.)

Above: Close-up (somewhat enlarged) of nest of the Red-eyed Vireo, showing compact but rather coarse construction. Nest held four eggs. Date: May 24. (Page 187.) *Right:* This nest of the Yellow-throated Vireo, also in a basswood tree, was much more firmly woven, more perfectly lined with finer materials, and much more artistically decorated than the Red-eyed Vireo's nests shown above. It was also somewhat more bulky. Date: May 21. (Page 187.)

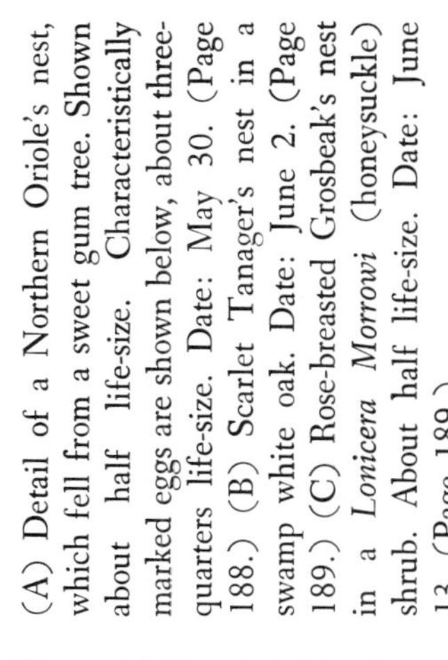

(A) Detail of a Northern Oriole's nest, which fell from a sweet gum tree. Shown about half life-size. Characteristically marked eggs are shown below, about three-quarters life-size. Date: May 30. (Page 188.) (B) Scarlet Tanager's nest in a swamp white oak. Date: June 2. (Page 189.) (C) Rose-breasted Grosbeak's nest in a *Lonicera Morrowi* (honeysuckle) shrub. About half life-size. Date: June 13. (Page 189.)

Shown exactly as it was found, with toes locked to a twig, this Hummingbird evidently died in its sleep. Slightly enlarged. (Page 190.)

Above: Saddled on a swamp white oak branch, this nest of the Ruby-throated Hummingbird was disfigured by a predator, as shown, who left one of the eggs. The birds deserted the site. Natural size. Date: June 1. (Page 190.) *Left:* Nest of the Wood Peewee, also saddled on an oak branch and decorated with green and gray lichens, rivals the hummingbird's in beauty. Shown natural size. Date: June 30. (Page 191.)

the crocodiles, the eight-foot flesh-eating monitor lizards of Komodo and other Pacific islands, scores of smaller lizards, gigantic pythons and other constrictor snakes, hundreds of smaller snakes (some poisonous and lethal), and a large number of turtles and tortoises.

And who could imagine the *human* vertebrate lasting for even another million years, when after, say, only two million as an erect tool-making tenant, man has already so abused and endangered the whole web of life that maybe most, if not all of it, including Homo sapiens himself, may well be exterminated long before A.D. 1,000,000 is even remotely approached.

The little turtle in hand was indeed thought provoking. Man, it would seem quite definitely, would never live to boast a lineage like that of the reptiles unless he mends his ways very drastically and promptly.

Many who read this book through, or have read this far, may realize for the first time how much life can be destroyed by the felling of a woodlot or a single half-acre of vegetation.

Six species of snakes occurred in and on the mini-wood borders during the fifteen-year study period, not a bad record for so small an area, knowing that in all of the United States and Canada east of the one hundredth meridian there are only 164 species and subspecies. The 6-snake record may reflect increasing and encouraging local concern with, and the movement against, dangerous pesticides, although the human population explosion will probably eventually doom many of our small reptiles locally.

The snakes which still choose to abide about our homes here in the East are seldom venomous species. All of them are beneficial in one way or another, but a difficulty in protecting them still exists because of parental fear of them, and the apparent inborn desire of many children to kill them on sight.

Any brown-colored snake, or any species that has a more or less triangular head, or which flattens its body and hisses, automatically suggests a copperhead, or, most "obviously," a venomous species. Believe it or not, the author's forty-seven years as a museum curator proved that many people still believe that a snake's harmless forked tongue is its venomous "stinger!" Even allowing a snake to tickle one with this soft sense organ or "feeler" failed to convince numerous nervous onlookers.

Our largest northern water snake, *Natrix sipedon,* is a non-venomous, semi-aquatic species frequenting the borders of ponds and pools and large lakes. Thus it has appeared from time to time on the border of the mini-wood, where a pool twenty feet in diameter harbors aquatic plants and many fishes and frogs, some of which the water snake takes for its food.

Robust, extremely fast moving, it is usually a dull brown snake with darker bands anteriorly, and with alternating narrower and lighter-colored bands between the darker ones. The broad bands become blotches posteriorly, one series on the back, another series on each side. The snake is hard to describe accurately because of the ground color, which may be pale brown, reddish brown, or a very dull shade, due as often as not in the latter case to mud dried on its scales which obscure the blotches and bands to such an extent that the reptile may seem to be of a solid, lusterless dark color. At other times, especially when the skin has been freshly shed, as it is every now and then in a single paper-thin piece, the snake may be quite striking, particularly in specimens in which the reddish scales are brightest.

On the underside, if one ever gets a long-enough look at a free-living specimen of this lightning-fast mover, one may glimpse red or black spots, half-moon-shaped markings, or little or no pattern of any kind. In other words, it will not be easy for the casual observer to identify this snake with certainty, but it is probably safe to state that, anywhere in New England, the only large dull brown or mostly mud-colored snake seen getting into the water in a flash as one approaches, will be this species, *Natrix sipedon,* usually called the northern water snake (plate 154A).

A favorite situation chosen by this snake is still water over which the branches of a fallen limb or tree trunk project a few feet above the surface, where it may climb to sun itself safely, and from whence escape from a hawk means simply an instantaneous drop into the water below.

Sometimes, after a good feed, a replete *sipedon* was seen sunning itself contentedly on the warm dead-leaf carpet on the mini-wood border and close to a good food supply, where it remained on occasion for two or three hours at a stretch in early spring when the sun was not too hot. This one helped itself to goldfish and frogs at times, and as these creatures were usually well able to maintain balanced numbers of their kind within this pool, an occasional water snake was allowed its rightful quota unmolested by the author. Within natural ponds, where it has dwelt through the past centuries, its status has doubtless been that of a normal, necessary predator in helping to maintain the normal balance of living things in such habitats.

These snakes are non-venomous, but will bite hard and quickly if cornered. They grow to three and a half feet in length, sometimes flatten the head when molested, and are well able to drive off most other animals.

This is a viviparous species. At the Bruce Museum some years ago, young to the number of thirty, were whelped in July. These looked nothing like the parent, being rather light gray, banded with black, and with thickly black spotted underparts, whereas the belly color of the adult was dark gray to almost a yellowish brown, with fewer dark markings. (Plate 154 shows the adult.)

The first warm days of April bring out the garter and ribbon snakes. They bask in the sunshine to limber their dozens of winter-stiffened joints, and are soon agile again and ready to defend themselves most valiantly against even the largest intruders, their sharp little teeth easily able to penetrate an incautious finger or the canvas of a sneaker. Garter snakes are the more pugnacious of the two; the ribbon snake is semi-aquatic.

When fresh from hibernation, their colors and scalations are often hidden or dulled by mud or earth; thus they may look much alike to the casual observer until this coating wears off. The eastern ribbon snake, *Thamnophis sauritus,* may be dark brown or almost black above, with three bright yellow stripes, one along the center of the back and one on each side, emphasized by the much darker ground color. These lateral stripes are always on the *third and fourth* rows of scales above the abdominal scutes or broad belly scales (plate 155).

The heavier-built garter snake, *Thamnophis sirtalis,* which is the commonest species in Connecticut and other eastern states, also has three more or less bright yellow stripes on a brownish to black ground color, the lateral two always on the *second and third* rows of scales above the abdominal scutes (plate 154). These belly scales may be yellowish or greenish white in this species, maybe more greenish in the ribbon snake and usually unmarked, but sometimes also marked with black quite distinctly in the garter snake.

Both species were found in or near the mini-wood from time to time, most often on its sunny borders, where matings took place. Both are viviparous, the garter snake producing young in astonishing numbers. An individual under observation by the author at the Bruce Museum gave birth to thirty young, a fair-sized litter in itself, on August 15, but five days later the same female whelped thirty-three more—sixty-three altogether in one great family. All but three were born alive, but two were curiously deformed, with half of their bodies formed at a right angle to the other half, so that, although alive, they could not move one way or the other. The mother, seemingly disturbed by her double achievement, began swallowing her offspring and had to be removed from the cage.

Garter snakes are great destroyers of insects, and like a great variety of other animals, are also fond of earthworms. They also eat toads and frogs, but in the experience of the author, less often. The slimmer ribbon snake, being at home in the water, dives after tadpoles and small frogs, small fishes, and possibly the larvae of newts, and on land takes earthworms and will even pursue peepers and wood frogs.

Occasionally found near and within woodpiles in the mini-wood, probably hunting there for shrews, but also very often attracted to rural and suburban outbuildings for the young and adult house and deer mice which constitute its other natural food items, was the eastern milk snake, *Lampropeltis doliata triangulum.*

The general ground color of this attractive, rather slender reptile is gray or ocherish tan with contrasting (rows) of reddish brown, to chestnut brown dorsal blotches or "saddles," and with smaller blotches below and alternating with the larger ones. Each "saddle" is narrowly bordered with a "saddle cloth" of black, while the underside in this species is distinctly checkered in black and white, which seems to be diagnostic of the species (plate 155).

Specimens actually captured within the mini-wood measured twenty-four and twenty-six inches in length, and were probably nearly full grown, although specimens thirty-six, and even slightly over forty-seven inches—the latter probably a record—have been recorded.

Naturally more docile than either garter or water snakes, the milk snake remains cool when confronted, and seldom tries to bite unless roughly handled. It becomes very tame in captivity and seems to enjoy the warmth of human hands, yet it crushes its prey ruthlessly in its coils.

Its food is swallowed whole, bones, hair, and all. The digestive powers exhibited by the rodent-consuming snakes of many species has always seemed astonishing to the author, for the bones of their victims often appear to be completely dissolved. Their feces include large proportions of white material along with some of the animals' hair, the former masses evidently the processed bones, thus such reptiles are important not only because they help to control rodents, but also because they return a certain amount of mineral matter in addition to nitrogenous material to the soil.

The milk snake lays batches of tough-skinned oval eggs in July or later. In the author's experience, these hatch in September, and the young which emerge are truly beautiful little creatures, their markings being distinctly reddish and becoming crimson as they grow somewhat older; but this attractive phase changes again as they become one or another shade of brown as adult reptiles.

We come now to two other species of snakes which the author had found to be rare at all times wherever a search for them was made in his area of Connecticut. It is therefore quite remarkable that specimens of each of these turned up in the area of the mini-wood and may therefore be included as temporary members of this always interesting and often surprising half-acre community.

Slender, about ten inches in length as found in the mini-wood (although they may grow to twice that size), bluish or grayish black above, yellow on the underside, and adorned with a narrow yellow to golden collar, the northern ringneck snake, *Diadophis punctatus edwardsi,* is strikingly colored and might be confused only with certain of the similar subspecies occurring in other parts of the country, in which more numerous black dots or other shaped markings on the belly scutes distinguish these closely related forms.

A frequenter of rocky woodland hillsides strewn with fallen bark or branches and containing other shelters such as a flat rock or perhaps an aging woodpile, specimens living in the mini-wood had found just such habitats, one in fact having crawled on April 15 from a woodpile in which it may have hibernated, its appearance of course delighting the author.

Feral house cats must be included among the mini-wood predators, and the ringneck snake shown (plate 155) had just been mauled by one of these nuisance animals.

Specimens found elsewhere in Connecticut by the author exhibited ground color variations ranging from dark bluish gray to slaty gray, but all had the characteristic, centrally located small black belly spots.

In captivity at the Bruce Museum, this species accepted earthworms for food. In the wild, in its natural woodland habitat, the ringneck also feeds on salamanders, an interesting interrelationship in the mini-wood definitely established as such, for when one of these little reptiles was suddenly grasped, it disgorged a partly digested red-backed salamander, a species regularly resident in the same habitat.

The species is oviparous, and as recorded by the late authority and long-time curator of reptiles at the New York Zoological Park, Raymond L. Ditmars, in his widely known *Reptile Book,* the embryos are already well developed within the eggs before they are deposited, thus the ringneck stands midway between other oviparous and viviparous species, for its young emerge from the eggs in about half the time usually required after deposition.

Young snakes examined by the author were very dark, almost black, about four inches in length, and with the yellow collar much paler than in the adults. They were found under a flat rock in the woods early in September.

On a hot day in mid-July, the author observed what from a distance appeared to be something like a length of green cord moving through the twigs and foliage of a highbush blueberry shrub at the wet lower border of the mini-wood. Actually, there was no mistaking the object, a real surprise that proved to be a beautiful specimen of the smooth green snake, *Opheodrys vernalis,* a slender sixteen-inch bright green reptile with a white belly tinged with yellow (plate 155).

Gliding with the greatest of ease, it moved up through the twigs and leaves of the blueberry, leaving the author astonished at its apparent effortless progress, and trying to imagine the motions of its belly scutes in making such an ascent.

As it turned out, this snake was not hunting its insect and spider prey, but simply moving to a well-shaded stopping place, thus it soon came to rest and at once blended almost perfectly into its leafy surroundings, coiled up to some extent and remained in the bush all that hot day.

It is one of the most docile and pleasantest snakes to handle. It never has even attempted to bite, in the experience of the author, and its scales are, as its name implies, smooth and satiny. *Vernalis* is a safe and interesting reptile, a splendid pet for any young collector or student of biology who is willing to tenderly care for it and handle it with gentleness, for gentleness is its nature.

Being strictly a feeder on insects and spiders, the only

problem involved in keeping the snake in captivity is inducing it to eat what is introduced into its cage, and its food, of course, must always be supplied *alive*. It is possible under some circumstances that green snakes may occasionally take small salamanders, but its habit of climbing bushes and even small trees indicates that arthropods form its main items of diet.

This species lays small batches of elongate eggs with thin parchmentlike shells. These lose their original shape, or become dented as development within proceeds. An adult, stupidly killed and brought to the museum on August 30, for identification as an imagined "poisonous" species, contained five eggs which contained small embryos on that date, but none lived to emerge. Others have reported young snakes normally hatching around August 25.

It is strange, when one stops to think about it, that so many turtles and snakes emerge so late in the year when the young must so soon prepare for hibernation, especially here in the North, where cold nights begin in October. No doubt there is some very intresting explanation for it.

CHAPTER XIX

BIRDS OF THE MINI-WOOD I: CAVITY AND BOX NESTERS

Probably most, if not all, who will be reading this volume will either own a Peterson *Guide,** or some other field guide to the birds of the eastern United States. They will know a large number of the commoner species on sight, and no doubt all of those included in these chapters.

No special attempt was made therefore, to include among the book's illustrations stills of living individuals, the author confining the photography almost exclusively to the nests and eggs of the species which actually bred within or on the mini-wood borders.

Strange as it may seem, there are still large numbers of bird enthusiasts who know dozens of species at a glance, yet have not had the exciting experience of personally finding the nests even of many familiar species, having missed thereby that sudden sensation of joy which comes with each such discovery and vis-a-vis view of these structures, especially when freshly completed and beautified by their complements of often almost gemlike eggs.

Nests and eggs of wild birds have always held a special fascination for the author. Nests are often so remarkably constructed, so cleverly hidden in their natural surroundings, and their contents of eggs so delightfully colored or patterned, or both, that chancing suddenly upon such a combination may actually result in a form of shock, if extreme pleasure might so be described.

Each spring and fall, the mini-wood had its migrant transients; each summer, its temporary visiting or settled

nesters; and always, its year-round resident hardy species.

During the warmer months, birds of all three categories were not only attracted by the area's insect and other food offerings, as well as the prospect of some shelter, but especially by a tiny spring-fed watering place which, like the great watering holes of Africa, in its miniature scale became a hublike lure not only for birds but a few mammals as well, some "dropping in" from territory well removed from the mini-wood proper.

The pool's soil-filtered trickling water, cool at all times in its niche beneath tall shading trees, became a drinking and bathing spot favored over all man-made pools and birdbaths and their often chlorinated offerings.

During the nesting season, many an "outsider" winged in from beyond the mini-wood borders to partake of the little pool's elixir, whose gentle flow cleansed it day by day of pollutants from the bathers' doused skins and feathers, and their unwittingly but frequently dropped excrement.

Robins and Wood Thrushes gathered sticky mud from the pool's edges, scooping it up in uncomfortable looking beakfuls, sometimes with bits of dead leaves and other debris. Robins, in particular, so burdened themselves at times that one or two stops for rest were necessary on their journeys to more distant nesting sites.

During migrations, especially in May, occasionally there were "waiting lists" for the use of the spring, with transient Warblers, a Vireo, perhaps; Cardinals; even a Scarlet Tanager—all awaiting the departure of the rougher and less mannerly Starlings and House Sparrows. But at times the tiny Black-throated Green Warbler became

---

*Roger Tory Peterson, *A Field Guide to the Birds* (Boston: Houghton Mifflin Company, 1934, 1939; rev. ed. 1947). Giving field marks of all species east of the Rockies. Illustrated.

amazingly bold, flitting nearer and nearer to the water, even making little darts in the direction of the larger birds as if trying to intimidate them!

The finest nests found within the mini-wood area were naturally those which were built in the "open," so to speak, under outdoor conditions, much greater care in their construction and placement of course being mandatory.

On the other hand, those species which used cavities or bird boxes seemed more occupied with filling the allotted spaces than in producing beautiful nests, some of these birds having definitely resorted to very careless nest building in changes that possibly evolved with the elimination of the danger of exposure, at least in certain cases. Examination of the plates illustrating this chapter (Woodpeckers, of course, excepted) will demonstrate these tendencies when these nests are compared with those photographed for the following chapter.

The most interesting of the box and cavity nesters were the White-breasted Nuthatches, *Sitta carolinensis*, cheerful permanent residents and dependable customers of the seed and suet feeders.

Nesting within the woodlands usually near to where they have been feeding all winter, often in a birdhouse placed there for them, their nests are often pilfered by raccoons, possibly by 'possums, and occasionally by deer mice, all of which predators seem too able to disrupt nests built in birdhouses and natural cavities. With dismaying frequency, the eggs and nestlings are eaten or otherwise destroyed as nesting materials are pulled through the entrance holes by the larger of the predators, or eggs are consumed within the nests by deer mice. Occasionally an egg or a nestling is overlooked and may fall to the ground; thus, at times, what has not been legitimately taken and devoured on the spot may be unhappily eliminated otherwise. Doubtless nestlings often reveal their whereabouts by their daytime peeping, and at night more by their odors.

Nestings in the mini-wood were sometimes interrupted by gray squirrels not seeking eggs or young, but "planning" to take over the houses by roughly enlarging the entrance holes to admit their own bodies, the birds in such cases, of course, abandoning the nests.

To outwit the larger raiding mammals, extra deep boxes were tried by the author, and while they were effective to some extent, the birds seemed to prefer the shallower type.

To afford protection from the squirrels, the author finally hit on the idea of surrounding the box entrance holes with metal in the form of angle irons screwed on in such a way that the rodents could no longer get at the wood. After some baffled attempted gnawing on the part of the squirrels, discouragement followed, and being mammals quick to learn, they soon left these birdhouses strictly alone.

The diamond-shaped openings resulting from the placement of pairs of the angle irons did not seem to bother the birds, even though the round form to which they had been long accustomed was no longer visible. (See plate 134, and the birdhouse at bottom right thus protected.)

Nuthatch nesting activities commenced very early in the spring. During the fifteen-year period, the earliest pairs to exhibit such intentions were observed within the mini-wood during the second week in March, after continued singing by the males. With the site finally selected, the birds then followed amusing, somewhat standardized routines.

First came the thorough cleansing of the box or cavity. If old nesting material or other debris was found therein, the fouled wherewithal was removed bit by bit, each beakful being subjected to much whacking and back and forth beating against the box or the nesting tree, and these discards, when having thus been sufficiently "punished," were dropped or stuffed into a bark crevice, or perhaps a crack in the birdhouse, if a convenient one happened to be present.

Once the site had been cleansed to the pair's satisfaction, a job which required from one to several days, a second and very curious routine followed.

Large strips of bark were now pulled from the dead limbs of various mini-wood trees, and, in most instances, flown directly into the nest, only occasional pieces being first subjected to the energetic beating and whacking processes. According to the nature of the pair, the bark-carrying routine was in order between the second week in March and the third week in April. Some pairs filled the whole bottom of the box or cavity with bark slabs to a depth of two or three inches!

Inner bark from a dead basswood tree, already well shredded and left by a nest-building squirrel, was much in demand by the Nuthatches. Carrying it to or into the nest box, it was then beaten and whacked and carried in and out of the nest, and thus more thoroughly shredded. Occasionally the whole thing was stuffed into a crack or left on top of the nest box after all the labor.

Some birds took bits of material from the ground just below the nesting place, and then, for no discernible reason, hammered it into cracks in a sort of "caulking" routine. Within the birdhouses there was often banging on the walls, and strange shuffling sounds by one or both birds in there together.

These are indeed strange little creatures, and one might almost believe at times that some of the routines were due to a showing-off complex, like a person saying, "Look! See what wonderful things I can do!" Such an anthropomorphic idea is absurd, of course, but the author was left wondering at some of these seemingly nonsensical routines of the Nuthatches.

Following the accumulation of bark and shredded material, many additional objects were added to the nest—pellets of dried hard mud, bits of cellophane, and always quantities of animal hair, the completed cup for the eggs being often lined with this, or with cottony material and finely shredded bark as well.

Seven to ten eggs constituted the sets. These were white or creamy white, thickly and finely spotted with rufous brown, and with specks of lavender scattered here and there (plate 156).

Occasionally the female would remain within the nest box or cavity for long periods, even at times before the first egg had been laid, the male bringing her tidbits. When fed thus outside of the nest, she crouched and quivered her wings as their well-grown nestlings and those of other species often do.

On the single occasion in which a setting of ten eggs was deposited, a careful check was kept on this nesting by means of a box with a hinged cover, made removable from the tree for photography. Working very rapidly, inspections could thus be made within a minute or two while the Nuthatches were both absent.

This nest was completed in eleven days. The first egg was laid on April 9, and the last on April 18. Incubation then required fourteen days, and surprisingly enough, healthy nestlings then emerged from nine of these eggs, and all were able to leave the box successfully, the last one observed as it made an awkward flight ending in a tumble into some bushes on May 27. The actual nesting period thus covered only about one month, despite the large number of ravenous nestlings to be fed. The great value of Nuthatches as destroyers of insects will be realized by this graphic illustration.

As these birds always raised large broods, the author tried supplying them with side-by-side two-hole nesting boxes, believing this would simplify matters at that time when all bird box nestlings are pushing and shoving to be the first fed at the usual single opening. But, strange to say, the project failed, *all* feeding taking place at one doorway in the age-old customary manner! Food items definitely identified included white geometrid moths, hairless caterpillars, spiders, bugs, and the butterfly *Euptychia cymela,* the little wood satyr (see plate 133M).

An astonishing discovery about the Nuthatches during the nesting season, was finding a pair plucking hair from a *living,* treed young raccoon! The little mammal seemed not to notice the birds thus helping themselves, their actions proving that mammals dead *or* alive may furnish the hair so often found in the Nuthatch nests.

At the approach of another bird species, or a squirrel into the vicinity of the nest, the Nuthatches performed a unique and remarkably graceful action, but one which, as an intimidating procedure, seemed to have little or no effect except in the case of birds.

The "defending" Nuthatch clung to the tree head down and within a foot or so of the intruder, and there, with wings spread to their full, swayed slowly back and forth in what the author has termed the Nuthatch "fan dance." This was performed so gracefully, and so persistently continued, that it was a truly fascinating reaction to observe. In order to record it in color motion pictures, a small stuffed bird was fastened to the nest box, and many feet of film were thus secured showing the "fan dance" in all its beauty. There was no need to resort to sixty-four-frame speed, for the dance was performed in slow motion by the Nuthatch (plate 158).

Comparatively recent as a species and nester within the mini-wood as well as in more open situations, where suitable birdhouses or natural cavities were available, was the Tufted Titmouse, *Parus bicolor.*

Previous to 1950, the author had never recorded this bird in this area, but by 1954 it had become a resident, and now here, as well as in other parts of New England, it is not rare as a breeding species. The author believes that its northern spread has been due to great numbers of people all over the East (and not only the fanatic bird watchers) who now winter-feed birds, rather than to any startling *natural* ecological factors or changes. It would seem that this generous supply of grain and seeds, especially sunflower seeds, in supply all winter, has made it possible for the Titmouse to extend its range northward. This available abundance of winter seeds has doubtless also had much to do with the rapid northward extension of the Cardinal's range; and now, for somewhat more cryptic reasons, the Mockingbird is moving north, and even nesting here and there in New England. It is the author's belief that these three species have always been hardy enough, and that it was only the absence of sufficient winter food in the northern states that prevented their moving into New England previously.

The Tufted Titmouse, the only resident small bird with slate gray upperparts, white underparts tinged with cinnamon-rufous (rust color) on the flanks, and with a tufted crest, has nested in the mini-wood, or close to its borders, on several occasions. Probably preferring natural cavities when such are available, it has nevertheless taken readily to bird boxes with entrance holes small enough to bar Starlings, and protected from squirrel intrusions by corner irons, as described earlier in this chapter.

The female Titmouse did the heavy nest-building, "assisted" on rare occasions by the male's picking up a bit of moss and usually dropping it again in favor of a burst of song. Once or twice, however, he *did* deliver a bit of something to the nest entrance, possibly moss, and must be given credit for at least trying.

Notwithstanding the fact that the male did none of the hard work of nest building, he was ever near the female on her many trips for material, cheering her on with his buoyant, persistnt, repetitious, and penetrating notes.

Large quantities of moss, pulled living and green from beds growing in flat mats on heavily shaded acid soil, were accumulated in the bottom of the nest box to a depth of three inches. Some of the moss had been pulled up with earth still adhering to it in considerable quantities, but such heavy loads were flown to the box only from five or six yards away, and taken from mats in the shade of a nearby garage. Upon opening the hinged top of this box for a look May 7, the author found that the layer of moss had been topped with quantities of mammal hair, some of it in odoriferous bunches, doubtless plucked from a dead animal. Three days later, more moss, interspersed with bits of

paper and dead grasses, formed the completed cup, in which four eggs were deposited—one a day until the small set was complete. The eggs were white shelled, but very thoroughly covered with pale brown markings (plate 156C).

Incubation began very promptly, and the parents were taking food to the nestlings on May 25. About two weeks later, the members of this brood were clambering to the entrance hole one after another in their eagerness to be out of the box. A few days later, these youngsters were observed being fed by both parents, and the brood remained in the trees near the nesting place well into June.

Tits had been seen in the mini-wood examining the boxes and occasional old Woodpecker holes as early as March 15. Accompanied by much singing on the part of the male, a site was finally chosen by another pair, and building begun March 18, with the customary carrying in of green moss, followed, by quantities of mammalian hair.

Again eggs were laid and successfully hatched in this box at fifteen feet above the ground. All went well until the brood of five were about half-grown, with their feathers normally expanding, and the parents feeding them large numbers of insects.

The time was May 28. On that night, a fatal one for most of the Titmouse family, a raccoon or some other predator pulled most of the nest through the entrance hole, in the act killing two of the nestlings, which were later found on the ground beneath the box. Of the five, two others had doubtless been devoured on the spot, but upon chancing to look within the box, the author found the fifth nestling beneath some refuse. It was numb with the cold, having been unprotected through a night with temperatures in the fifties.

First warmed by hand, then in a teacup set in warm water, the little bird soon revived. In a few hours it was feeding greedily upon hard-boiled egg yolk made into a paste with milk. (This, by the way, makes an ideal food for foundling young birds, for if one stops to think for a minute, young birds are initially eggs, and while embryos, lie floating and feeding upon the yolk, by which they are mostly nourished before hatching.)

Before the destruction of most of this brood, the parents had been observed feeding them numbers of canker moth caterpillars. This was during the year of the explosive increase of the elm looper caterpillars described in chapter XIV.

Left: The Redstart's nest, beautifully constructed of tightly wound fine grasses and silvery plant fibers, was saddled on a red maple branch. Shown reduced about one-third. Date: May 25. (Page 191.) *Right*: The Yellow-throat's nest, placed solidly on the ground, as shown, may therefore be much more loosely constructed except for the lining. About natural size. Date: May 25. (Page 192.)

Plate 163

A mini-wood ornithological-botanical study. Both birds chose plants of the same genus in which to place their nests. *Left:* The Cardinal's rather loosely constructed but nicely lined nest, shown half natural size, was built in a tangle of Japanese honeysuckle, *Lonicera japonica;* that of the Catbird, shown with its complement of four deep greenish blue eggs (one-fourth life-size) in a bush honeysuckle *Lonicera Morrowi.* Dates: Cardinal, April 15; Catbird, June 1. (Pages 192, 193.)

Above: A typical Robin's nest, with compact mud walls maintaining its very neat form. Date: **May 9.** (Page 193.) *Below*: Wood Thrush's nest in a green-brier. Date: June 8. (Page 194.)

Above: An exhibit mounted for the Bruce Museum by the author shows a Robin on its nest, placed at the corner timbers of an old building. Thick mud walls of the nest are plainly visible. (Page 193.) *Below*: A Wood Thrush takes cotton waste for its nest. Date: June 9.

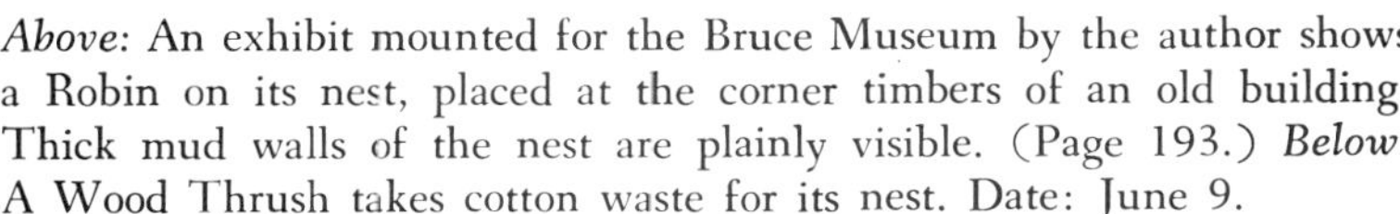

Above: Sharp eyes or a magnifying glass are needed to discover the Brown Thrasher sitting on her nest in a greenbrier in this photograph. Thrasher eggs are bluish or grayish white thickly speckled with cinnamon and rufus brown (see plate 159). Date: June 17. (Page 194.)

Below: The bulky but very well made nest of the Blue Jay. The well-spotted eggs may be pale olive green or brownish gray in ground color. Date: May 19. (Page 194.)

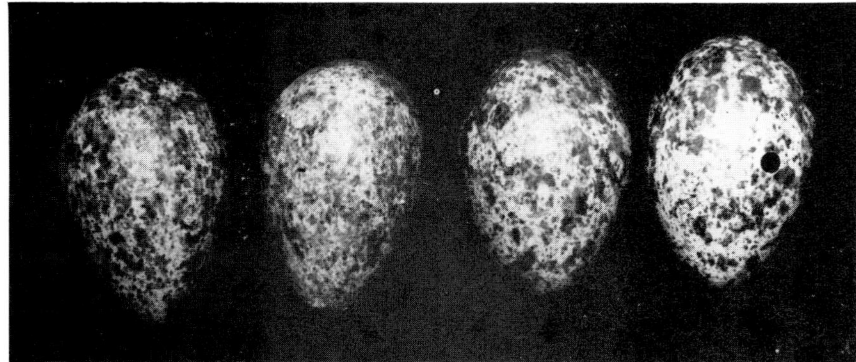

Crow's eggs, about natural size. Varying considerably in color, these were bluish green thickly marked with various shades of brown. Dates: April 10, April 25. (Page 195.)

The Virginia opossum (only species of marsupial inhabiting the United States), here shown near a woodpile (at the entrance to the mini-wood) under which opossums dwelt at various times. (Page 197.) *Above:* A litter of opossum pups, as they lay within the mother's belly pouch, with their mouths tightly holding her nipples. While shown well beyond the half-inch embryolike stage in which they enter the pouch, they were still weak and naked. Shown about life-size, but exact age unknown. (Pages 197, 198.)

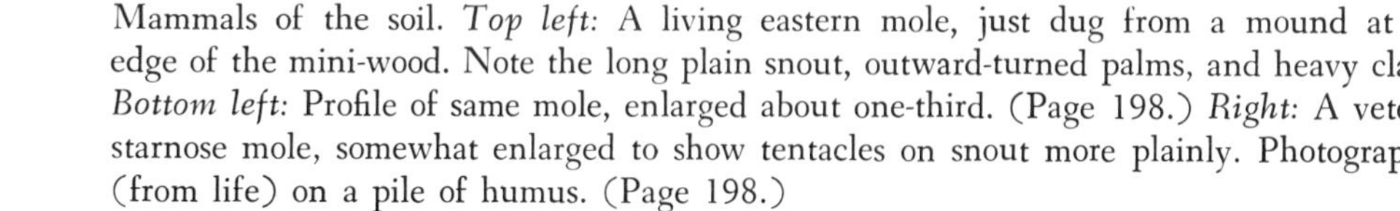

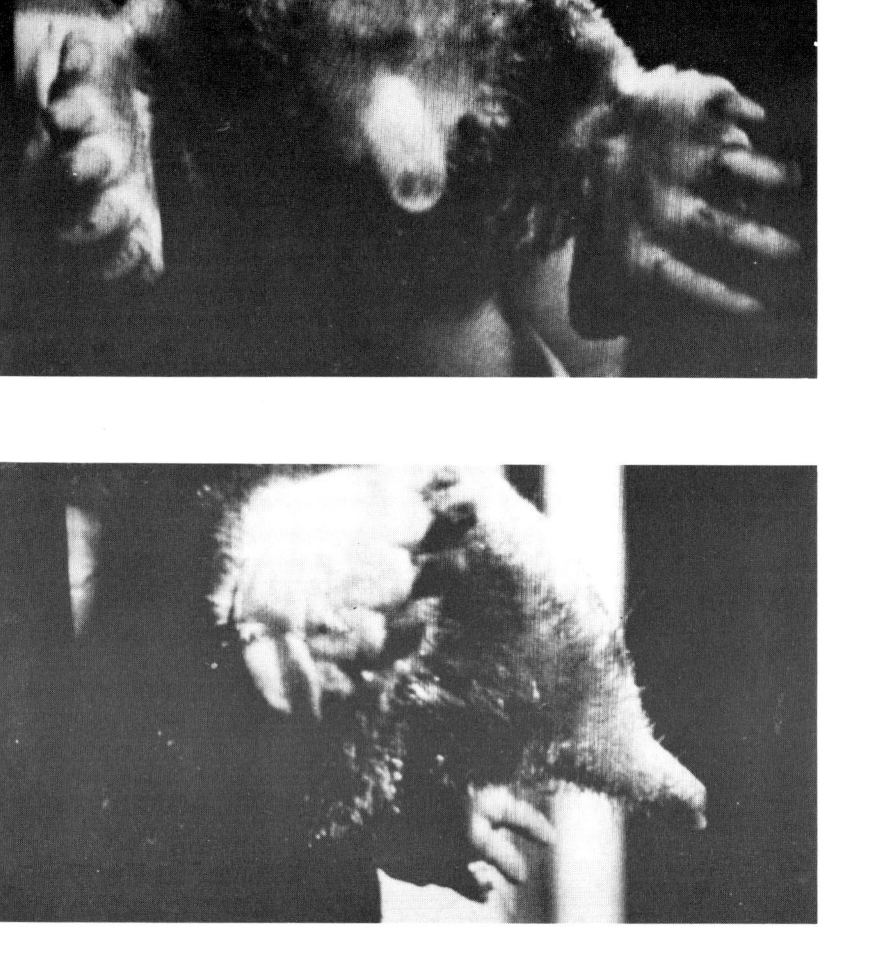

Mammals of the soil. *Top left:* A living eastern mole, just dug from a mound at the edge of the mini-wood. Note the long plain snout, outward-turned palms, and heavy claws. *Bottom left:* Profile of same mole, enlarged about one-third. (Page 198.) *Right:* A veteran starnose mole, somewhat enlarged to show tentacles on snout more plainly. Photographed (from life) on a pile of humus. (Page 198.)

PLATE 168

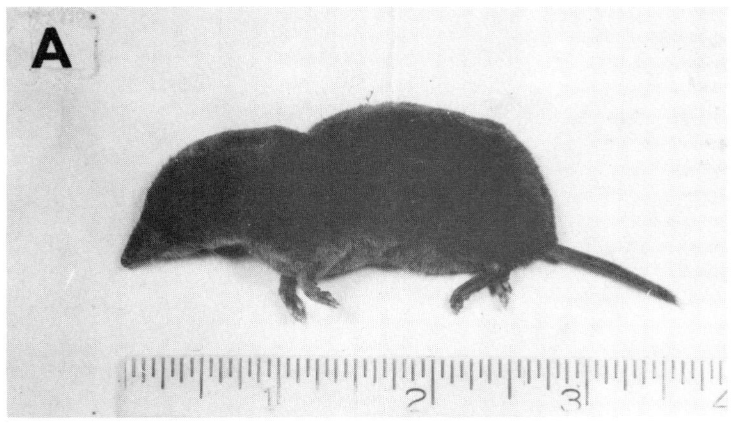

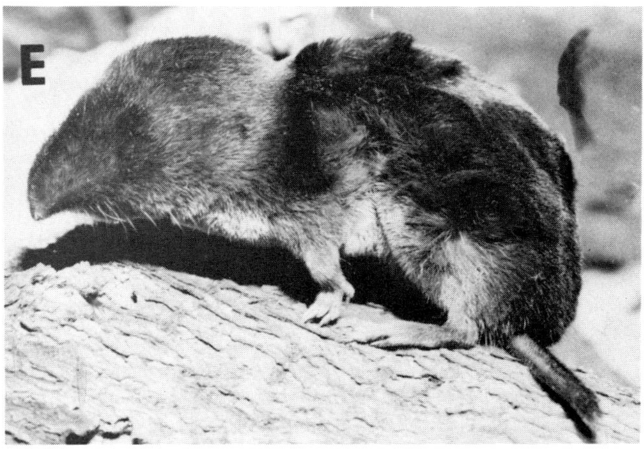

The short-tailed shrew, diminutive carnivore of the mini-wood often mistaken for a herbivorous rodent. (A) Actual size. (B) A good view of the short thick pelage with its beautiful sheen. (C) Rummaging among dead leaves for worms. (D) Enlarged view of the curiously underslung mouth. (E) A shrew badly roughed up by a cat—but still alive—that was rescued by the author. (Page 198.)

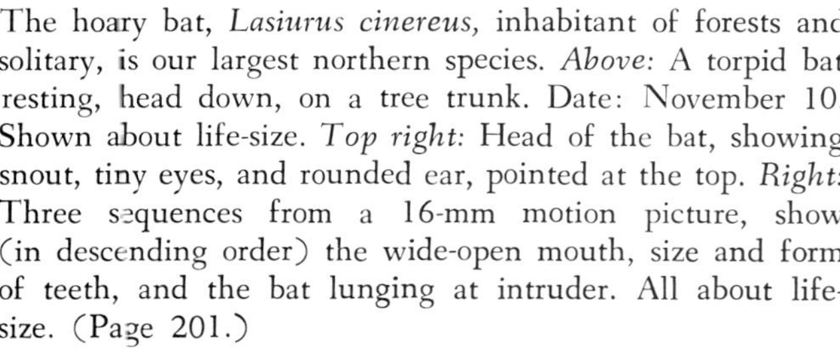

The hoary bat, *Lasiurus cinereus,* inhabitant of forests and solitary, is our largest northern species. *Above:* A torpid bat resting, head down, on a tree trunk. Date: November 10. Shown about life-size. *Top right:* Head of the bat, showing snout, tiny eyes, and rounded ear, pointed at the top. *Right:* Three sequences from a 16-mm motion picture, show (in descending order) the wide-open mouth, size and form of teeth, and the bat lunging at intruder. All about life-size. (Page 201.)

Unwittingly believing that almost all smaller, hairless moth larvae would be good safe food for a vigorous, half-grown nestling, the author offered small elm loopers to the young Tit. These were gobbled down with avidity, and the bird's powerful digestive system appeared to be in first-class working order, as evidenced by the passage of large amounts of fecal matter, usually directly after feeding.

In the case of young Tits, as with many other species of nestling birds, nature evolved a cloacal sack with which to enclose the feces, which are defecated wrapped in a shiny film which holds the whole mass together. These sacks are carried away or swallowed by the adult birds, the whole strange process serving to eliminate soiling of the nest materials. The young Tit in question ejected these sacks from its upturned vent. They measured an inch or somewhat over in length, and five-eighths of an inch in width. White at one end, dark at the other, these deposits, containing, among other things, urea and other nitrogenous matter of high value, were doubtless returned to the woodland in one way or another, thus to enter into the recycling process.

Sad to relate, the young Tit died within a few hours after being fed upon the brown and blackish elm looper caterpillars, and it was realized soon afterward that the adult birds of many kinds which regularly feast on the cankerworm and other caterpillars were leaving the elm loopers strictly alone. It seemed obvious therefore, that it *was* the elm loopers that had killed the young bird, and that in all likelihood they contained some substance which prevented adult birds from devouring them. It is a subject that would bear further investigation.

The hungry nestling Tit, with its wide, bright yellow gape shown in plate 156, was photographed only four hours before it suddenly died after avidly consuming a number of the elm looper larvae. The author had fed many young birds before, but nothing like this had ever happened.

Belonging to the same family as the Tufted Titmouse (the Paridae) the Black-capped Chickadee, *Parus atricapillus*, needs no further description. A greatly loved little species, a steady customer at the bird feeders, especially where sunflower seeds are offered in abundance, pairs will often remain in the vicinity of one's feeding station to nest and rear their large broods of young, provided rotting natural tree trunk stubs are available. Chickadees will occa-

sionally nest in an artificially hollowed-out log, or a birdhouse, but despite their tiny, and one might imagine weak, little mandibles, they much prefer to *excavate* their own nesting holes in soft wood.

It is truly curious what unprotected and seemingly bad choices they sometimes make when the nesting season early in April comes around.

On one occasion a pair chose the decaying stub of a small tree that had been cut off square, about four feet above the ground, and on which a plant box had been fastened. This stump was so close to the author's driveway that cars passed within a foot or two of the entrance to this cavity which the Chickadees had excavated. Both birds labored incessantly on this project for many days, and color movies show the birds bringing out slivers of rotten wood held crosswise in their tiny mandibles and lending them a comical mustachioed appearance.

This pair had started nesting in late May, no doubt having had an earlier attempt broken up. The long, tedious excavating job was finished by May 31, then came a protracted period of nest building, both birds bringing quantities of *yellow* mammal hair and occasional small soft feathers. Some time between June 9, when the nest-lining process began, and June 28, the eggs were laid and hatched, and the nestling Chickadees very quietly fed and cared for. Then, on July 9, without the author having been present to count them, all had left the stub. Later, however, seven young were being fed by the adults near the nesting site.

In another curious case, Chickadees chose an absurdly exposed, vulnerable, and almost completely decayed stub at the base of a much branched "tree hydrangea," growing close to the author's house and some thirty feet from the mini-wood border. The birds had made this very poor selection a few days after having abandoned a much better, taller stub within the woods.

And now hear this!

Long broken off, this second stub was only twenty inches in height and three inches in diameter, while the entrance hole and the beginning of the cavity were located only ten inches above the ground! (Plate 157.) What happened in this odd case was all but inevitable. As the birds dug down, they kept getting closer to one side of the stub, and finally, having excavated but five inches below the entrance, broke through the wood into the open.

Dismayed, the pair kept examining the ruin of their plan. A piece of building paper was neatly wired over the offending hole by the author in the belief that the excavation might be continued, but to no avail. The birds gave up and left the site for good, but later selected one of the author's birdhouses.

An interesting nesting, constructed almost entirely of green moss and yellow mammal hair, is illustrated in plate 157, its odd form being due to the narrow, oblong shape of the birdhouse.

Chickadees lay from five to nine eggs, large sets being deposited even in the cramped quarters of the self-excavated nest cavities. Eggs are white, well speckled with cinnamon and brown, mostly around the larger ends. As most of these hatch, one may visualize the vast number of insects necessary as food for the success of such broods, which remain in the nests for about two weeks.

Chickadees are energetic gleaners of leaves, stems, twigs, and bark. In a single year, a single Chickadee must account for thousands of aphids alone, as well as countless other small insects. One may imagine the importance of this species in the mini-wood economy and its usefulness to arboreal foliage in general.

The author has also observed the parent birds bringing large numbers of both the plain green and striped canker moth caterpillars to their offspring.

In May, usually between the fourth and fifteenth of the month, the Crested Flycatcher, *Myiarchus crinitus*, arrives in our woodlands fresh from its winter home in Florida and Central America. A noisy bird, it at once announces its presence with loud, rather raucous calls—almost shrieks—uttered singly, or several times repeated. Again these loudest calls may be preceded by as many as twenty or thirty run-together short "cheeps" or clearly uttered "wheeps." A distinctly crested species nine inches in length, with a yellow belly, gray throat and breast, and a rufous tail, it could not be confused with any other bird inhabiting our New England summer woods.

All of the calling and chattering of these Flycatchers seemed to have had much to do with the territory selected, the males thus trying to attract the females to a chosen birdhouse or cavity. Sometimes, however, couples appeared to have been already mated upon arrival, in which cases inspection of the available sites was commenced at once.

Most of the boxes and cavities were occupied at dates between May 15 and the first two weeks in June, and some of the sites were used up to four years in succession. During other seasons, no nesting birds occupied either the mini-wood or sites on its borders, and few males were heard calling elsewhere.

Once established, no other hole- or box-nesting species was more interesting to observe, and the nests were usually put together rather rapidly, considering the amount of material used and the variety selected. A nest examined as this was being written was mostly constructed of coarse and also much finer grasses, and long thin strips of grapevine bark. To these basic materials were added dry and green needles from a tall white pine tree near the author's house, dried bunches of oak tree catkins, fragments of cast-off snakeskin, and bits of cellophane, doubtless mistaken for the former. Snakeskin in birds' nests is not unusual, especially in those of these Flycatchers, and its significance is anyone's guess.

The actual cup for the eggs averaged two and a quarter inches in diameter, and was neatly lined with fine grasses, grapevine strips, white pine needles, and a few soft wild bird feathers.

The four to five eggs in a set were curiously patterned, being *longitudinally* streaked with chocolate brown on a creamy white ground (plate 157), reminding the author strongly of nutmegs.

Most nests were completed between June 2 and 19, the nestlings breaking out of the eggs thirteen or fourteen days after incubation was begun.

And now we come to the most interesting part of the Crested Flycatcher's life-history, the rearing of the nestlings on a progressively coarser diet, including food items unique in the experience of the author during over fifty years of observing our New England birdlife.

For the first six or seven days, these young birds were fed conventionally enough, the items including small, soft-bodied, hairless caterpillars in considerable numbers, occasional white "millers," and other unidentified small insects. Beginning with the nestlings' second week, however, and in three cases observed in detail from hideouts or blinds close to the nesting boxes, a complete dietary change took place.

Fat short-horned grasshoppers, those spiky-legged, when mature brittle-winged, always succulent insects, were brought to the youngsters, and the diet was varied by occasional large dark-winged, scarey-looking insects three inches in length, with a pair of crossed jawlike scimitars protruding from the head (in the males), useful in clasping the females when mating. These grotesque, unpalatable-looking creatures were dobson flies, *Corydalis cornuta*, whose aquatic larvae are the predacious animals called *hellgram-*

mites, known to every fisherman as excellent live bait for black bass.

Strange as it may seem, grasshoppers and dobsons were swallowed *whole* by the nestlings, legs, spines, wings, "jaws," and all!

With the author's lily pool and a neighbor's pond conveniently located in relation to the mini-wood and the nesting Flycatchers, many of our largest northern dragonflies (aquatic in the early stages) were captured while skimming over the water and promptly fed to the young birds.

A common species with a large dark brown marking on each of its four wings, technically called *Libellula pulchella,* became a frequent victim, and even the giant plain-winged species, *Epiaeschna heros,* with a huge head, three-inch abdomen, and a four-inch wingspread, also entered the young Flycatchers' menu.

All of these bulky insects were swallowed whole! If one of the largest of these creatures could not be well managed by a nestling on the first try, the parent bird would pull it from the youngster's gape and then stuff it farther down the baby's throat, until somehow, wings and all, it would vanish amazingly into the little bird's anatomy.

A number of interesting relationships were here revealed, the nymphs or immature dragonflies (within the pond) preying upon other aquatic insects, tadpoles, and occasional minnows; the winged adult dragonflies preying upon other airborne insects rising from the pond or elsewhere; the Flycatchers preying upon adult dragonflies and, sometimes, on blowflies attempting to lay their eggs on weakling nestlings within the bird box.

Still more unusual from the author's point of view was to find that certain pairs of these Flycatchers preyed regularly upon large butterflies as food for their young during the last half of their occupancy of the nests. Few birds feed on large butterflies, what with their comparatively small bodies, and bulky, brittle, and seemingly unpalatable wings.

Watched at close range on many occasions, as stated, the birds were observed bringing their broods familiar species of butterflies, namely, the big yellow and black tiger swallowtail, *Papilio glaucus,* with a wingspread of up to almost four inches; the red admiral, *Vanessa atalanta;* and the mourning cloak butterfly, *Nymphalis antiopa.* Of the three, the largest or swallowtails were captured and fed most often. Red admirals were apparently the second choice, and the purple and yellow mourning cloaks were fed the least.

As with the dobsons and dragonflies, the nestlings swallowed these butterflies *whole!* Incredulous at first, the author opened the boxes many times while the old birds were foraging, and directly after such feedings, but no insect debris was ever found in the nests. Somehow, these little Flycatchers ingested these big insects successfully, discarding neither wings nor wing fragments, or other awkward parts, and it required a lot of watching and many such examinations to convince the author that there had been no mistake about this conclusion.

Experiments were then carried out with another pair of these birds nesting in a box placed ten feet up on the trunk of an oak tree, and which could be watched from a clump of closeby bushes. A dry, mounted fritillary butterfly, *Speyeria cybele* (with pin removed), was fastened with collodion to a leaf on a twig just above the box, which the parent birds often used before feeding their young. A monarch butterfly, *Danaus plexippus,* a species said to be nauseating to birds if eaten, was also fastened to an oak leaf close to the nest, and two freshly caught mourning cloaks were suspended on shoe threads, so adjusted that the butterflies bobbed about in the slightest breeze about twenty-four inches in front of the entrance.

Reactions were interesting and varied. One bird swooped upon the fritillary at one, breaking off its wings, but otherwise leaving it alone. The monarch was subjected to the same rough treatment, and having long been dried, it fell to pieces, but the two fresh mourning cloaks were quickly pulled from their suspending threads and *fed at once* to the nestlings.

In later years, much to the author's gratification, and with a motion picture camera documenting some of the events, another freshly caught monarch, red admirals, tiger swallowtails, and more mourning cloaks were observed being fed to young Crested Flycatchers. (See Plate 133 for illustration of these butterflies.)

It may interest some uninitiate to know that monarch butterflies in the larval stage feed upon various species of milkweed. Certain birds in captivity that were fed these caterpillars and, the author believes, a butterfly, were recorded as having experienced nausea, followed by actual vomiting. That the young Flycatchers just described were not so affected would indicate that Connecticut monarchs in the larval stages are nourished by milkweed species whose latex, bitter as it is, leaves no such disagreeable properties in the butterflies employed in the nestling Flycatcher diet.

During daylight hours, these young birds gave away their whereabouts to cats and other predators by their loud and continual peeping whenever hungry, which appeared to be almost all of the time between the visits of the adults. The voices of the nestlings sounded very much like the calling of the spring peeper tree frogs, especially so in the tape recordings of these sounds.

The parent Flycatchers attacked unoffending gray squirrels if they approached the nesting tree, while Grosbeaks, Woodpeckers, and other birds which by chance came too near were also attacked, and sometimes violently driven to the ground, after which the Flycatcher would fly off in triumph, uttering harsh shrieks and other notes. Strange to relate, however, a pair of Catbirds that had built their nest in some shrubbery but a few yards from the Flycatchers' box were never molested, for reasons known only to these other aggressive individuals.

After the breeding season, the Crested Flycatchers, like most other summer residents, disappeared from the scene, and probably by mid-September had all left for their winter homes far south of New England.

The much loved House Wren, *Troglodytes aëdon,* was a dependable early May arrival in and out of the mini-wood, as well as a regular nester throughout the fifteen-year study period covered in this book. From long association with human beings, it now seems to prefer man-made boxes to natural cavities, doubtless for the reason that the holes in such boxes are usually made too small for any other birds but Chickadees to enter, so that, in these houses, they may usually raise their broods more or less in safety.

The birds spend a great deal of time trying to manipulate long awkward twigs through these entrance holes, and many, despite the persistence and patience of the Wrens, are finally dropped. The nests examined over many years, in and out of the mini-wood, revealed a similar basic foundation which in all consisted of small dry twigs piled in the bottom of the box or cavity in varying depths up to about three inches. Some of these twigs were so long and awkwardly bent that they recalled to the observer on sight the patient struggles of these little birds trying to manipulate such unwieldy and irksome burdens into the one-and-an-eighth inch entrance holes.

The foundations also often included the silken egg cocoons and the wintering cocoons of spiders stuffed into the twigs at random, the birds having first eaten their contents, then taken the white material along as desirable nest additions.

Various grasses and long strips of bark peeled from grapevines (a favorite building material of many woodland and other birds) occupied varying regions in the nest box or natural cavity above the basic foundation, while the neatly formed egg cup, averaging two inches in diameter, was lined with fine grasses, very narrow grapevine strips, sometimes a few dry or green white pine needles, and horsehair (formerly used by many birds when it was everywhere available), and usually the cups also had a few soft feathers worked in.

According to the whims of these odd little birds, various odds and ends were often added to the nests—fragments of snakeskin, cellophane, and cotton; bits of wire probably mistaken for twigs; and, in one unique instance, several dozen inch-and-a-half rusty brads!

The six to ten delicate little eggs making up the completed sets varied considerably, but most were so thickly covered with brown and pinkish brown specks that in most of the nests they appeared to be of one solid color, although they were really spotted on a whitish ground (plate 158).

As far as could be ascertained by careful watching at several nests, the incubation period lasted for thirteen days. Once out of the shells and dry, the nestlings were fed at short periods from dawn to dusk with a great variety of small insects, including many small white moths, and caterpillars. Spiders were also tirelessly hunted and fed to these young Wrens, and as they became older and stronger, the adults arrived at the nest boxes carrying what appeared to be small grasshoppers, according to the legs sticking out of their beaks.

The young birds remained in the nests about two full weeks, sometimes the last one in a brood lingering another day or two.

At this juncture, the author wishes to repeat the description of an episode already written in the preface, an important part of every book, which, however, many people do not read.

Birds, of course, have been great if not the greatest sufferers from our own past stupidities. The author admits to his share of guilt, for in years gone by "tree men" in and about the mini-wood and its surrounding area were allowed to spray the trees and shrubbery with chemicals then thought best to protect the foliage. Sprays lethal to small birds that ate the sprayed caterpillars were unwittingly applied to the trees at the author's place and to those of neighbors on either side of the mini-wood, and at a time when the inpouring summer birds and transients were depending upon cankerworms and other small hairless caterpillars, according them top priority in their own diets and in those of their nestlings.

All at once we found to our dismay that some of our most valued songbird species were dying. We found Wood Thrushes and Robins, certain Warblers and Scarlet Tanagers lying quivering on the ground and quickly succumbing. Vireos disappeared from our woods and shade trees. So did Chebec (the Least Flycatcher) and some other small species.

Even today, many of them have not returned to their former numbers despite the fact, thanks to Rachel Carson, that lethal sprays are now banned by states and communities, and that most people no longer employ dangerous pesticides of any kind.

It was during these years of stupidity on the author's part, and on the part of his neighbors of that period (when most spraying was done with improperly understood chemicals) that cessation of all activity around the boxes was noticed, but knowing House Wrens to be fickle little birds, it was believed that they had simply found and moved to other sites more to their liking, so for the time being no more was thought about the matter. A short time later, however, Wrens were found dead, one of them already merely feathers and bones, and another one, not long dead, lying with outstretched wings, as at times we had found other kinds of birds poisoned by DDT.

The boxes were now opened, and their dismaying contents graphically revealed what poison sprays had been doing to our nesting songbirds, a tragedy that had probably been duplicated all over the country.

In one nest the Wrens had evidently hatched their brood successfully, but the young had either died, like the adults, from poisoned food or, what was just as likely, from starvation when food was no longer being brought to them. Both of the old birds found dead were on the ground within a few feet of the pole and nest box.

In the cup of this first box opened lay the skeletons of the nestlings, one almost intact. All bones in the nest were clean and white, probably left thus by blowfly larvae that consumed all of the fleshy parts after death, as is seen in plate 158. In the other Wren nest that year, a set of seven eggs had dried up or rotted, indicating that the adult pair had died (plate 158).

In these and other equally sad instances, we had had a graphic lesson which ended our spraying with all sus-

pected materials from that day onward, and it was suspended by our neighbors of that day as well, all of us convinced of the damage we had been inflicting upon our wildlife.

Normal male House Wrens must sing at least a hundred thousand times in a single nesting season, then they suddenly cease altogether and are seldom seen again thereafter until another spring.

At the time of this book's completion, which was not until some time after the fifteen-year study period, some species of birds seemed to be holding their own, but the House Wrens, Vireos, Robins, and Crested Flycatchers had not regained their former numerical status.

Three species of Woodpeckers nested within the mini-wood habitat during the fifteen-year study, and two of these were permanent residents, feeding all winter at the suet supplied by the author.

The Downey Woodpecker, *Dendrocopus pubescens*, and the very similar but much larger Hairy Woodpecker, *D. villosus*, drilled their own nesting cavities in dead limbs or old stubs, usually in wood somewhat softened by decay and fungus growths. On one unusual occasion, the former species excavated a cavity on the underside of an almost horizontal limb, weakening it to such an extent that it broke—fortunately, after the nesting season—one end hanging down from just beyond the bottom of the cavity until winter winds blew it to the ground.

Both of these Woodpeckers deposited sets of four pure white eggs on beds of wood chips in the bottom of the cavities. Except for size, the nests were very similar. A typical one of the downey species is shown in plate 159.

While Woodpeckers are typically gleaners of bark insects, borers, and insect eggs, they often brought their well-feathered young to the suet feeder and thus supplied them with additional easily obtained animal food. Wild berries were other foods taken by them.

Arriving from the South early in April, the familiar mating calls of the male Flickers, *Colaptes auratus*, were heard daily in the mini-wood, and the courtship antics of these big Woodpeckers were most amusing to observe. Sometimes they encircled the tree trunks, male following female, or the male crouched on a limb close to his intended mate, uttering strange sounds. Even more amusing was his circling of the female (when both birds were on the ground), following this action by sudden pauses during which the male assumed ridiculous poses and attitudes.

Mostly insectivorous, Flickers also eat wild berries, as do the smaller Woodpeckers mentioned above. Ants are a favorite food item of Flickers, and they were often seen on the ground inserting their long, sticky tongues into an anthill and drawing out the adhering inmates.

At other times they were seen making another, curious use of these insects, "anting," as this is referred to by ornithologists. Here the bird lies on the ground in a sunny spot beside an anthill and deliberately transfers *living ants* into its plumage. It has been supposed that the odor of the ants possibly helps in delousing the bird. Perhaps the ants catch the lice and carry them back into their nests, or maybe, like many human beings, Flickers just like to be tickled. The reader's guess will be as good as any other!

As mentioned, all Woodpeckers lay pure white eggs. A Flicker clutch may number as many as nine, but in the author's experience six or seven is more normal (plate 159).

Flickers nested in the mini-wood for several seasons in a cavity high in a decaying stub on a large black oak tree, returning to it each spring until a winter storm blew it down. Their deep cavities may be drilled high up, as this one was, or sometimes only two or three feet above the ground, if a stub happens to be suitable.

Once the nestlings are old enough to thrust their heads out of the entrance hole, feeding is most interesting to watch. Accomplished by the process of regurgitation, the young bird inserts its bill and part of its head well into the mouth of the parent, then, in a rapid back-and-forth motion of both heads, the partly digested food is transferred, and not forgetting the large broods, in what would appear to be rather an exhausting process for the parent by the end of a full day of such feeding.

Two of the most uplifting sounds of spring are the loud, repetitious calls of the Flickers, and the far-carrying drumming of the smaller Woodpeckers on dry dead limbs and the sides of empty birdhouses. Flickers drum also; in one case, the bird selected the TV aerial on the author's house, which transmitted a most startling sound into the room below.

Two other nesters, in bird boxes or in natural cavities and crannies, wherever found, were the House Sparrow, *Passer domesticus,* and the imported English Starling, *Sturnus vulgaris.*

For those who are not familiar with the nests of these species, it may be mentioned here that both build rather unkempt, trashy types, carelessly put-together affairs (especially is this true of the Sparrows), and composed of grasses and straws, shreds of bark and other vegetable fibers, bits of string, paper and cellophane and other odds and ends, and feathers (of which the Sparrows use as many as they can find, and the Starlings only one or two or, just as often, none).

House Sparrows have two broods a year, laying from four to six eggs in a set. These are variously spotted and otherwise marked with gray and brown on a whitish to greenish ground color.

Those Starlings' nests that were examined contained from four to seven eggs; the shells were unmarked and of a beautiful burnished pale blue shade most attractive to the human eye (plate 158).

Much as these two bird species are disliked by some people, it must be said that both were found to be highly beneficial insect destroyers throughout their long (double-brooded) nesting seasons. Like all species introduced from foreign countries, they have their serious faults also, chief among which in these two birds are of course their aggressiveness and tendency to usurp the available nesting sites formerly used by more timid native species. House Sparrows and Starlings may be discouraged, however, by repeatedly destroying their nests, the Starlings deserting a site under such treatment more readily than the Sparrows.

The Starlings ad-lib amid more standardized themes, chatterings, gratings, and whistles, some perhaps in imitation of other species, and they sing cheerily on and off even in winter. When nesting time comes around in early April, the male often sits close to the nest site, his head tilted upward in song, at the same time expanding his wings to their full and fanning them rapidly, a display which, to the thinking of the author, spells *ecstasy*, perhaps not as strongly as we ourselves experience it, but ecstasy, nevertheless.

And what better way could there be to express it? Would that we too had wings!

CHAPTER XX

BIRDS OF THE MINI-WOOD II: THE OTHER NESTERS

Known to people more by its repititious summer-long notes than by sight, notes which sound to the human ear somewhat like its last name, the Red-eyed Vireo, *Vireo olivaceus,* is a slim little bird somewhat over six inches in length. Its upper parts are light olive green, without wing bars; its underparts, white; its crown, gray with black borders; and over each eye there is a conspicuous narrow white line. As for its eyes, it is the *iris,* not the pupil, that gives the bird its name.

Spring migration records formerly kept by the author for thirty-nine consecutive years at Stamford, Connecticut —only a few miles from the mini-wood's location—revealed that this Vireo always arrived from the South between May 4 and 15.

Breeding began soon after the birds' arrival. Nests observed in and around the mini-wood area were placed from four to twenty feet above ground, in basswood and oak trees. Invariably pensile and firmly attached to the twigs of a crotch, a typical nest was constructed of fine grasses and thin strips of grapevine bark bound together with spider webbing and decorated on the exterior with wasp-made paper fragments and bits of birch bark. The compact cup for the eggs was usually lined with delicate strips of grapevine bark, apparently especially shredded for the purpose, as indicated by the deckled edges of these strips. Fine plant tendrils were also used for the lining of the nest's two-and-a-quarter-inch (in diameter) cup, in which four white eggs lightly spotted with black and umber were deposited (plate 160).

Another particularly interesting nest was found in the crotch of a basswood sapling, four feet above the ground, and placed in such a position that a single large leaf three inches above the cup acted as a parasol, perfectly shading the eggs or the sitting bird below. After this leaf was lifted very carefully in order to take color motion pictures of the nest interior, it would flip back exactly into place when released, to become the perfect sunshade once more. Mere chance? Or did the female plan it so?

Unfortunately, a Brown-headed Cowbird, *Molothrus ater,* broke up this nesting, laying three of its brown and umber spotted eggs in the cup, and their size and weight broke those of the Vireos, causing the pair to abandon the nest. This common parasitic Cowbird seemed to search out Vireos in particular for its victims, at least in the area of the mini-wood.

Much rarer as a nesting species was the Yellow-throated Vireo, *Vireo flavifrons,* which usually arrived from the tropics sometime between May 7 and 14.

Not observed some years even as a transient migrant, it nested on one occasion in the mini-wood, placing its pensile cup in a small crotch formed by twigs of a small basswood branch fifteen feet above the woodland floor. This nest was easy to observe, constructed as it was on a so-called sucker branch growing out from a side of the main tree trunk (plate 160). A nest almost identically placed had been found outside of the mini-wood area several years previously.

Somewhat more heavily constructed than the nest of the red-eyed species, this Vireo also employed grasses and strips of grapevine bark for the main body, binding them with spider webbing and then decorating the exterior with bits of birch bark, and in this case, the empty egg cocoons of spiders. The nest measured two and three quarters inches in width, and two inches in depth. The four white eggs, sparingly spotted with brown and black, lay upon a lining of very fine encircling grasses. Freshly finished and containing the eggs, this nest was almost startlingly beautiful when suddenly looked into on a brilliant May morning.

For those who do not know this Vireo, it may be recognized by its olive-green upper parts, white wing-bars, bright yellow throat and breast, and this color combination coupled with its "Vireo" notes. To the author's ear, the bird's song is richer than that of the Red-eyed Vireo, and not so incessantly repeated.

Like most Vireos, the two which nested in the mini-wood seemed to be strictly insectivorous. Both species stayed all summer, thus aiding greatly in the control of foliage-consuming caterpillars, aphids, and other arboreal insects.

By the third week in September, the red-eyes had departed for the South, and it was assumed that the yellow-throated species had left somewhat earlier, as they were not long seen after completing their nesting activities, whereas the former were heard singing until late in the summer.

The Baltimore Oriole, *Icterus galbula*, needs no plumage description.* Its loud, seemingly joyous songs, varying greatly in different males, yet ever unmistakably "oriole" in quality, are first heard issuing from the treetops (where the birds are feeding on arboreal insects) during the first week in May. These songs continue for several weeks, or until the parents are feeding their well-feathered young outside of the nest.

In the experience of the author, Orioles may choose the site for the nest as early as May 15, the female beginning the long, tedious work at once. Sites have been an elm branch overhanging the road leading to the author's property; an outer branch of a very tall tulip-tree, and, on one rare occasion, in a mature sweetgum somewhere, a nest blown down and later brought to the author by children, and shown in plate 161. In a fourth instance, the site was a wildly swaying tip-end beech branch high above the ground on the mini-wood border. Most often, however, Orioles choose the drooping and long-favored elm branches.

---

*Henceforth, in the official A.O.U. check list, to be known as the Northern Oriole, in one of the most annoying decisions ever made, in the opinion of large numbers of bird lovers.

Why these birds, which spend the greater part of each year settled comfortably in the tropics, mostly far from people and traffic, so often hang their woven nests in elm branchess *directly over* northern concrete highways, with thousands of automobiles and trucks roaring along just below, is anyone's guess.

The nest in the beech on the mini-wood border was already partly woven when first seen on May 15. Swinging wildly in the breezy spring weather, it was a revelation to observe the female at the tip end of the branch weaving and sewing this deep cup together, with the male watching and often singing close by. Working downward with each beakful of grass and plant fibers, and stopping now and then to wind extra supporting strands about the supporting twigs, she gradually formed the sides and finally brought them to a rounded bottom and soft cup for the eggs.

In some cases, the nests have been less expertly constructed, less neatly bag shaped, and shallower than others, and which may have represented the work of younger females constructing their first examples.

Even against a strong light, the average Oriole's nest may appear solid and airtight, but actually it is loose enough for fresh air to pass through very readily and yet keep the nestlings warm.

When the nest in the beech was nearly completed, this pair suddenly discovered an outdoor rotary clothes drier, and finding the dangling loose ends of the cotton cords ideal for their purpose, proceeded to shred out large fluffy masses with their sharp bills, and which they carried off as the lining material for the nest. It was one of the few instances in which the male took part in actually carrying material for the female.

This was an attentive male otherwise, often seen bringing food to his mate in the nest and later laboring strenuously to feed the brood after they were able to fly. The newly fledged birds sat indolently in various trees, uttering a curious and characteristic call to reveal their whereabouts to the foraging, hard-working parents. The fully feathered young were fed thus for several days more before learning to care for themselves.

The Orioles do not stay very long once their nesting activities are finished. One wonders why they bother to come North at all!

In these deep nests the Orioles deposit from four to six very interestingly patterned eggs. They are white or off white, or slightly bluish in ground color, variously spotted with black, and curiously scrawled with threadlike black lines (plate 161).

The nest in the beech tree, which the young birds left on June 18, then withstood the following winter, another summer, and a second winter before its last few shreds had disappeared.

In the author's experience, no Orioles were ever observed returning to, repairing, and again using one of these tough old nests.

The Rose-breasted Grosbeak, *Pheucticus ludovicianus*, and the Scarlet Tanager, *Piranga olivacea*, two always gratefully welcomed migrants, arrived in the mini-wood area around the first week in May, and not later than the twenty-first, except in those fortunately infrequent years when long cold rains in spring delay these tender tropical species, and sometimes cause the death of many by starvation due to the absence of normal insect food.

Both birds are well known by the striking plumages of the males, the Grosbeaks flashily black and white, with rose-colored breasts, and with heavy, thick bills; the Tanagers in fiery scarlet with jet black wings, longer-type bills, and of lighter construction.

Although of different genera, and differing widely in form, they have been included next to each other in this chapter only for the reason that there are striking similarities (except for size) in their eggs, and in some characteristics of their nests.

Both species occasionally bred in the mini-wood or on its borders, and over the years there were very probably still other nests undiscovered by the author.

Those of the Scarlet Tanager, closely observed, have been situated from twenty-five to thirty-five feet above the ground, and saddled among leafage on more or less horizontal branches of swamp white oak and pignut hickory trees, while the Grosbeaks, on the other hand, have usually selected sites from ten to fifteen feet from the ground in shrubs such as the winterberry (partly overgrown with Virginia creeper), or in arrowwood bushes or honeysuckle vines, and, once, thirty feet up in a young red maple growing in a crowded group of others in a great confusion of upward shooting branchlets, where the female bird experienced much difficulty getting the nest foundation to stay put.

Both the Grosbeaks and Tanagers used twigs and what appeared to be weed stems and undetermined shreds of some other form of vegetation in building their nests, but the main bulk of the nests examined in the mini-wood or close by have consisted of very special rootlets and tendrils, from what plants or vines was undetermined.

The Grosbeaks' nests were much bulkier than those of the Tanagers, but they looked alike otherwise, and the linings of the cups for the eggs consisted of the finest of the tendrils and rootlets mentioned above (plate 161).

Four eggs seemed to constitute the normal set in the nests of the Tanagers; four or, rarely, five in those of the Grosbeaks. Freshly laid, they were pale bluish white (those of the Grosbeak sometimes tinged with green) and in both species, spotted and marked with reddish brown, with some olive-brown on the latter's eggs also. Generally speaking, the eggs of these two widely differing species do indeed look very much alike, an observation, however, with no deeper significance than that it illustrates an interesting coincidence.

The Scarlet Tanager is a bird of the woodland treetops, where throughout the summers it devours countless insects, including many caterpillars, and no doubt as many harmless spiders. It may be regarded as one of the extremely valuable avian species helping in the control of foliage consumers, especially those of the canopy, which are quite inaccessible to many other animals that by habit are strictly terrestrial.

The Tanager's rich, rhythmical, penetrating song, slightly reminiscent of a Robin's, but never as prolonged, typifies spring at its best in the woodland.

The Rose-breasted Grosbeak is generally more omnivorous than the Scarlet Tanager, feeding on seeds as well as on insects and some fruits.

During May, when the elm trees put forth their abundant clusters of flat seedpods (usually during the first ten days of that month), this botanical event is nicely synchronized with the arrival of many of the Grosbeaks from the South, who forthwith feast on the soft forming seeds enclosed within each of these curious receptacles as a welcome seasonal variation in their diet. Nesting over, both species depart for the South early in the fall, the male Tanagers having replaced all their scarlet feathers with dull green plumage above, dusky wing feathers, and yellowish plumage below. The loud, vigorous song of nesting time is also replaced with a simple, double-call note which sounds to the author like *"Chick-anh,"* while the delightful and often extended warbling song of the Grosbeak males will not be heard again until May.

Two species which built their exquisite little nests within the mini-wood, but only rarely, were the Ruby-throated Hummingbird, *Archilochus colubris,* and the Wood Pewee, *Contopus virens.*

Both arrived from their southern winter homes the first or second week in May, only the Pewee delaying in unseasonable weather with a scarcity of flying insects, as these little Flycatchers seem to depend mostly upon food caught on the wing. Likewise, the Pewees' departure in fall was governed by the available winged insect supply, whereas the Hummingbirds were often found feeding late in the season at the still blooming jewelweeds and late-blossoming cultivated flowers.

In the northeastern states, these Hummingbirds could not well be mistaken for any other species, what with the male's brilliant red throat, the female's white throat, and the iridescent green overall plumage and long needlelike bills of both.

The Wood Pewee is a sparrow-sized Flycatcher with somber olive brown plumage above, shading into white below the breast, and with two distinct pale wing bars, but no white ring around the eye, as in some other small Flycatchers. Its plaintive, drawn out song—"Pee-weeeeee,"—is delivered at intervals from the shadiest arboreal realms throughout the summer.

Both species have two habits in common. They saddle their little nests, the Hummingbirds choosing somewhat sloping (either upward or downward) branches of small diameter, and the Pewees seemingly preferring somewhat larger, near horizontal, sites (their nests being of course much broader than those of the Hummingbirds, this is probably a necessity).

Both species "decorate" or camouflage the nest exteriors with fragments of gray and green lichens, at times so thickly and cleverly applied and bound with spider webbing that the nests blend very well with their natural surroundings.

Hummingbirds in the mini-wood constructed their nests mostly of yellowish to cinnamon-colored hairy or scaly material gathered from the mature stalks of cinnamon ferns, binding it together with spider's silk. The nests were lined with the same material neatly shaped to the bird's body form. The example shown in plate 162 had been only partly camouflaged with lichen fragments when it was

partly disfigured and one of the eggs taken by some predator, something which all too often happens to songbirds' nests and eggs and young. Every nature photographer well knows that it is best to make one's pictures immediately upon discovering such subjects, and the author has only himself to blame for this unfortunate photograph.

The walls of this nest were a quarter of an inch in thickness, and the cup for the customary two eggs measured just under one inch in diameter. The single plain white egg (all Hummingbird eggs are this color) lying against the yellow tinted fern down, was deserted by the birds, just as shown in the picture.

Nests built in the identical area were not identical in construction. Two females, for instance, that chose and built in the mini-wood, took the soft woolly material which coated the stems of the sterile cinnamon fern fronds as well as the cinnamon-colored similar stuff from the fertile fronds, but one of the birds used only the light-colored wool, while the other (the female which built the nest in the illustration) employed much of both kinds.

Collecting was accomplished on the wing by both birds, and it was most interesting to watch them gathering the soft bunches bit by bit as they slowly rose and descended in front of the frond stalks until their little bills were well loaded. Dozens of trips to the fern beds were made by both females, but the males were never seen during these activities.

The first of these nests was saddled on a slender swamp white oak branch fifteen feet above the ground, and the second, somewhat higher, on a sloping branch of a white oak on the mini-wood border.

In about thirteen or fourteen days after the eggs were deposited in one of these nests, two diminutive Hummingbirds were successfully hatched, being smaller than many common insect grubs—and as naked. The author watched the feeding process from a distance through binoculars so as not to frighten the adults away. The process was seen to be by regurgitation of a liquid substance, during which the adult's long tongue was much in evidence. Run out and in many times, it aided the delivery of the food well into the throats of the nestlings, which were fed for nearly three weeks before they left the nest.

Few people realize that most species of perching birds which sleep nestled on small twigs and branches require a device which will prevent them from falling off such perches when in this state. Nature has supplied such a safety measure in the form of locking bones; in other words, the bird has a built-in device which locks its toes around the branch or twig.

That the device is efficient to an astonishing degree is illustrated in the case of a Hummingbird found by the author, which had died in its sleep in an epiphytic vine. On its death from some undetermined cause, the bones in its feet remained locked as it fell forward, and it remained thus upside down! The bird had been dead for perhaps two days when discovered, but the toes were still firmly gripping the stem, as shown in the photograph in plate 162.

The typical nest of the Wood Pewee found in the mini-wood was saddled on a white oak branch about twenty-five feet above the ground. It was mostly constructed of fine grasses and silky vegetable material, and bound with spider webbing to some extent. It was lined with bits of tendrils, and very fine grass, some of the stems still bearing the seed spikelets. The Pewee also camouflages the exterior of its nest with lichen fragments to produce an altogether extremely beautiful object and example of bird-building skill (plate 162).

The cup, when photographed, contained two white eggs wreathed with dark reddish brown specks; there were four when the set was complete, and they just about filled the space (only two and an eighth inches in diameter). Exteriorly, the nest measured an inch and a quarter in height.

The diet of the Wood Pewee seemed to consist mostly of insects caught on the wing. While the diet of the Hummingbirds was mostly nectar taken from a wide variety of flowers, they no doubt also consumed some of the smallest insects and spiders found within the nectar cups, those which are sometimes drowned therein and taken up automatically with the nectar.

Among the endless associations to be found between woodland organisms, we have witnessed in the latter two bird species alone their use of such plants as lichens and ferns, their relationship with spiders and use of their silk as nest-binding material; how the Hummingbirds feed on nectar and its minute insect corpses and how, in turn, these birds aid flowers, cross-pollinating them with the pollen grains which adhere to their long bills, while both species (the Wood Pewee and the Hummingbird) doubtless help in the distribution of ferns and lichens by carrying spore-bearing fragments of them to somewhat distant nests.

For those who might like to lure Hummingbirds by natural means, the author recommends the planting of butterfly bushes (*Buddleia*), coral bells, and jewelweed or touch-me-not, which, once started in reasonably wet ground, propagates itself lavishly from year to year by means of its "explosive" seed dispersal capsules (see plate 75).

When the migrant Warblers are beginning to appear in the trees in May, one of the first to be noticed is the brilliant American Redstart, *Setophaga ruticilla*, the males being black above and white below, and with bright orange-red patches at the bend of and across each wing and on the outer tail feathers. The somberer females are olive brown above, white below, and with yellow patches on the wings and tail. Drooping the wings and spreading the tail as they dart about after insects, the birds seem to flash through the trees.

Redstart (Warblers) could not be mistaken for any other if markings given are carefully noted, together with the male's song, consisting of five notes—"*tsee tsee tsee tsee tsee-o,*" or "*teetsa teetsa teetsa teetsa teet*"—so well described in the Peterson *Field Guide*.

Male Redstarts were sometimes observed in the mini-wood traveling back and forth over the same short course some ten to fifteen feet above the ground and frequently darting into the air. Thus covering about fifty feet horizontally, the back-and-forth trips were repeated for an hour or more at a stretch. Close inspection, although it was only early May, revealed what appeared to be a marriage flight of termites, the insects rising at intervals from the woody debris on the mini-wood floor.

Once in a while a pair of Redstarts remained to nest within the mini-wood, two very different sites having been selected. One was a small crotch twenty feet up in the only pin oak in the wood; the other, an almost horizontal branch fifteen feet above the ground, upon which the nest was saddled among the leaves of this red maple tree (plate 163).

This nest was mostly constructed of fine grasses and finely separated strips of grapevine bark, and was thickly wound around on the exterior with that threadlike and strong *silvery* plant fiber that so many small birds (Warblers in particular) find and use for their nests, and which still remains unidentified by the author.

The nest cup, one and a half inches in diameter, was lined with delicate bits of brown and light-colored grasses. To secure the nest on this limb, much of the silvery plant material mentioned above had been worked into the exterior, then wound around two small side twigs (hidden in the photograph), these strands holding it securely in place in a horizontal position.

The four eggs were bluish white, rather sparingly spotted and blotched with reddish brown and darker shades, mostly concentrated about their larger ends.

As nearly as could be ascertained by observing the other Redstart's nest (in the pin oak), thirteen days were required for its completion. Hatching probably occurred thirteen days after incubation had begun, and the nestlings were fed for fifteen days before they were able to fly.

One regrets the disappearance of these colorful and lively Warblers when their nesting season is over, but as these local individuals depart for the South, many others that nested much farther north come hurrying southward, thus our woods and shade trees may be populated with them for many days in early fall after the others have gone. Even at this season, when most bird plumages seen are duller after molting, or are juvenal plumages, the Redstarts retain enough of their colors so that they still seem to flash as they dart about the foliage.

A red-letter day for the author came when a nest of the Maryland Yellow-throat, *Geothlypis trichas,* was discovered well hidden in grasses and jewelweeds on the wettest borderland of the mini-wood, with a large patch of skunk cabbage close by in a fully shaded niche of its own.

The Yellow-throat, as it is usually called, is a little Warbler that arrives each spring in May.

The male, with its bright yellow throat and a very distinct black mask across its face and eyes, is easy to recognize, while the female, minus this mask but with the yellow throat, may be less so.

The species frequents low vegetation in wet woodland habitats or along its borders, feeding among the skunk cabbage patches and fern beds, from where the male's rollicking song is often repeated, and more often than the bird is seen.

The nest shown in plate 163 was an interesting example of a firmly situated one, and accordingly more loosely constructed than Warbler nests saddled on branches. It was formed mostly of coarse grasses, dry wed stems, and grapevine bark (the thirteenth species of mini-wood organisms to employ the latter material) and the cup lining was of very fine dried grasses. The four white eggs were lightly spotted with rufous brown and umber.

Found on May 25, 1964, the eggs were taken by some predator the same night. This, of course, was a great mis-

fortune from a photographer's point of view, but as a natural woodland community event, had to be accepted and discounted, although a rare opportunity to study the Yellow-throat's home life with the camera had been lost.

Prior to 1946, the Cardinal, *Richmondena cardinalis,* had not been recorded by the author, but from 1950 onward, its spread northward from its former normal range included Connecticut, and eventually the species nested in the mini-wood area, in the shrubbery close to the author's and neighbors' homes.

Now a welcome resident in many northeastern localities, the Cardinal has become the aristocrat of the bird feeders, and a most welcome and colorful songster that sometimes begins to tune up as early as late January, a holdover perhaps from its long lineage in warmer climates and earlier nestings.

The uplifting quality and the variety in the Cardinal's singing proclaims its success as the northward movement continues, success no doubt made possible, at least to some extent, by the thousands of winter feeding stations now maintained. In the opinion of the author, the Cardinals might soon disappear from the North were it not for these stations.

Cardinals are rather aloof and gentlemanly birds, preferring to arrive late at the feeders, after the common herd of House Sparrows, Starlings and Blue Jays have departed toward dusk. The author keeps the feeders filled all summer, and the Cardinals enjoy this service, often feeding in comfort right up until darkness, and sometimes bringing their fully fledged young with them.

These offspring were often fed sunflower seeds and, normally, also many kinds of insects, and berries. Not infrequently, the parent birds were observed extracting the soft immature seeds (cut out whole) from the ovaries of the Asian day flowers (see plate 74), then feeding them to the still inexperienced young.

Around the mini-wood borders, the Cardinals selected various sites for their nests, usually only five or six feet above the ground, in honeysuckle vines and, when close to the house, in syringa and other shrubs such as the mountain laurel, in which nests were built within a foot or two of living room windows.

Building commenced as early as April 5, but at other times might be delayed until well into May. Sometimes, when the nest had been completed very early, before the leaves of the chosen shrub had expanded, predators easily found and destroyed the eggs or young; but in such cases, second nestings were the rule, with these extending into June and July or even early August.

The female carried on all of the nest-building, first bringing rather long, often *bowed* twigs for the foundation, then coarse grasses, rootlets, and weed stems, and while the growing nest became somewhat bulky and rather loosely formed, it stayed in place very well within the honeysuckle vines or among the small branching twigs of such shrubs as the mountain laurel and syringa, favored sites already mentioned (plate 164).

Nest cups were neatly lined with fine rootlets and tendrils or thin strips of honeysuckle bark. Some nests were so thinly woven on the bottom, however, that the eggs could be dimly seen through it.

Four eggs completed an average set. They were white, and when freshly laid, sometimes had a bluish hue; they were always variously sprinkled with olive brown, grayish brown, and cinnamon markings.

Both birds fed the young, and it was a common sight to find the adult pair feeding late, fully fledged broods (from the second nestings mentioned above) on sunflower seeds first opened for them.

Three species of birds which nested fairly regularly in or on the mini-wood borders, although one or the other occasionally skipped a season, were the Catbird, *Dumetella carolinensis,* the American Robin, *Turdus migratorius,* and the Wood Thrush, *Hylocichla mustelina* (plates 164 and 165).

Catbirds, those "mewing," mostly slaty gray denizens of heavy shrubbery, woodland thickets, and brier patches, are of course good singers, their catlike harsher utterances being merely their call or warning notes. Some individuals possess astonishing repertoires, and all of them sing in interesting variety, often mimicking other species.

Within the mini-wood, Catbirds built their nests within a few feet of the ground in greenbrier vines and in the honeysuckle shrub, *Lonicera Morrowi* a plant shown in plate 63. Those nesting around the woodland borders preferred the arrowwood shrubs.

A typical, well-made nest was constructed of many twigs, dried weed stalks, and shreds of bark from the Japanese honeysuckle, and was bound around on the outside with strips of grapevine bark, dozens of these shreds adding bulk to the nest. The cup measured only an inch in depth, and three inches in diameter. Lined with leaf midribs and fine

black rootlets, the deep greenish blue eggs lying on this dark compact carpet were displayed to their best advantage. Suddenly coming upon such a nest always startled the author with its beauty, for there is no other species laying more exquisitely colored eggs (plate 164).

Nests were built at various dates in May and June, and second nestings, as in the case of the Cardinal, much later in the year, when the first settings failed because of predators.

The nestlings were fed a wide variety of insects and occasionally, suet from the feeders; the adults often partook of foods supplied by the author, including oranges. They arrived during the first week in May and remained until well into September.

As a rule the Robins arrived in the mini-wood area early in March, but seldom commenced nesting before April 15. Thereafter, their activities were observed well into the summers. Whether or not two broods are regularly raised, or the late nestings represent disturbed earlier attempts, the author cannot say for certain, but newly "weaned" Robins in their juvenal plumage with speckled breasts were often seen feeding themselves late in September.

The heavily mud-walled nests, with more or less purposely or accidentally enclosed reinforcing grass, leaf fragments, and stems or other debris, were constructed in white oak trees, beeches, and red maples, in climbing rose vines, in arrowwood shrubs, and for several years in a great white pine close to a window of the author's house.

These were comparatively simple nests, often incorporating bits of paper, and lined with fine grasses. The four or, rarely, five "robin's-egg-blue" settings possessed a beauty all their own, although very much paler than those of the Catbird, and without their greenish tinge (plate 165).

Representing still another of the numerous animal forms depending to a considerable extent upon earthworms, the adult Robins also fed them to their young in great quantities when these broods consisted of four ravenous nestlings.

After each frequent feeding, most nestlings elevate their bottoms and vents. As the old bird waits, one or more of the youngsters excretes a single lump of feces enclosed in a membrane, which is swallowed or carried away and dropped by the parent. In this species, as in so many others, this interesting provision of nature serves to keep the nest clean until the young birds are able to sit on the rim of the nest, or eject their feces over its walls.

When Robins are seen hopping over the ground and stopping to listen and observe their surroundings, they are not only feeding on earthworms, certain beetles, and bugs, but also on insect larvae, which they are able to locate and exhume, including the grubs of the Japanese beetle.

Rarely, a Robin or two remained in the mini-wood area all winter, and how they maintained themselves was a mystery, as there were few wild berries then available, and no patches of sumac (on whose hard seeds Robins often feed) were growing nearby, nor were these birds ever observed feeding on suet, always in ample supply.

The return of the Wood Thrushes in early May was ever an event keenly looked forward to. Suddenly, some morning, the exquisite and somewhat varied songs could be heard with great dependability, the arrivals sleek in their fresh rich brown plumage, with black-spotted white breasts, and their big black eyes glistening. Continuing their lush phrases for long periods at a stretch, day after day until late in the summer, they made life happier, always. Typical of shady woodlands and moist thickets, they seemed to sing at their best when answering each other after summer showers.

Nests were less bulky than those of the Robin, and with less mud employed for their walls. Most of the nest was made up of dead leaves, weed stems, and grass, often with bits of paper, cellophane, or even cottony material tucked in here and there, and the cup for the eggs lined with fine rootlets (plate 165).

Situated from two to fifteen feet above the ground, nests were found in the crotches of sapling beeches and sugar maples, but more often among vines overgrowing old stumps. Others were in greenbrier and climbing rose vines, and in arrowwood shrubs. One pair of Thrushes had a nest in exactly the same spot in a vine-covered stub for three years in succession.

Four blue eggs, somewhat darker than the Robin's but much paler than the Catbird's, usually completed the sets, most of which were being incubated from the third week in May, but others not until the first two weeks in June. On one occasion, the Thrushes and a pair of Robins had nests only fifteen feet apart in climbing rose vines on separate supports, a very unusual case.

While the female Thrush sat patiently brooding, day after day, the male nearby sang for long periods, his almost liquid phrases echoed by other males singing in their respective territories. Such rivalry brought forth the very best that these birds had to offer. A replayed tape record of an individual's *own* song seemed somewhat puzzling or disturbing to the bird, who came at once trying to locate the intruder in his domain!

Adult Thrushes used the birdbaths a great deal, especially a bath constructed on the top of a glacial boulder at the edge of the woods. Late in summer they brought their brood of four, and sometimes the whole family of six bathed together. At such times, many seeds of fruits or berries eaten were left in the pool in excrement, again illustrating how many plant species are thus distributed (see plate 53). The Thrushes departed for their winter homes in Central and South America in September, the author having once observed them in the forests of Panama in February.

Appearing within the mini-wood in late April or early May, but singing its mockingbirdlike repertoire for only a short period each year, was the Brown Thrasher, *Toxostoma rufum,* often mistaken by the uninitiate for a very long-tailed, black-streaked (on the underparts) brown-backed Thrush with white wing-bars. Employing the simple description just given, it would be hard to confuse the Thrasher with any other bird in the area under consideration.

While the species undoubtedly nested nearby, it was never found building within the mini-wood or on its borders; but having often found the nests elsewhere, usually in tangles of greenbrier, the author has included an illustration taken at Stamford, Connecticut (plate 166).

The nests observed were constructed of twigs, and leaves, tendrils and rootlets, and well lined with finer rootlets, and the four or five eggs were bluish or grayish white, extremely well covered with minute cinnamon specks and spots (see plate 159 for eggs).

The birds sat very tightly once the eggs were laid, and only flushed when the author was about to touch their feathers. It is strange and really too bad that the Thrasher's season of song is so brief, for the odd succession of notes are wonderful to listen to, and their variations seemingly endless.

Blue Jays, *Cyanocitta cristata,* nestled quite regularly in shrubs on the mini-wood borders and in ornamental bushes close to the author's house. The bird needs no description, but not everyone sees their nests and eggs (plate 166).

The nests found were constructed of numerous long twigs, some of which were broken from branches by the birds, who seemed to enjoy this extra labor. These foundations were overlaid with more and finer twigs nicely interwoven, and all nests examined were less than twenty-five feet above the ground. Four or five eggs completed the sets, whose ground color was either pale olive green or brownish gray, heavily marked with shades of brown and cinnamon. Full sets were recorded on May 6 (in a hydrangea bush) and on June 9 (in a nest on a horizontal beech limb); on May 29, the nest in the photograph in plate 166 also contained a fresh set, while in other cases beyond the mini-wood area, well-fledged noisy nestlings in nests were recorded on May 30 and during the first week in June.

On one occasion a pair of crows, *Corvus brachyrhynchos,* were seen carrying materials into a lofty crotch of a black oak within the mini-wood, and during another spring a pair raised a brood in a similar tall oak somewhat beyond the author's study area.

The first nest, commenced about March 23, appeared to have been completed by April 1 as a bulky mass of mostly large twigs, but was probably lined and had a neatly formed egg cup of cedar or grapevine bark, as in nests examined elsewhere, in which four to six large eggs were deposited. These were bluish green, heavily marked with various shades of brown (plate 166).

A great deal of activity followed nest construction, and during the incubation period, other Crows tried to raid the nests, causing battles, accompanied by much raucous shouting, always won by the rightful pair.

The young Crows were extremely noisy from the time they were about half-feathered out, and even noisier after leaving the nest, in this case on June 5. Later, they could still be heard squawking from various perches in the woods, and occasionally making curious vascular gurgling sounds indicating that they were swallowing large food items brought by the parents—all while trying to continue their hunger cries at the same time!

Now that people who dwell in suburbia no longer shoot at Crows, they have lost most of their former fear of all human beings, and are often seen walking about on well-kept lawns, and not infrequently coming down to take hunks of suet from the author's feeders.

They may take a few eggs or nestlings from smaller birds also, but Crows have always been a part of the ecosystem; they have a natural right to such food items. The author's attitude being "live and let live," their nesting presence was welcomed as a rare chance to study them right at home.

CHAPTER XXI

MAMMALS OF THE MINI-WOOD

Surprisingly enough, small as the area under observation was, it nevertheless proved roomy enough to satisfy and support four small year-round resident mammal species, each of which was perhaps more interesting than some other irregular, larger, and merely chance visitors to the mini-wood.

The most primitive of these resident species (but not confined to the mini-wood) was the wandering Virginia possum, *Didelphis virginiana*, a marsupial (pouched mammal), and the only North American member of the family Didelphiidae north of Mexico, a very nonchalant species; its members dwelt at various times in brush and woodpiles in the mini-wood.

About twenty inches in length when full grown, with mixed, rather loosely placed gray and blackish hair, a white face behind a long pointed snout, black ears, and a ratlike and prehensile tail, the little beast could be mistaken here for no other (plate 167).

An omnivorous feeder, these coolly indifferent little mammals often wandered about the author's house and grounds at night, feeding on the scraps thrown out for any wild takers that might come along. Not infrequently they lifted the lids of garbage cans and became trapped therein before these were enclosd in a protective house by themselves. 'Possums raided birds' nests, on several occasions consuming both eggs and young. They also relished insects, wild berries, fresh or rotted fruit thrown out for them, and as mentioned, were scavengers of garbage wherever it was available.

As we well know, many opossums, like other small mammals (skunks in particular), are killed on the roads when panicked by automobile lights, yet the former species seems to hold its own even in heavily populated suburbia and in the presence of a large population of dogs from which skunks are mostly immune.

On the female opossum's belly there is a fur-lined pouch within which her nipples are situated. Young are born in an extremely undeveloped state, and might in fact be easily mistaken for aborted embryos. Although only a half an inch or less in length, they are able to crawl and reach the pouch by means of their forelegs, which at birth are the only pair of limbs sufficiently developed for the purpose. Here we see a special provision of nature to assist these tiny creatures, which must make blind journeys from womb to the warmth of the sheltering pouch and the mother's comforting milk.

Once safely within the pouch, each young one grasps and *retains* a nipple in its mouth, and thus remains hidden and safe from predators and also assured of nourishment, whenever it is required. Thus crowded in with its brothers and sisters, it so remains until partly furred and strong enough to leave the pouch and crawl upon the mother's back.

It is doubtless the opossum's prolificacy in giving birth to two or more litters a year that has made it possible for the species to exist as it continues to, and possibly even increase in numbers, although now living in the midst of civilization. In an animal at the bottom of mammalian classification, a creature having a comparatively small and sluggish brain for its cat-sized body, and accordingly slow in its motions, it seems the more remarkable that the 'possum, a representative of so ancient a mammalian family (the

Didelphiidae), is, nevertheless, still here, living, one might say, right in our own backyards.

These mammals are expert climbers, the five toes on each foot acting almost like hands, the hind foot in particular, because on this, the inside toe, much like our own thumb, is opposable. When an opposum climbs after fruit or to a bird's nest, it can hang by his prehensile tail and thus reach to branches that would not bear its weight.

As many as ten young may be born to the 'possum at one time, but six has been the number in the author's experience. As litters may come quite close together, a mother may be found carrying one litter on her back, and another still younger litter in her belly pouch at the same time. In such cases, the prehensile tail of one or more of the riders may be wound around the lower part of the parent's tail, while the rest of the litter hold onto her fur.

Up a rung or so in the mammalian ladder, we come to the moles, two species of which occurred on the borders of, but rarely within, the mini-wood, while both were also occasionally in evidence by their mounds in lawns and swampy areas nearby. They were the eastern mole, *Scalopus aquaticus,* and the less often seen starnose mole, *Condylura cristata.* Both have forelegs with claws strongly adapted for deep digging and forcing their way even through tough sods.

The eastern mole, averaging about five inches in length when full grown, possesses beautiful gray fur with a silvery sheen, reminding the author of very fine sealskin. The tail is short and *naked,* and the palms and claws are turned curiously outward in this species (plate 168).

In the starnose species, the fur is much darker, ranging from brown to almost black at times. The animal is about an inch shorter than the common mole; its tail, bearing hair, is much longer, and is thickest near the center and constricted near its base. About this mole's nose, twenty-two curious, fleshy tentacles extend in a circle, a feature which gives the species its name and unmistakably identifies it (plate 168). The snout tentacles are no doubt useful in detecting subterranean food items such as beetles and their larvae (grubs) and earthworms, organisms relished by both moles, and prey sought by a great variety of animal forms, as we have seen in these chapters. (Indeed, earthworms seem to be almost a universal food, and of enormous import to ecosystems in many parts of the world.)

To a considerable extent aquatic, the starnose mole's occasional presence on the mini-wood borders (and doubt-less its penetration of the woods, at times) was probably due to portions of the area remaining wet or even saturated for considerable periods each year, and with actual swampy areas nearby. The starnose mole being an excellent swimmer—and often a burrower in mud—such spring-saturated land would be much to its liking. The ample supply of earthworms to be found in this habitat constituted an additional lure which no doubt also had much to do with the mini-wood visits of both species.

Unlike marsupials, of which family the opossum is a typical representative, moles are *placental* mammals like ourselves, creatures fed and to a high degree developed in the mother's uterus before being given birth. Although these moles may be delicate and hairless at birth, as are rodents such as mice and rats, they are *not* hardly more than embryos at birth, as are the opossums.

Never having been fortunate enough to find a nest of either species of mole, the author cannot say how many there are in a litter, or exactly what they look like.

Intriguing little residents of the mini-wood and areas beyond it were the short-tailed shrews, *Blarina brevicauda,* diminutive placental mammals. Though common enough, they are rarely seen by the casual observer, and their presence is probably not even imagined by the majority of human beings. Difficult to maintain in captivity or study intimately, a special effort was made to obtain detailed photographs of these curious little mammals (plate 169).

Three to four inches in length and, as the animal's name implies, with a tail adding but an inch or less to this overall dimension, this species possesses fine dark gray fur resembling that of the moles, smooth and soft, with a sheen that changes according to the light, causing silvery waves on the pelage (plate 169).

Easily mistaken for mice by the uninitiate, shrews are classified next above the moles. Compared with mice, the dentition (character of the teeth, in form and arrangement) is radically different, the common shrews having thirty-two teeth that differ greatly in form and arrangement from the sixteen of mice. The very obvious chisellike incisors of the latter rodents are conspicuously absent in the shrews. It was also noticed that the teeth were stained black. Their tiny eyes and lack of visible ears will also help to identify them (plate 169).

Living mostly under logs and woodpiles, or under woodland debris and rock piles, and traveling in little passages beneath the dead-leaf carpet, they may be seen occasionally in daylight scampering from one shelter to another, but generally they come out to feed only at night.

They are fiercely carnivorous little mammals, feeding upon insects, including large beetles, and also earthworms, young mice, and probably an occasional nestling bird. One individual kept for observation and photographed in a terrarium, and given a large supply of earthworms, was so ravenous and greedy and ate so many at a single "sitting" that it suddenly just keeled over and died an hour later!

Shrews caught and held in long forceps fought valiantly to be free, at the same time emitting an intensifying and rather disagreeable musky odor which might be a protective measure.

One clever short-tailed shrew appeared on many nights during July and August to feed on moths and beetles that had dropped to the ground after hitting the three-hundred-watt electric light bulb outside of the author's laboratory. Watched from the doorway, it would scurry for shelter at the slightest human move, surprising behavior when remembering the shrew's tiny eyes, usually obscured by the fur. When things seemed safe again, the little mammal would hurry forth, pounce upon a moth or beetle or other flopping insect, and then either consume it on the spot or, for greater safety, carry it to shelter.

In the economy of the mini-wood, where many shrews dwelt, they were important in helping to control boring beetles, the larvae of some small foliage-consuming moth larvae, and, possibly to some extent, the meadow vole and its young, a species occurring only rarely in or near the mini-wood but a nuisance mammal in some more open meadowlands. The shrew descends into the vole's burrows and nests, attacking and devouring both young and adults.

Owls seemed to be the only predators taking these shrews (possibly other than domestic cats), at least within and about the mini-wood, although snakes, weasels, foxes, and other mammals doubtless prey upon them elsewhere. Pellets containing fur and parts of shrew skulls, mouth-ejected by owls, were found on rare occasions actually within the mini-wood.

Never hibernating, active all winter under and on top of the snow, shrews turn to vegetable food at such times. The author often found little round holes from which shrews had emerged to travel *over* the snow to reach weed stalks, bushes, and the bird feeders. Their neat trails could be identified by the fact that they were rounded and smooth on the sides, little half-tunnels made as the mammals crept through the fresh snow, quite unlike the regular series of footprints left by winter prowling deer mice. Under one suet feeder, these shrews also ate the crumbs dropped by the birds.

No nests were found in the mini-wood by the author, but a litter of two-thirds-mature young shrews, which were brought to the museum for identification, measured just under two inches in length, not including their tails. This was on June 15, at which date their bodies were very slender, and their snouts less pointed than when adult. Their fur was already very smooth and beautiful, but of a much lighter shade of gray than that of mature individuals. Probably not fully weaned and still unable to hunt and feed properly by themselves, all died soon after being dug from an old celery bleaching embankment from which the nest was not recovered.

Shrews, in general—and there are numerous species and subspecies in various parts of the United States and the world—have greatly interested zoologists, one good reason being that they are living representatives of ancestors which were among the very earliest placental mammals whose physical aspects and modes of life were very possibly much like those of our "backyard" shrews of today.

Other stirring facts came quickly to mind when holding one of these tiny creatures (with such a venerable lineage) in one's hand. It was born warm blooded, from the womb of a warm blooded mother; it was suckled in its infancy with *milk,* as human babies are, facts which clearly point to our kinship with these little mammals. Recognized also was the pleasurable truth that the tiny shrew and the comparatively gigantic mammal holding it were both vivid manifestations of the eternal energy and warmth of our sun. It was such thoughts as these about the small fry of the mini-wood and its surroundings which brought great satisfaction to the author.

We come now to those curious mammals, the bats, creatures disliked instinctively by many people, and feared by those who still look upon them as the malevolent animals of fiction.

Warm-blooded mammals minding their own business, and with every right to occupy and breathe the same air that we do, they really far surpass us in possessing *natural* wings, due to which they are not only able to catch, eat, and thus rid us of innumerable detrimental airborne insects, but may when convenient, carry their babies aloft, suckling at their breasts, while catching this food for themselves! In addition to these accomplishments beyond human ability, bats avoid aerial collisions with others of their kind and also with branches and wires and other obstructions by means of *built-in radar* equipment, seemingly situated within their ears.

In view of the countless flying insects and the thousands of species of birds, and not forgetting that there once were flying reptiles of great size, it is strange indeed that alone among all the mammals, only bats evolved wings, leaving proud mankind dependent upon machines for flight!

Four species of bats occurred within, above, or around the borders of the mini-wood during this fifteen-year study. The largest and rarest is illustrated in plates 170 and 171.

Of the smaller species, the little brown bat, *Myotis lucifugus,* was the most frequently seen. Recognized in hand by its glossy-tipped brown dorsal fur, paler hair tinged with yellow below, and a ten-inch spread of nearly hairless wings, the species indeed became a tiny mammal when its wings were tightly folded. As with other bats, little hooks for climbing occupied the position where thumbs would be in much higher mammals.

Little brown bats were mostly observed at dusk coursing back and forth along the mini-wood borders, or swooping over the author's aquatic plant and goldfish pool in quest of gnats and mosquitoes and other insects flying over or arising from the water. Since they also feed on moths and flying beetles, these bats should be encouraged, not "broomed" out, or killed when found roosting or sleeping the daylight hours away in the garage, or behind some screen or shutter, or maybe in little groups tucked away in one's attic, which they might enter by way of a broken louver.

Bats are very clean little animals personally, often licking every part of their rubbery wing membranes, twisting this way and that, and even occasionally stretching a wing tightly over the head for some reason unknown to the author. One will seldom if ever get lice from handling bats, but they do carry very curious and tiny parasitic flies of the family Streblidae, which have become adaptively flattened for life in the mammals' fur, but these insects do not crawl upon and bite human beings. It may be that the bats *eat* some of these tenants during the grooming process.

The author once observed a little brown bat making **sorties** and returning to a certain small tree and what were **two little** restless dark objects clinging to a twig. Suddenly gone, after one of the flying bat's visits, these had undoubtedly been "parked" young that had finally been taken "aboard" again. This trait of leaving the offspring by themselves as with growth they become heavy burdens for the female probably is common to other species, including the big brown bat, *Eptesicus fuscus,* also observed about or over the mini-wood.

With a wingspread of twelve inches, this bat's fur was dark brown and its wing membranes very black; otherwise it seemed to be simply a larger edition of the little brown bat, except for its facial expression. But it does, of course, belong to a different genus.

On hot summer nights, the big brown bats were occasionally seen flying at dusk high over the woodland canopy, where their erratic courses, their rapid turns, and sharp dives and rises probably indicated that there were marriage flights of ants or termites upon which the bats were feeding.

All of our eastern bats require water daily. In very dry weather they go straight to ponds (probably located earlier) where they fly down, crawl to the water's edge, and lap the fluid with a very rapid motion of the tongue. In normal and wet weather, it is the author's belief that they lap the raindrops which accumulate on the leaves, an observation, however, that was only verified with a bat in captivity.

The third small species seen in the area under observation was the red bat, *Lasiurus borealis,* a species of still another genus, with a wingspread of about eleven inches, and the most distinctively colored of all the eastern species.

In the male, the fur varies from rusty red to an almost orange red at times, but lighter on the ventral side and with the hairs more or less generally tipped with white, thus giving the animal a well-termed "frosted" appearance. Lighter fur occurred on the shoulders of specimens examined at the Bruce Museum, and in living specimens actually handled, other white-tipped hairs formed one or more bands in the area of the nape. Females were duller colored, but still exhibited the "frosted" appearance.

The author's pool, often alluded to and situated on the edge of the mini-wood, occasionally attracted a red bat which came for the insects which habitually fly over water, as when adult, winged mayflies transform from the aquatic nymphal stage and issue into the air in hundreds or thousands all at the same time.

*   *   *   *   *

By great good fortune, four living examples of the largest and rarest of our eastern bats came under the author's observation. These were hoary bats, *Lasiurus cinereus.*

One of these found within the mini-wood was hanging by its feet, from the twig of a pin oak tree, head down, in the customary bat manner, on a cold day late in November.

Stiff from the cold, it was gently removed from its perch and taken indoors, where, after an hour and a half, it had not only revived but had become quite pugnacious. Later in the day, having warmed up considerably, this bat was taken out and released, and was watched as it flew out of sight southwestward.

Experimenting with the angry little beast before letting it go, its defensive actions were seen to consist of displaying its very pink gape and rows of interestingly shaped, very sharp white teeth, while at the same time performing a series of motions that could only be described as "push-ups." While resting, either on horizontal or oblique surfaces, with wings folded, it raised and lowered itself over and over again on its elbows, strongly reminding the author of a weary recruit endeavoring to do the required number. It made no sound at any time as it opened and closed its mouth, but no doubt it would have severely bitten any living thing that came too close to those strong double teeth (plate 170).

During this display, the edges of its mouth were seen to be outlined with black hair, which served to emphasize the pink interior and the white teeth. A sudden confrontation with this colorful gape, tongue, and dental equipment might very easily and promptly frighten off any would-be small predator, as the three pictures in plate 170, enlarged from 16-millimeter motion picture films, plainly attest.

The fur of the hoary bat, according to one recent *Field Guide,* might be "yellowish brown to mahogany brown with hairs tipped with white over most of the body," and in another *Field Guide* it is described as "dark brown to light yellowish brown, more or less tipped with silvery white."

In the specimen captured within the mini-wood (see plates 170 and 171), as well as in others mounted by the author for the Bruce Museum, where they are on display, the dorsal fur was streaked with black and white and pale gray, and there were heavy ruffs of fine grayish white, black-tipped hair standing up conspicuously around the posterior part of the head, while some of the hair on the forehead was distinctly buff, and some of it black with white tips. The throats of these bats were buff, their prominent ears outlined with black hair, and the upper side of their rather sharply pointed tail membranes well furred.

In addition to the clawed feet, all bats are equipped with short, sharply hooked wing-claws or "thumbs," which are very necessary assets used in climbing and shuffling over the ground. The finger bones other than in the thumbs have evolved into long slender supports for the wing membranes, but are true finger bones nevertheless, the joints of which may be readily seen. The various wing bones and the clawed thumbs may be seen in the illustration of the hoary bat in plate 171.

Within the ears of all bats there are more or less leaf-like membranes arising externally from their bases, one such projection in each ear, and varying in form according to the genus or the species. This fleshy object is called a *tragus,* and is part of the sensitive sonar equipment of these flying mammals which makes it possible for them to avoid objects in pitch darkness. It is probably also useful to some extent in detecting food.

The hoary bat is apparently migratory, but quite hardy, as was attested by the four specimens with which the author has had first-hand experience, one of which was found clinging to the branches of a shrub late in November, as mentioned above and one earlier in that month.

A very interesting fact in regard to one of these late migrants was that its stomach contained cankerworm moths, an insect species which flies in November and December, when most all other insects are absent, and which are thus available to late transients such as these hoary bats. We have here another suddenly revealed interrelationship, nicely timed to be of benefit to the bats—and thus to the trees on which the moth's larvae feed—since in some small measure the bats help control these insects by reducing the number of males flying to the wingless females.

Certain people who once brought a hoary bat to the author for identification warned that such animals were "crawling with lice."

Now, the author has collected and handled large numbers of bats of many species, at home and in Central and South America, and while it is true that they may have certain ectoparasites,* belonging to the order Mallophaga,

---

*Ectoparasites (species living externally) do not suck blood, but feed on feathers, skin scales, and hair.

at times, and that most bats, as already stated, support curious tiny flies belonging to the families Streblidae and Nycteribiidae, which live in their fur, these insects stay very close to their specific hosts as long as the hosts remain living. When the hosts die and their bodies become cold, such parasites naturally move out, seeking warmth elsewhere, but soon die themselves unless again reaching specific hosts. In the long experience of the author, bat parasites never adopted *him!*

On the other hand, some birds—the House Wren and Barn Swallow, as examples—are often, as the people mentioned above said of the bats, "crawling with lice." Especially may this be so during the nesting season, when many other kinds of birds also become heavily infested with Mallophaga (lice) belonging to the family Philopteridae, which will crawl in dozens, and at once, upon anyone handling nests or nestlings of species so infested.

Only two representatives of the order Carnivora, or true flesh-eating mammals, occurred in the mini-wood during this study. These were the raccoon, *Procyon lotor,* and the striped skunk, *Mephitis mephitis.*

Both mammals being great roamers, neither was confined to the mini-wood portion of the author's property, but both were found actually established therein at various times, that is to say, having there made their dens. Raccoons are distributed over most of the United States, and skunks being even more widely distributed, are doubtless the best-known little animals of all.

Most of us know that raccoons stick their sharp snouts into everything. They turn over logs and rocks, dabble in water and mud, unfasten hooks to get into garbage huts, go into buildings (even down unused chimneys)—anywhere, indeed, in search of food.

One aged and fat monster raccoon broke through the roof of the author's hut in which the garbage cans were kept, and later made a hole into the structure through the siding after the roof had been repaired. As time went on, this 'coon discovered the large population of Japanese snails that inhabited the lily and goldfish pool. All at once the author found small groups of cracked and empty shells on the top of the wall surrounding this pool where, during the nights, this individual fed to repletion on *escargots* of a sort. Something had to be done before all were consumed, for it had taken much time to establish the snail colony from the few original breeders.

A good sized "have-a-heart" trap, baited with snails, did

the trick, and on the first night thereafter, a patriarch 'coon was captured (plate 172).

About thirty inches in length, not including the rather sparsely furred tail, he very nearly filled the trap. Grizzled and gray elsewhere, his fur was nearly worn off in places. Unlike a trapped gray squirrel, he was philosophical and calm. Exhibiting no sign of anger or fright, he looked out at the author as if to say, "So what?"

It was too bad to take him away from the environment in which he had doubtless been born and had lived all his life, but the snails and frogs were not in the pool in unlimited numbers, and in this case had to be preserved even though the raccoon's raids on them were simply in the nature of things. His sentence: release near a large natural pond and nearby wooded ridges thirteen miles north of the mini-wood. When let out of the trap, this old raccoon ambled slowly toward the woods without a backward glance, and without hurrying in the least.

The author and some of his neighbors often fed smaller 'coons, throwing out scraps and steak bones and parts of chicken carcasses; thus the raccoons could not be blamed for their other raids.

Becoming tame accordingly, they sometimes brought their litters with them, and young raccoons were occasionally found wandering in the mini-wood alone.

Frequently these animals got into fights at night, perhaps over a steak bone or the trimmings from a roast, or an especially relished ear of sweet corn-on-the-cob tossed out for them, or again, maybe over a fat frog captured at a neighbor's pond.

At such times, their odd, high-pitched chatterings and grating screams over the possession of food were intensely interesting to listen to, sounds which again brought the pleasant realization that wild things still choose to roam and fight and breed, even here on the borders of megalopolis.

A raccoon once took up its abode within the only large, partly hollow tree within the mini-wood, an aging beech in which the cavity later became so much enlarged by rot and the additional excavations of carpenter ants, that a raccoon could no longer stay put in the resulting central perpendicular shaft.

Where the many individual raccoons which still come to houses in highly populated areas hang out in the daytime, remains a mystery to the author, but they must have dens somewhere nearby in which to safely bear their litters, for there are probably more of these animals now than there were fifty years ago, when most people hunted them. Long live the raccoons! (Plate 172.)

And as for the striped skunk, the little beast is familiar to all of us in one way or another. The air of many a spring and summer night, ambient with the skunk's strange and characteristic scent, the product of specialized glandular activity in these viable little mammals, is a unique protective provision of nature, this "personal" smell which not only announces the owner's approach well in advance but quickly puts all other animals, including man, on guard. Held in instant readiness in solution, the contents of the gland may be sprayed as a foul mist, and with devastating effect upon any unwitting or venturesome enemy. It is an odor that easily penetrates doors, windows, solid walls, and is, without much doubt, among the most long lasting, once it is sprayed upon a victim.

The author once saw a tiny skunk that had lagged behind a party of several other pups following the mother. With its tail erect and stamping its little feet in warning, it stood confidently, backside facing a huge old cow in a pasture. A naturally curious animal, the cow seemed to be fascinated, standing still, its head lowered and almost touching the baby skunk invader in its meadow.

But nothing more happened, that is to say, nothing of a more serious imminently expected nature, for after all, a skunk may be reasonable if an opponent behaves itself. The comparatively tiny mammal simply outstood the cow, who suddenly turned away and hurriedly made off toward less odoriferous companionship.

Skunks are great destroyers of insects, fattening up in the fall on grubs, and on succulent members of the grasshopper families. They also devour frogs and snakes, and turtle eggs which they dig up. Birds' eggs and nestlings, young mice and rats, and meadow voles make up other portions of their diets, while scraps of bread, meat or vegetables, purposely thrown out for whatever nocturnal animals came along, were selectively consumed by the skunks, raccoons and 'possums. Really omnivorous, both skunks and raccoons will also avidly lap milk, when offered to them.

In May, or later, skunks give birth to litters of four or five pups, and no sight was ever more intriguing than a mother leading her babies, still clumsy and stumbling behind her, but already confident in themselves, as the little parade passed through the mini-wood near a brush pile in which she may have whelped them.

* * * * *

The order Rodentia, or gnawing mammals, was represented within and about the borders of the mini-wood by six species, five of them more or less regularly. Rodents are characterized by having two powerful, curved, chisellike incisor teeth above, and two shorter ones below. When the animal gnaws, the pairs move against each other in such a way that the movement tends to keep them sharpened. One may visualize this dental action by recalling the appearance of a feeding rabbit or squirrel. The teeth apparently keep growing from above as their ends are worn down.

The familiar woodchuck, *Marmota monax*, famous for its appearance or non-appearance on Groundhog Day, strangely enough came on one or two occasions to live for a time under the same brush pile within the mini-wood in which the skunk (already mentioned) may have whelped a litter earlier. In the later years of the study, however, *M. monax* went elsewhere to dwell, probably to share in some more lucrative area near a large garden.

With a heavily built body on short legs, a woodchuck (weighing up to ten pounds) can consume a lot of garden truck. The biggest specimens may measure nearly twenty inches in length, not including the six- or seven-inch bushy tail. Two color phases occur, the common one, yellowish gray or yellowish brown, and the rarer, *melanistic* phase, in which the pelage may be almost black. The woodchuck's feet are always black (plate 173).

Often excavating a deep burrow near where there is an ample supply of clover and other grasses, except for a few daily sorties to feed, and sunbathe in fair weather, much of this rodent's time is passed below ground, sleeping. The winter is passed obliviously in hibernation, no doubt until Groundhog Day or very much later.

It has been stated in one book on North American mammals that, as soon as weaned, young woodchucks are driven from the burrow by the "hard-hearted" parents, and must thereafter fend for themselves. While this may be true usually, the author has observed a mother woodchuck followed by four young ones, though not as closely as young skunks follow theirs. The five woodchucks were traveling along the base of a stone wall bordering an old field which was still rich in weedy hay and some clover as well.

Three species of squirrels inhabited the mini-wood for varying periods during this fifteen-year study.

Occasionally, the eastern chipmunk, *Tamias striatus,* made its home among rocks on the edge of the woodland. One chose a crack in stonework backing the author's cinnamon fern display, and employing this fissure as a main entrance, continued its burrow well into the earth back of this wall.

For those few who do not recognize the chipmunk on sight, this species inhabiting the eastern United States may be identified by its small size—five to six inches in length—and by its bright stripes—a dorsally centered black one, with a buffy stripe between two black ones on either side of the body (plate 173). The little mammal runs very fast, often with its tail held straight up.

T. striatus is a rodent typically inhabiting hardwood deciduous forest land, but it is also found in more open brushy areas, and as we know, very often about country homes where bird feeders are maintained, especially homes with woodland backdrops.

It is interesting that these little rodents, although constructing deep tunnels, one of which ends in the grass-lined sleeping chamber, and others in hollows for food stores and for excrement, never accumulate piles of earth around the burrow entrances. Habitually, chipmunks, when collecting and carrying foods for winter consumption, pack these materials in their cheek pouches, and it is logical to suppose that they also remove the dirt from their excavations in this same way, dumping it at a distance and thus eliminating telltale mounds near the burrow openings. These little squirrels hibernate in winter, occasionally coming out of this deep sleep to feed on their accumulated stores safely and snugly below the frost line.

Active again early in April or early May—whenever the weather remains warm—it is no doubt at such times that pregnancy occurs, as the author has observed the beautiful tiny young chipmunks scampering playfully about near the burrow entrance by mid-June.

These squirrels eat many kinds of wild and cultivated berries and their seeds, including the bright red beads of the wintergreen, and also the pits of wild cherries. In spring and summer they take large grasshoppers and other succulent insects, and like most other squirrels, sometimes birds' eggs and embryos and tender nestlings, natural and rightful acts which for centuries never upset the woodland scheme of things as man has upset the environment so grossly.

As an amusing experiment, the author placed a whole flat of decaying and fermenting domestic grapes (weighing several pounds) within a short distance of a chipmunk's mini-wood burrow among the rocks. The provident little animal was not long in locating the offering, and although carefully observed for many hours, it was never seen stopping to eat any of the decaying pulp of the fruit, but with enormous patience it gathered seed after seed into its cheek pouches, only leaving for its burrow when appearing as if it might have a bad case of the mumps. Making journey after journey, it collected every last seed, accomplishing the task, with unexplained intermissions, in about a day and a half.

The chipmunk of course has its own enemies, but being so alert and so rapid in its movements, probably may often escape even Cooper's and other small swift Hawks. Squirrels in general set up warning chatterings and special cries the second a natural predator is sighted in their territory. Chipmunks follow this rule, all other small woodland denizens benefiting therefrom. Where the weasel is a member of the wild community, however, the chipmunk and its family doubtless suffer accordingly, but to such juicy morsels, this ferocious streamlined little predator has a natural right, as "chippy" has to a nestling bird.

About the ubiquitous eastern gray squirrel, *Sciurus carolinensis,* little need be said here. Everyone knows it by sight; everyone who has a bird feeder feeds it whether he or she likes it or not. One just cannot outwit a gray squirrel, once it makes up its mind to outwit *you.* In one way or another it will get at your wild bird food, but as the author has stated many times before, every animal organism, large or small or microscopic, has a right to eat and live as best it can until its own natural predator exerts control. (Plate 174.)

While the usual color of this squirrel's fur is gray, or yellowish gray with partly black hairs, not everyone realizes that the very dark brown to black individuals occasionally seen here are melanistic squirrels of the same species, a form occurring commonly or predominantly in other parts of the country.

There seem to be two litters of young, one of them born in spring or early summer, and then again one sometimes sees very small squirrels playing and chasing one another in the fall. Possibly all of these simply mean there are single annual litters born to different mothers at widely different times, but the author cannot answer this question with certainty. Most males, however, are seen persistently

Left: Hoary bat, dorsal view, somewhat enlarged, showing fur patterns and fur ruffs at back of head (at bottom of picture). Fur blends well with bark on which bat is resting. *Below:* An example with a fifteen-and-a-half-inch wing spread, mounted for the Bruce Museum by the author. Note the wing-hooks, or "thumbs," with which bats are equipped for climbing. (Page 201.)

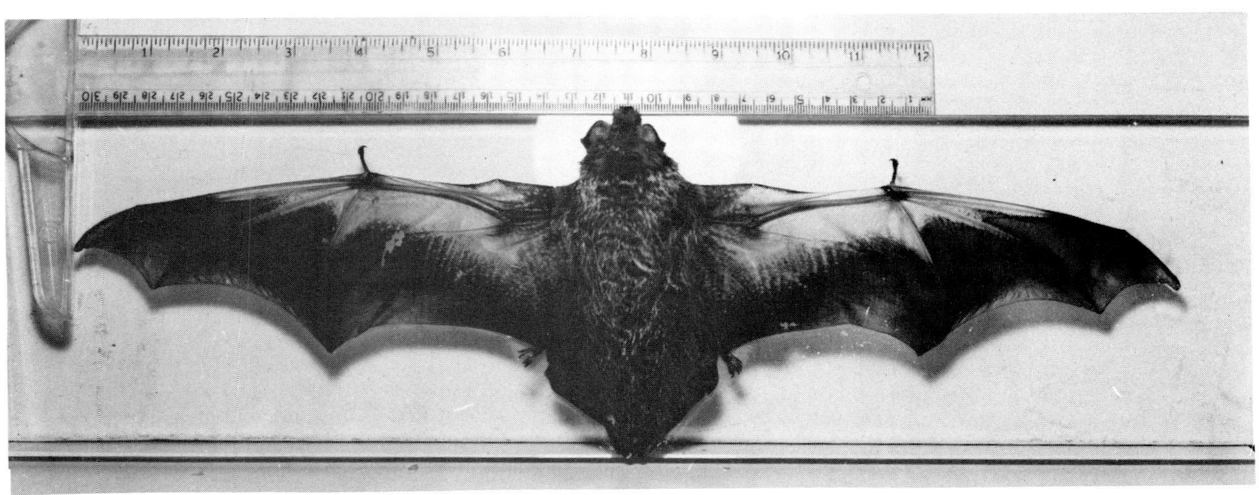

Baby raccoon gets cuddled. The lady is Olivia St. Germain, secretary and head of the clerical staff at the Bruce Museum. (Page 202.)

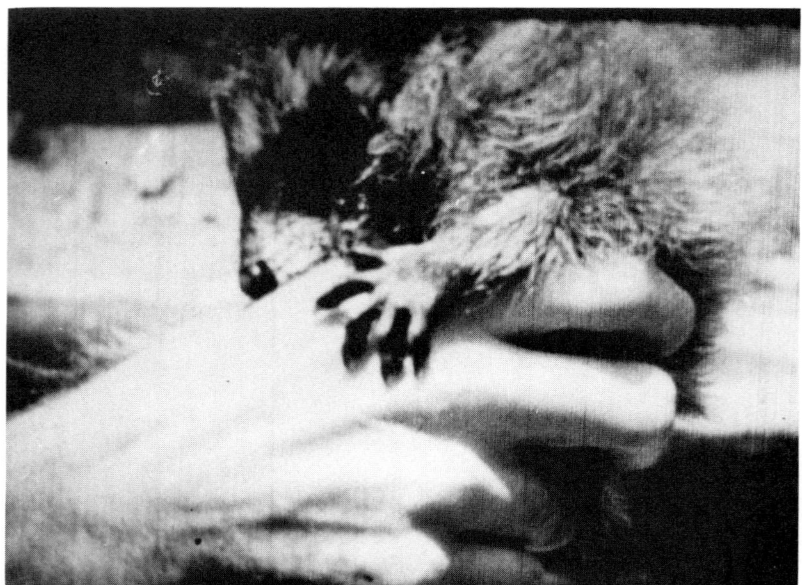

Above: A baby raccoon suckles the hand of Richard Crawford, head custodian at the Bruce Museum. Note the opposed thumb of this animal, which gives the raccoon such great capability with its hands. Date: May 18. *Below:* The patriarch raccoon that tore a hole in the roof of the hut for the author's garbage cans looked like this old-timer (mounted for the Bruce Museum), except that its fur was faded, thin, and absent in places. (Page 202.)

Left: The woodchuck moves with a curious undulating gait, alternately humping and stretching out its broad body, as shown by these two frames from a motion picture. *Below:* An adult female and her young at a burrow entrance, in a diorama by the author. (Page 203.)

Author holds half-grown opossum by the tail, designed by nature as a useful prehensile appendage. This treatment did not hurt the animal. Date: May 29. (Page 197.)

Above: The thinner portion of the mini-wood. Chipmunk had a burrow beside the large angular boulder. An occasional opossum wandered in, gray squirrels made many visits for the acorns, as did deer mice for the wild cherry pits, and three flying squirrels had a nest in the box at arrow. *Right:* Gray squirrel just after leaving nest. Date: August 1. (Pages 204, 205.)

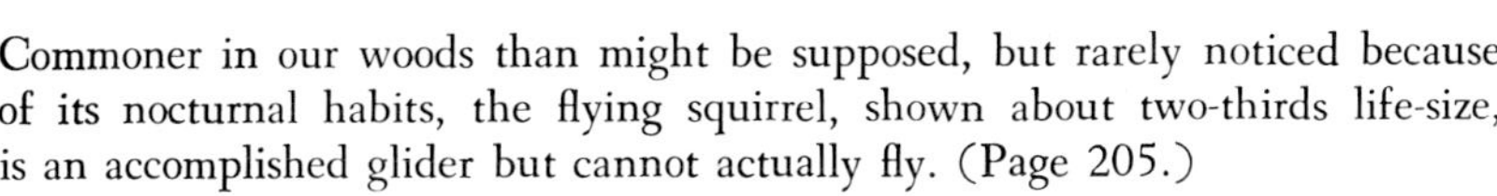

Commoner in our woods than might be supposed, but rarely noticed because of its nocturnal habits, the flying squirrel, shown about two-thirds life-size, is an accomplished glider but cannot actually fly. (Page 205.)

Flying squirrels kept snug and warm together all through one winter in the mini-wood in this nest, shown natural size. The squirrels constructed it entirely of shredded inner bark taken from a dead basswood tree. (Page 205.)

PLATE 175

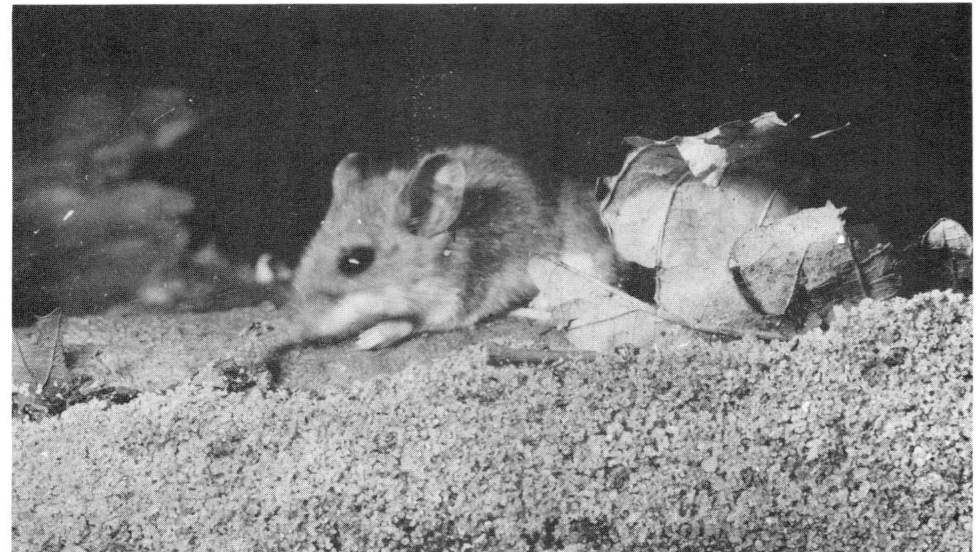

Deer mouse cautiously comes out to feed at night. All mice illustrated on this plate are shown about two-thirds life-size, although species may grow (not including tail) to four inches in length. (Page 206.)

Deer mouse may easily be mistaken for the white-footed mouse, but the former (shown here) has a bicolored tail and white undercoat with dark fur above, whereas the latter has a more uniformly dark tail. (Page 206.)

Pregnant deer mouse, found injured, died the day after giving birth to three robust young. (See plate 177.) (Injured small rodents are shown to emphasize constant predator problem.) Date: May 24. (Page 206.)

Below: Deer mice born May 24 of injured mother shown in plate 176. Young were hairless, with grayish pink skins. Enlarged three times life-size. (Page 206.) *Right:* Barely weaned deer mouse found wandering alone in the mini-wood, being warmed in the author's hand. Date: June 15. (Page 206.)

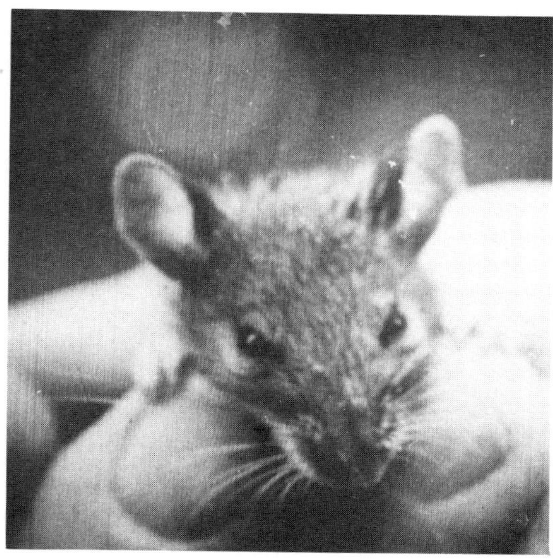

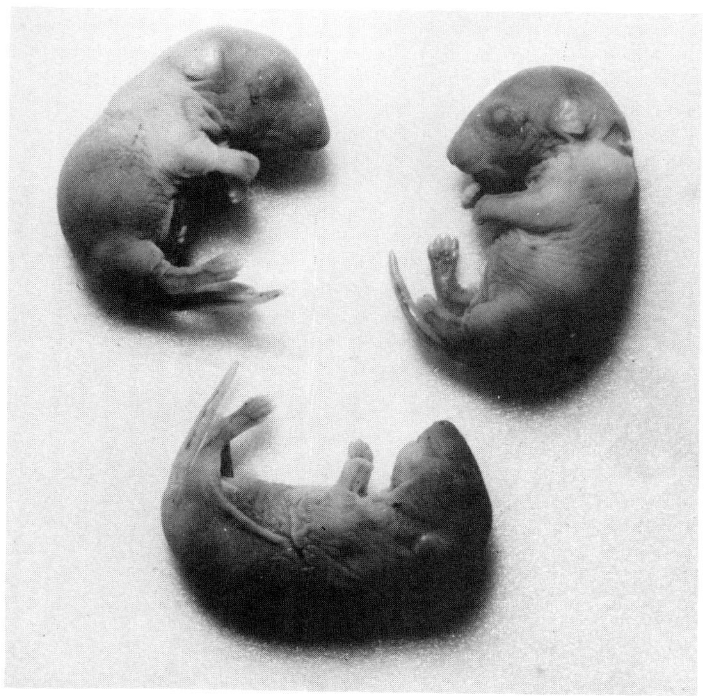

Above: A white-footed mouse feeds on a winged maple seed. (From an exhibit mounted for the Bruce Museum.) *Left:* A badly mauled meadow vole hides in the grass. These little mammals are the natural prey of many other mammals and of hawks, owls, and black snakes. Shown about half life-size. (Page 207.)

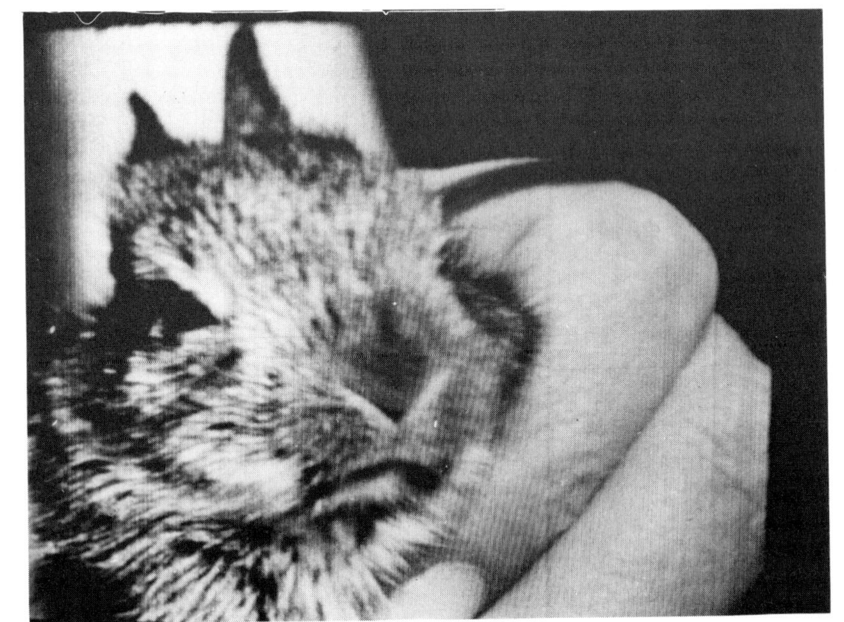

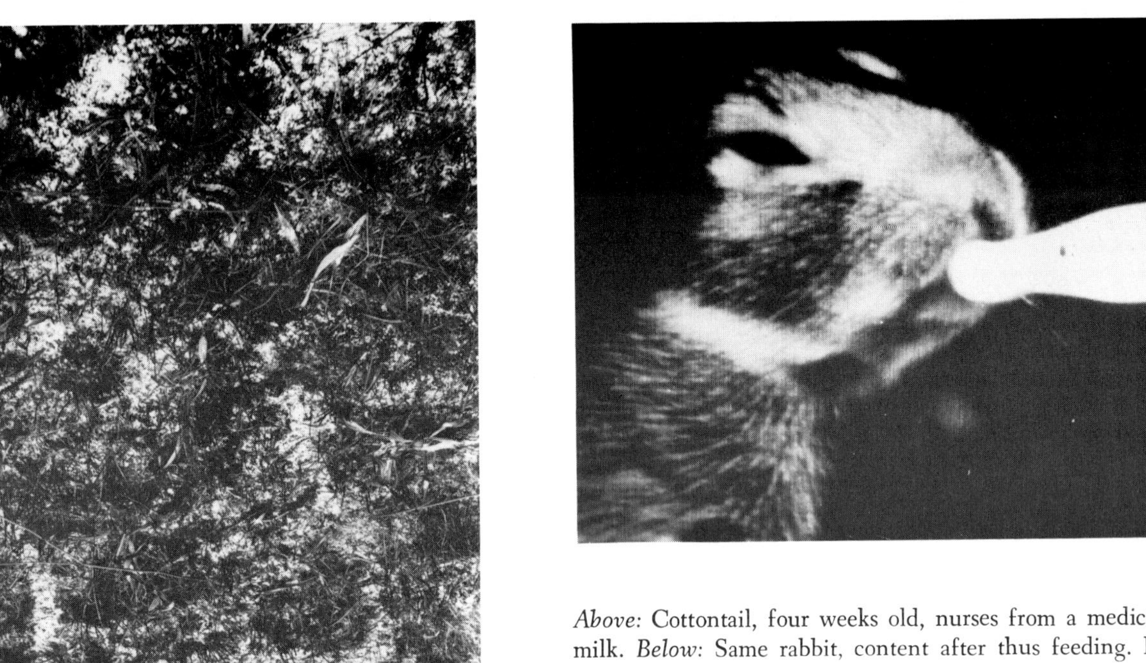

Above: Cottontail, four weeks old, nurses from a medicine dropper of diluted milk. *Below:* Same rabbit, content after thus feeding. Date: June 15. (Page 208.)

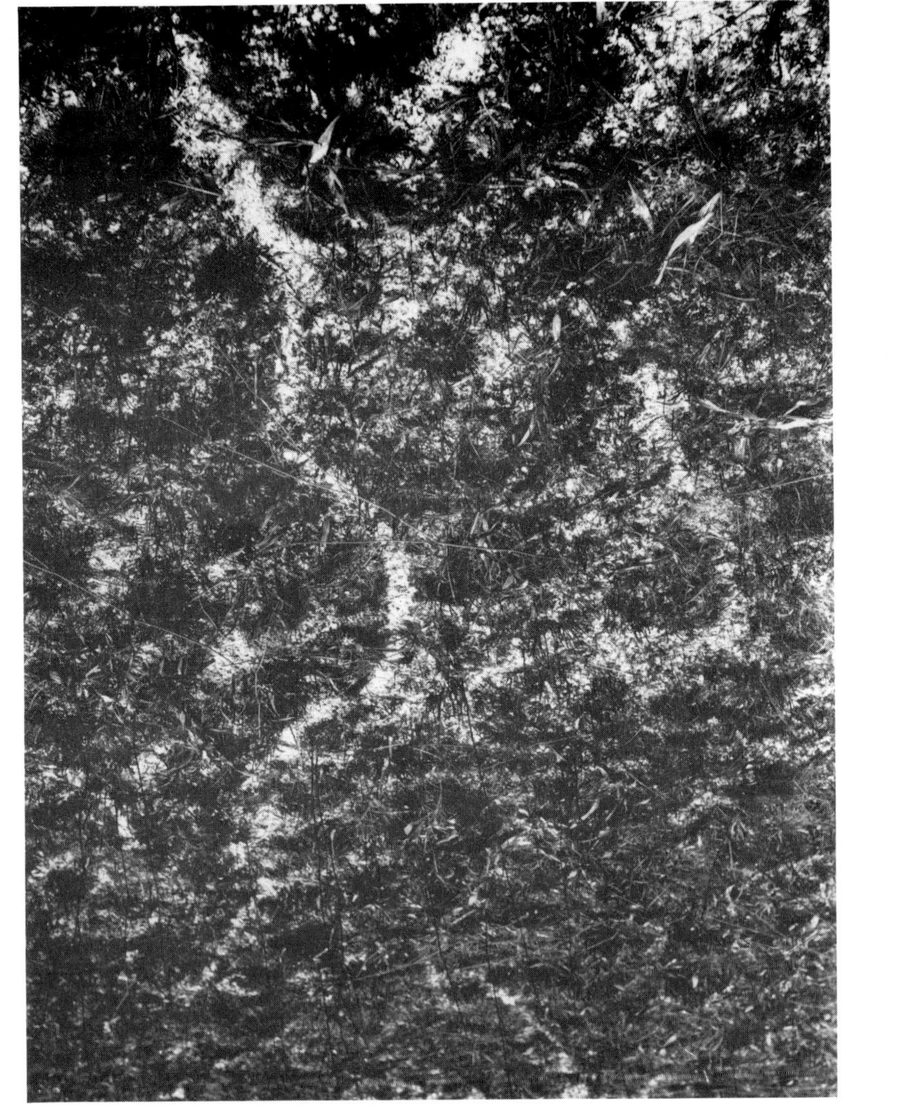

Grass, eaten down to its roots, reveals resulting roadways of the meadow voles in burned-over land. (Page 207.)

following and "hounding" the females in spring, and seeming to show little interest later.

Within the mini-wood, and especialy along its borders, and often in isolated trees in the open, these squirrels build their bulky nests of dead leaves and twigs, usually high up in the taller trees. Where birdhouses have been put up in numbers, as about the author's grounds, they also try to appropriate some of these for nesting purposes, and would succeed in enlarging the entrance holes with their chisellike teeth, were the boxes not protected with metal, as described elsewhere in this book (see plate 134).

Gray squirrels build their nests in spring or fall. As late as November, the author has witnessed such activities in full swing by tardy individuals heeding the sudden warning of cold nights. In one case, *two* gray squirrels were observed building a nest together, carrying great mouthfuls of dead leaves to a crotch high in a mature beech tree. When the bulky affair was nearly completed, it was snugly lined with shredded basswood bark prepared in this soft form by one of the squirrels. In this observation we have one more example of the uses to which dead material is put by animal organisms of the mini-wood.

Other, rather odd, observations on the gray squirrel include an individual building a nest of *wet* dead leaves in the middle of May, and a group of six quite small ones playing about the trunk of a tree and chasing one another over the crusted snow on January 31!

A year-round resident of the mini-wood and most woodlands in this area, yet seldom seen by most people, is perhaps the most beautiful, docile, and secretive of all our native rodents. This is the flying squirrel, *Glaucomys volans*, actually commoner than is generally realized because of its mostly nocturnal habits. Spending the daylight hours tucked within a natural cavity in a tree, in a birdhouse, or an abandoned Woodpecker's excavation, it comes forth at dusk, climbs higher, if need be, and then, by stretching its four limbs widely and thus pulling taut the ample skin between them, forms a plane by means of which it can sail or glide (downward only) for considerable distances between trees, especially where these are large and unencumbered by smaller growth about their bases. "Flying" squirrels cannot actually fly, bats being the only mammals in the world that can. But when one has studied these little rodents and observed their enormous energy and love of gliding, as when in a zoo, under infrared lighting, which they seem to assume to be the equivalent of darkness, one

gains the impression that they are actually *trying* to fly. Under normal conditions, at dusk in the woodlands, they are among the most active of its small mammals, and being gregarious, two or three sometimes issue from the same nest.

In color, this (southern) flying squirrel is olive brown dorsally, glossy, and sometimes tinged with russet. On the ventral side it is pure white. Its fur is short but thick, and beautifully smooth and soft to the human touch, and the squirrel's big black eyes lend it a confiding and appealing expression. The prominence and size of the eyes indicate their efficiency after dark.

Overall lengths of head and body in a number of individuals measured by the author varied only from five and a half to six inches, but the rather wide *flat* tail added from three and a half to four inches more (plate 175). The mammal's folded layer of rubbery skin between the legs on both sides, the flattened tail, and the lightness of the body (a mature sample from the mini-wood weighing only two and a quarter ounces) all contribute to the efficiency of the animal's gliding ability.

Every now and then, during the author's inspections within the mini-wood, a tap on the bottom of a box might result in a flying squirrel's hurried exit and a glide to a nearby tree. Scrambling a few feet upward, it would remain there for a short time, its body flattened out and its four legs stretching the skin between them taut, and thus, motionless, its fur blending very well with the bark. At such times the squirrels were not overly timid, evidently informed instinctively that their natural camouflage offered security of a sort, but being impatient for darkness, they would soon glide back to the nest box or cavity and scamper into the hole with great rapidity.

A nest examined on August 1, during the squirrel's absence, was found to consist entirely of *fresh green* leaves taken from the white oak to which the box was attached. Another, larger box, with an entrance hole large enough to admit animals up to the size of a gray squirrel, and painted red, was occupied so consistently by three flying squirrels that the paint on one side (left of the entrance) was entirely worn away, and had dozens of fine claw marks proving the route habitually followed in and out of the box. In this case, a *winter* nest had been constructed entirely of shredded inner bark taken from a dead basswood tree (plate 175). In still another instance, a flying squirrel lived for a long time in a two-hole bird box intended especially for White-breasted Nuthatches to facilitate the feeding of their large broods, and which these birds had utilized prior to the squirrel's occupancy.

If taken from a nest in infancy (in one case young were born in May) and carefully brought up on diluted warm milk until weaned, they are easy to maintain. Under such conditions, their behavior is sweet and serene. Their cleanliness at all times, and the absence of any odor offensive to human beings, make these little squirrels ideal pets for children or adults.

Specimens kept by the author at home and at the Bruce Museum almost always deposited their feces in one spot in a corner of their cage, thus making cleaning very simple. Being very active, they require a wheel on which to exercise, and this they will use very often, apparently enjoying an audience.

In the wild, there is little doubt but that these squirrels will occasionally eat birds' eggs and maybe nestlings. They certainly eat insects also, but as they thrive in captivity on rolled oats, wheat grain, corn, sunflower seeds, apple halves, bits of orange, and mixed wild bird seed, one may say that they are mostly herbivorous. A squirrel that lived in a birdhouse within a few yards of a well-stocked bird feeder (as this was being written) certainly picked an ideal situation for a rodent species which apparently does not always hibernate.

Of the great family of rodents known technically as the Cricetidae, which includes the mice, rats, voles, and lemmings, only two species were found within the mini-wood or on its borders during the fifteen-year study period.

One of these, the deer mouse, *Peromyscus maniculatus* (plate 176), often confused with the white-footed mouse, has been a regular resident, but in some years much rarer than in others, fluctuations doubtless being due to the available food supply, especially in winter, as this mouse does not hibernate, nor deep-sleep, except periodically during the cold months (plate 176).

Beautiful little creatures, with big bright black eyes, the pelage varies in different individuals from distinctly grayish buff to a rather rich reddish brown. This coloring is attractively contrasted with the white underparts, while the tail is always characteristically bicolor, dark above and white below. The white-footed mouse, *Peromyscus leucopus,* already mentioned, is so similar that only a specialist could distinguish it from a brown-phase deer mouse, the species usually found here in the northeast. Possibly the two interbreed.

Finding two mice in the same nest, one in each color phase, long baffled the author when he was just a budding naturalist years ago, it being then assumed that they were always mated pairs, whereas the gray specimens were usually young mice, which seem to stay gray for some time.

Formerly, the author found many rather bulky nests of the deer mice situated some distance above the ground in grapevines, in which cases the nests were constructed of grapevine bark and also of bark brought in from nearby honeysuckle vines.

Now, however, deer mice living within the mini-wood seemed to prefer bird boxes, or natural cavities, when available, or, occasionally, some cranny or receptacle, or even a tool-cabinet drawer in an outbuilding or garage. One deer mouse constructed a snug ball-like nest in the author's tool house, using various colored material plucked from an automobile duster. Another one commenced a nest under the seat of the snow plow stored in the same building.

Some of these mice cover over old birds' nests, using them mostly for summer shelter and food storage. Being naturally of a provident nature, these little rodents establish caches of food—goodly accumulations of seeds and cherry pits—which they tuck away for use when other food is scarce in winter. The author has found their stores in unused drawers, in holes or hollows of most any sort, such as in the odoriferous bowls of the author's old pipes, and often under discarded lumber (see plate 53 for a deer mouse cache).

With prominent ears and long whiskers, these mice are very alert, yet they seem naturally trusting, seldom attempting to bite if gently covered by the whole hand when first picked up. They do not cry out like some mice, but if held too firmly, they utter barely audible little chattering sounds. Young mice, when first on their own, seem slow and almost half asleep at times, and as they sometimes wander abroad at this stage of their lives, they become easy prey for Hawks and owls, shrews, and larger mammal predators (plate 177).

Baby mice, born on May 24, were minus fur on their soft grayish pink bodies, which measured less than an inch in length, and their tails about a half inch more (plate 177).

The mother deer mouse nurses her litters of from three to five young, lying partly on her side, with the babies lined up at her nipples. As the little ones become well furred, which is quite soon, when seen thus nursing in a row, they remind one strongly of so many miniature puppies, what with their stubby heads, prominent ears, and their little tails sticking out behind.

Deer mice are omnivorous. They eat some birds' eggs and nestlings at times, and being very fond of meat, they will usually take animal food of any kind, especially in winter. They will scrape the last remaining shreds of meat from bones thrown out for whatever takers might come along, and then gnaw the bone itself for needed mineral matter. In the experience of the author, a beef bone left in the woods was sooner or later certain to bear the telltale marks of their teeth. They also consumed suet and drippings, and it was all but impossible to keep a cake of Ivory soap in the garage, if left where they could get at it. They were just as avid for seeds, such as sunflower and millet, and for peanut crumbs, as they were for animal food. In summer their diet included insects and berries, and annually they hoarded large numbers of wild cherry pits, later laboriously gnawing a single small hole in each pit to obtain the meat, as so graphically illustrated in plate 53.

Some of such hoarded seeds were lost, still undamaged, through the bottoms of nests placed in vines, or beneath woodland debris. Thus have these mice aided in the distribution of wild cherry trees and probably such other woodland species as spicebush and winterberry, or black alder shrubs.

Deer mice remained active all winter. In and out of the mini-wood, their tracks in the snow often led to brittle weed and flower stalks, to which they had resorted in search of a few remaining seeds. In severe winters they sometimes became hungry almost to the starvation point. In one case, when the author had set out a number of live-traps in the woods to ascertain what might still be active, two deer mice had fallen into a can-trap sunken in the snow and baited with cooked fish. These two had not only eaten all the bait, but one mouse had killed the other and had eaten part of the victim—no doubt an exceptional occurrence, it was very likely an act of desperation, which, let us hope, these little rodents seldom need resort to.

Other species of the genus *Peromyscus* inhabit many parts of North America. In fact, these little mice may be the most abundant of the wild species, and only a specialist or mammalogist could distinguish which was which in some cases. For instance, the deer mouse and the white-footed mouse, as already mentioned, are species both of which inhabit wide areas of the United States, and here in the northeast are so alike that it is often a question as to which mouse the uninitiate may be examining.

The deer mouse usually has a more sharply defined bicolored tail, and a plainly seen area of dark dorsal hair along the back, but not forming a distinct stripe. The pregnant female deer mouse shown in plate 176 was dark reddish brown, with the darker dorsal area very distinct.

The meadow vole, *Microtus pennsylvanicus*, known to many as the meadow mouse, is not strictly of the woodland habitat. Preferring moist meadows or even drier fields, especially abandoned and thinning hay fields, it does occasionally occur under the woodland carpet and in the more open habitat on its borders.

It is a four- to five-inch stubby, thickly furred rodent, mostly grayish brown, with a bicolored tail up to two and a half inches in length. There is considerable variation in color of the fur, however, many individuals showing little gray on the upper side, and with black hair so intermixed that it looks dark brown. Underparts show gray or buff, the fur lightly contrasting with the darker pelage, and with a sheen, if seen in a bright light (plate 177).

Voles make their networks of passages at the bases of weeds and grasses. These become distinct, well-trodden roadways, the rodents eating as they extend each branch road, but often leaving many uneaten pieces of the stems, and also occasional accumulations of their droppings, which may have some nutritional value for the grasses. The extent of these networks in many fields is seldom realized because of their position. In days gone by, after a winter with heavy snow cover, we used to burn over the fields in the spring. Afterward, sometimes a late light fall of the white stuff would outline every winter-made vole passage, thus revealing in certain fields wide-spreading reticulated patterns like huge road maps, and thus graphically illustrating the extent of this rodent's winter feeding (plate 178).

From the borders of the mini-wood, occasional meadow voles made passages into a one-time extensive vegetable garden, where they made one nest in a small corn patch, a rounded affair constructed of shredded corn leaves, and conveniently located near the ripening ears, but from which the vole was promptly evicted by the owner of said garden. While predominantly herbivorous, voles doubtless eat some animal matter (insects) taken with their other food. They also gnaw the bark from trees, unseen under the snow.

A nest accidentally dug up on the edge of the mini-wood had been constructed one foot below the surface. Made of dry grasses, it held five minute, naked, pinkish voles with their eyes still closed and still looking like embryos. The time being early April and the nights still frosty, this nest with its contents was carefully replaced, but somewhat nearer the surface, and with a man-made entrance passage composed of a cardboard tube, around which the earth was repacked. Two days later the young voles were gone, whether removed by the mother or a predator, who could say?

Probably no other small mammals are more hunted as food items, being preyed upon by Hawks both large and small; by Owls and Crows; also by shrews (suspected of devouring young voles); by weasels, foxes, raccoons, cats, blacksnakes; probably by skunks; and maybe by even the sluggish 'possum, who may occasionally eat a young vole.

In other parts of the country, there are doubtless many other hunters for which countless voles are constantly being born as nourishment.

Meteorological data may sometimes be supplied by these rodents. A past severe winter's snowfall, for instance, may be indicated in inches by the height to which the bark of many small trees will be seen eaten away when the snow melts.

The eastern cottontail, *Sylvilagus floridanus,* known to most at sight, is the little rabbit identified here in New England by its conspicuous cottony white tail as it scampers down the country lane in front of us, or springs from its bed almost under our feet as we disturb long grasses backed by shrubbery.

Seen close up, its fur is a mixture of brown, gray, and sometimes russet hairs. It is grayer toward the rump than anteriorly, and white on the underside. There is a more or less distinct brown band across the breast, and the ears are edged with dark hairs.

Several times a cottontail startled the author as it jumped from under the edge of a brush pile purposely accumulated in the mini-wood for whatever wildlife might care to use it. These little rabbits bound away, but by a protective habit suddenly "freeze" in their tracks when a few yards away and seem to fade into their surroundings at once, especially against a woodland carpet of dry dead leaves. Often completely baffling some predators by this

maneuver, the cottontail, which is a very important food animal, often falls victim to Hawks and mammal predators.

Cottontails were active throughout the year, in and out of the mini-wood area, their characteristic tracks in the snow indicating their year-to-year fluctuations numerically.

In the spring the female prepares a nest of grasses under a bramble or other tangle of vegetation, then snugly lines the nest with fur from her own body. This acts as a blanket for the young rabbits when they are still very tender, especially if the litter comes in early spring, before the nights are very warm.

Newborn rabbits found by the author have had their eyes closed, and at this time they are curious looking for the reason that at this stage their large ears lie flat back against the nape.

Well-furred by the time they are a few weeks old, they make delightful pets for a time, readily taking diluted cow's milk from a medicine dropper, and soon thereafter learn to chew commercial rabbit pellets and small seeds and other solid vegetable foods (plate 178).

Usually, and for quite some weeks, young cottontails may become playful and amusing, but strange to say, as they get a little older, there is a tendency to revert to the wild state. Playfulness quickly diminishes, and they no longer care to be fondled even by the human being that may have rescued them and brought them safely to near maturity.

We have seen that meadow voles seemed destined to become food items of importance to many predators, and probably young cottontails fill a similar role in the normal nature of things. But notwithstanding these obvious truths, both mammals have been always capable of holding their own, doubtless their prolificacy being mainly responsible for their success.

Living about our homes, feeding on our lawns, in gardens, and within the woodlands, it has always been a pleasing experience to find a cottontail bobbing along the road to one's home, or nibbling grass and clover on the lawn toward evening in summer. Better yet is to know that they whelp their litters successfully, often, so to speak, right under the noses of our cats and dogs!

And now that the reader has come to the end of this book, it might be well to stop and think for a few minutes about the astonishing variety of wildlife that is still to be encountered even in a mini-wood and on its borders. And when you have done this, try then to comprehend the numbers of living things in a hundred acres, a thousand acres, and in a square mile of forest land. Impossible, of course, because their numbers are astronomical.

Finally, think of the loss of life when even a single acre of forest is felled, or destroyed completely by even one small forest fire.

RECOMMENDED BOOKS
FOR FURTHER READING

Braun, E. Lucy. *Deciduous Forests of Eastern North America.* Reprint of 1950 ed. New York: Hafner Publishing Company, 1967.

Burch, John B. *How to Know the Eastern Land Snails.* Dubuque, Iowa: Wm. C. Brown Company, 1962.

Cobb, Boughton. *A Field Guide to the Ferns.* Boston: Houghton Mifflin Company, 1956.

Collins, Henry Hill. *Complete Field Guide to American Wildlife: East, Central & North.* New York: Harper & Brothers, 1959. A very handy guide for the student, as it includes the fishes, amphibians, reptiles, birds, and mammals, as well as invertebrates.

Elton, Charles S. *The Pattern of Animal Communities.* London: Methuen and Company; New York: Barnes & Noble, 1966.

Gleason, Henry A. *The New Britton and Brown Illustrated Flora of the Northeastern United States and Adjacent Canada.* 3 vols. Rev. ed. New York: Hafner Publishing Company, 1968. Originally published for the New York Botanical Garden, 1952.

Grout, A. J. *Mosses with Hand Lens and Microscope.* Augusta, W. Va.: Eric Lundberg, 1972.

Hale, Mason E. *How to Know the Lichens.* Dubuque, Iowa: Wm. C. Brown Company, 1969.

Holland, W. J. *The Moth Book: A Popular Guide to a Knowledge of the Moths* . . . New York: Doubleday, Page & Company, 1908. Republished several times. Always valuable for its color plates. Rare-book dealers can usually supply some edition.

Klots, Alexander B. *Field Guide to the Butterflies.* Boston: Houghton Mifflin Company, 1951. Covers North American species found east of the Great Plains.

Lutz, Frank E. *Field Book of Insects of U.S. and Canada.* G. P. Putnam's Sons, 1948. The best all-around field guide.

McIlvaine, Charles, and Macadam, Robert K. *One Thousand American Fungi.* Barton, Vt.: Something Else Press, 1972; New York: Dover Publications, 1973.

Meglitsch, Paul A. *Invertebrate Zoology.* New York: Oxford University Press, 1967.

Peterson, Roger Tory, and McKenny, Margaret. *A Field Guide to Wildflowers of Northeastern and North-Central North America.* Boston: Houghton Mifflin Company, 1968.

Petrides, George A. *A Field Guide to the Trees and Shrubs.* Covers northeastern and north-central United States, and southeastern and south-central Canada. Boston: Houghton Mifflin Company, 1958.

Rickett, Harold William. *Wildflowers of the United States, vol. I: The Northeastern States.* New York: McGraw-Hill Company, 1966. Publication of the New York Botanical Garden.

Smith, Alexander H. *The Mushroom Hunter's Field Guide.* Rev. and enlarged. Ann Arbor, Mich.: University of Michigan Press, 1963.

Weygoldt, Peter, *The Biology of Pseudoscorpions.* Cambridge, Mass.: Harvard University Press, 1969.

Wilson, Carl L.; Loomis, Walter E.; and Steeves, Taylor A. *Botany.* 5th ed. New York: Holt, Rinehart & Winston, 1971.

INDEX